IDENTITIES AND FORMULAS

Basic Identities [5.1]

	Basic Identities	Common Equivalent Forms
Reciprocal	$\csc \theta = \dfrac{1}{\sin \theta}$	$\sin \theta = \dfrac{1}{\csc \theta}$
	$\sec \theta = \dfrac{1}{\cos \theta}$	$\cos \theta = \dfrac{1}{\sec \theta}$
	$\cot \theta = \dfrac{1}{\tan \theta}$	$\tan \theta = \dfrac{1}{\cot \theta}$
Ratio	$\tan \theta = \dfrac{\sin \theta}{\cos \theta}$	
	$\cot \theta = \dfrac{\cos \theta}{\sin \theta}$	
Pythagorean	$\cos^2 \theta + \sin^2 \theta = 1$	$\sin^2 \theta = 1 - \cos^2 \theta$
		$\sin \theta = \pm\sqrt{1 - \cos^2 \theta}$
		$\cos^2 \theta = 1 - \sin^2 \theta$
		$\cos \theta = \pm\sqrt{1 - \sin^2 \theta}$
	$1 + \tan^2 \theta = \sec^2 \theta$	
	$1 + \cot^2 \theta = \csc^2 \theta$	

Sum and Difference Formulas [5.2]

$\sin(A + B) = \sin A \cos B + \cos A \sin B$

$\sin(A - B) = \sin A \cos B - \cos A \sin B$

$\cos(A + B) = \cos A \cos B - \sin A \sin B$

$\cos(A - B) = \cos A \cos B + \sin A \sin B$

$\tan(A + B) = \dfrac{\tan A + \tan B}{1 - \tan A \tan B}$

$\tan(A - B) = \dfrac{\tan A - \tan B}{1 + \tan A \tan B}$

Double-Angle Formulas [5.3]

$\sin 2A = 2 \sin A \cos A$

$\cos 2A = \cos^2 A - \sin^2 A \qquad$ First form

$\quad\;\;\, = 2 \cos^2 A - 1 \qquad$ Second form

$\quad\;\;\, = 1 - 2 \sin^2 A \qquad$ Third form

$\tan 2A = \dfrac{2 \tan A}{1 - \tan^2 A}$

Cofunction Theorem [2.1]

$\sin x = \cos(90° - x)$

$\cos x = \sin(90° - x)$

$\tan x = \cot(90° - x)$

Half-Angle Formulas [5.4]

$\sin \dfrac{A}{2} = \pm\sqrt{\dfrac{1 - \cos A}{2}}$

$\cos \dfrac{A}{2} = \pm\sqrt{\dfrac{1 + \cos A}{2}}$

$\tan \dfrac{A}{2} = \dfrac{1 - \cos A}{\sin A} = \dfrac{\sin A}{1 + \cos A}$

Even/Odd Functions [3.3]

$\cos(-\theta) = \cos \theta \qquad$ Even

$\left.\begin{array}{l} \sin(-\theta) = -\sin \theta \\ \tan(-\theta) = -\tan \theta \end{array}\right\}$ Odd

Sum to Product Formulas [5.5]

$\sin \alpha + \sin \beta = 2 \sin \dfrac{\alpha + \beta}{2} \cos \dfrac{\alpha - \beta}{2}$

$\sin \alpha - \sin \beta = 2 \cos \dfrac{\alpha + \beta}{2} \sin \dfrac{\alpha - \beta}{2}$

$\cos \alpha + \cos \beta = 2 \cos \dfrac{\alpha + \beta}{2} \cos \dfrac{\alpha - \beta}{2}$

$\cos \alpha - \cos \beta = -2 \sin \dfrac{\alpha + \beta}{2} \sin \dfrac{\alpha - \beta}{2}$

Product to Sum Formulas [5.5]

$\sin A \cos B = \dfrac{1}{2} [\sin(A + B) + \sin(A - B)]$

$\cos A \sin B = \dfrac{1}{2} [\sin(A + B) - \sin(A - B)]$

$\cos A \cos B = \dfrac{1}{2} [\cos(A + B) + \cos(A - B)]$

$\sin A \sin B = \dfrac{1}{2} [\cos(A - B) - \cos(A + B)]$

Pythagorean Theorem [1.1]

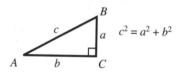

$c^2 = a^2 + b^2$

The Law of Sines [7.1]

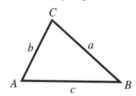

$\dfrac{\sin A}{a} = \dfrac{\sin B}{b} = \dfrac{\sin C}{c}$

or, equivalently,

$\dfrac{a}{\sin A} = \dfrac{b}{\sin B} = \dfrac{c}{\sin C}$

The Law of Cosines [7.3]

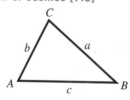

$a^2 = b^2 + c^2 - 2bc \cos A$

$b^2 = a^2 + c^2 - 2ac \cos B$

$c^2 = a^2 + b^2 - 2ab \cos C$

or, equivalently,

$\cos A = \dfrac{b^2 + c^2 - a^2}{2bc}$

$\cos B = \dfrac{a^2 + c^2 - b^2}{2ac}$

$\cos C = \dfrac{a^2 + b^2 - c^2}{2ab}$

Trigonometry

Third Edition

Trigonometry

THIRD EDITION

Charles P. McKeague
Cuesta College

SAUNDERS COLLEGE PUBLISHING

Harcourt Brace College Publishers

Fort Worth   Philadelphia   San Diego   New York   Orlando   Austin
San Antonio   Toronto   Montreal   London   Sydney   Tokyo

Text Typeface: Times Roman
Compositor: Progressive Typographers
Acquisitions Editor: Jay Ricci
Developmental Editor: Laurie Golson
Managing Editor: Carol Field
Project Editor: Laura Shur
Copy Editor: Ellen Thomas
Manager of Art and Design: Carol Bleistine
Art Director: Christine Schueler
Cover Designer: Lawrence R. Didona
Text Artwork: Techsetters and Grafacon
Director of EDP: Tim Frelick
Production Manager: Carol Florence
Marketing Manager: Monica Wilson

Cover Credit: © Paul Silverman, 1989/Fundamental Photographs

Printed in the United States of America

TRIGONOMETRY, 3/e

ISBN: 0-03-96562-4

Library of Congress Catalog Card Number: 93-085367

6789   032 987654

Contents

| | | | | | | | | |

v

Preface to the Instructor

ı ı ı ı ı ı ı ı ı ı

This third edition of *Trigonometry* retains the same format and style as the first and second editions. It is a standard right triangle approach to trigonometry. Each section is written so that it can be discussed in a 45 to 50 minute class session. The text covers all the material usually taught in trigonometry. In addition, there is an appendix on logarithms.

The emphasis of the textbook is on understanding the definitions and principles of trigonometry and their application to problem solving. However, when memorization is necessary, I say so.

Identities are introduced early in Chapter One. They are reviewed often and are then covered in more detail in Chapter Five. Also, exact values of the trigonometric functions are emphasized throughout the textbook. There are numerous calculator notes placed throughout the text.

Organization of the Text

The textbook begins with a preface to the student explaining what study habits are necessary to ensure success in mathematics.

The rest of the book is divided into chapters. Each chapter is organized as follows:

1. Introduction   Each chapter begins with an introduction that explains in a very general way what the student can expect to find. The introduction also includes a list of previous material that is used to develop the concepts in the chapter.
2. Sections   The body of the chapter is divided into sections. Each section contains explanations and examples. The explanations are as simple and intuitive as possible. The examples are chosen to clarify the explanations and preview the problems in the problem sets.
3. Problem Sets   Following each section of the text is a problem set. There are five main ideas incorporated into each of the problem sets.

a. *Drill:* There are enough problems in each problem set to ensure proficiency with the material once students have completed all the odd-numbered problems.

b. *Progressive Difficulty:* The problems increase in difficulty as the problem set progresses.

c. *Odd-Even Similarities:* Each pair of consecutive problems is similar. Since the answers to the odd-numbered problems are given in the back of the book, the similarity of each odd-even pair of problems gives your students a chance to check their work on an odd-numbered problem and then try a similar even-numbered problem.

d. *Application Problems:* Application problems are emphasized throughout the book. There are application problems in most problem sets.

e. *Review Problems:* As was the case in the previous edition, each problem set after Chapter One ends with a number of review problems. When possible, these problems review material that will be needed in the next section. Otherwise they review material from the previous chapter.

4. Chapter Summaries   Following the last problem set in each chapter is a chapter summary. The summary lists all the properties and definitions found in the chapter. In the margin, next to most of the topics being summarized, is an example that illustrates the kind of problem associated with that topic.

5. Chapter Tests   Each chapter ends with a test designed to give the student an idea of how well he or she has mastered the material in the chapter. All answers for these chapter tests are included in the back of the book.

Changes in the Third Edition

New Section Openings   Many sections now open with an application problem or historical information. The purpose of these openings is to motivate the topics under consideration with a real-life application or from a historical perspective. Whenever possible, these introductions are expanded on later in the section and then carried through to topics found further on in the book.

New Topics   In this edition, Section 4.4 is new. It shows students how to find the equation of a graph from the graph itself. Also new is Section 7.5 which extends the topic of vectors to include vectors written in the form $\mathbf{V} = a\mathbf{i} + b\mathbf{j}$.

Research Projects   Scattered throughout this edition are problems for students to research and then report on. Although they appear at the end of a number of the problem sets, they are not intended to be part of the student's daily assignment. In my classes, these are the problems I use for extra credit. In most cases, I require students to type their reports, just as they would type an essay in their English classes. I have designed these problems so as to appeal to students with a variety of interests. There are problems for students studying English, history, philosophy, religion, and map making. Do not be concerned if you are not familiar with the topics shown in the research problems: The idea behind these problems is to have your students do the research, and then tell you what they have learned.

Inverse Functions Emphasized   The section on inverse trigonometric relations has been eliminated. The inverses for the trigonometric functions are now covered in just one section, with the emphasis placed on inverse trigonometric functions. Also, I have dropped the convention that uses capital letters to denote inverse trigonometric functions and lowercase letters to denote inverse trigonometric relations. The notation $\sin^{-1}x$ now stands for the inverse sine function only, which is the way students will see it if they go on in mathematics.

Circular Functions A third definition for the trigonometric functions has been added to this edition in Section 3.3. It defines the trigonometric functions as circular functions associated with the unit circle. The emphasis in the book is still on the definition that gives the trigonometric functions in terms of a point on the terminal side of an angle in standard position. I have included three definitions in this edition in order to show how they are related to one another.

I think you will find here, in this third edition of *Trigonometry,* a book that will assist you with your endeavors in the classroom. In particular, this text, together with the supplements listed below, form a package that will allow you to implement the Curriculum Standards for Trigonometry as set forth by the National Council of Teachers of Mathematics.

Supplements to the Text

A comprehensive collection of ancillary resources accompanies *Trigonometry,* 3rd edition, both for the instructor and for the student. New to this edition are the *Instructor's Solutions Manual;* the *Transparency Package; ExaMaster+*tm, including *ExaMaster+*tm *Computer Testbank, ExaMaster+*tm *Printed Testbank,* and *ExaMaster+*tm *RequesTest*tm; and *MathCue*tm *Interactive Software,* including *MathCue*tm *Tutorial* and *MathCue*tm *F/C-Graph*tm. Supplements for students using graphing calculators are also available.

For the Instructor

All ancillaries for instructors are free to schools who use *Trigonometry,* 3rd edition.

Instructor's Solutions Manual *New to this edition* Included are instructor-appropriate solutions for every exercise in the book, and answers for every even-numbered exercise.

Prepared Tests Included are 13 sets of ready-to-copy tests: one set for each chapter, one set for each odd/even pair of chapters, and one set for the entire book. Each set comprises 2 multiple-choice and 4 show-your-work tests. Items for half of the tests are ordered according to the sequence of topics in the book; items for the other half of the tests are in mixed-up order. Answers for all tests are included.

Transparency Package *New to this edition* Roughly 50 excerpts from the book, including graphs, worked examples, and key theorems, definitions, and properties, are provided in a full-color transparency format suitable for overhead display.

ExaMaster+tm *New to this edition* A flexible, powerful testing system, *ExaMaster+*tm offers teachers a wide range of integrated testing options and features.

- Using **ExaMaster+™ Computerized Testbank,** in either IBM or Macintosh format, teachers can select, edit, or create not only test items but algorithms for test items as well. Teachers can tailor tests according to a variety of criteria, scramble the order of test items, and administer tests on-line. *ExaMaster+™* also includes full-function gradebook and graphing features.
- In printed form, **ExaMaster+™ Printed Testbank** arranges *ExaMaster+™* databank test items by chapter.
- Using **ExaMaster+™ RequesTest™,** teachers can select *ExaMaster+™* test items or algorithms, or specify criteria for test items, then call the Saunders Software Support Department (1-800-447-9457) who will generate, print, and mail or fax the *ExaMaster+™* test within 48 hours.

For the Student

Videotapes and *MathCue™ Interactive Software* are free to schools using *Trigonometry,* 3rd edition.

Student's Solutions Manual   Included are detailed, annotated solutions for every other odd-numbered exercise in the book's Problem Sets and for every exercise in the Chapter Tests.

Videotapes   The videotape package comprises 9 VHS videotapes, one for each chapter and one for the appendix on logarithms. Each tape is approximately one hour long and is divided into lessons that correspond to sections of the chapter. In each lesson I work selected problems from the text.

MathCue™ Interactive Software   *New to this edition*   Available in IBM and Macintosh formats, *MathCue™ Interactive Software* for students affords opportunities for learning, reviewing, and practicing skills, discovering and exploring concepts, and pinpointing and correcting weak areas of understanding. MathCue™ is designed by George W. Bergeman of Northern Virginia Community College.

- Tailored to every section of the book, **MathCue™ Tutorial** presents students with problems to solve and tutors students by displaying annotated, step-by-step solutions. Students may view partial solutions to get started on a problem, see a continuous record of progress, and back up to review missed problems. Student scores can also be printed.
- **MathCue™ F/C-Graph™** allows students to graph and analyze any trigonometric, logarithmic, polynomial, or exponential function or conic equation they choose. *F/C-Graph™* can zoom, trace, display function values or coordinates of selected points, graph up to four functions simultaneously, and save and retrieve setups. Students can use *F/C-Graph™* to relate algebraic and visual forms of functions and conic sections, and to explore how changing parameters affects the graph. A printed collection of exercises and projects using *F/C-Graph™* accompanies the software.

Supplement for Graphing Calculators   Trigonometric concepts are explored using graphing calculators. Projects and exercises suitable for group or individual work are included.

Acknowledgments

A project of this size cannot be completed without help from many people. In particular, my thanks go to Laurie Golson for managing all the pre-production elements of this revision, to Jay Ricci for coordinating and approving all aspects of the revision, to Christine Schueler for the design of the book and cover, and to Carol Field for supervising the production process, making the changes I requested, and for doing so in the most efficient, professional, and pleasant manner.

One of the most difficult tasks in producing a book on trigonometry is checking the accuracy of the examples and the answers to the problems in the problem sets. This task was handled very competently by Kate Pawlik, Bob Martin, Cheryl Roberts, and Patrick McKeague. Their attention to detail and promptness in working the problems is unmatched, and I am particularly lucky to have been able to work with these four people.

Thanks also to Stacey Lloyd for her wordprocessing skills and for her help with the index, and to my wife Diane and my daughter Amy for their continued encouragement with all my writing endeavors.

My son Patrick assisted me with this revision from the beginning to the end. I am pleased with the way the book has turned out and much of what I like about it is due to his influence.

Special thanks to the following people for their reviews of the book. Their comments and suggestions on this revision were very detailed and extremely helpful.

Mack Hill
Worcester State College

Melba Morgan
Northeast Mississippi Community College

Rafal Ablamowicz
Gannon University

Frank Garcia
North Seattle Community College

John Spellman
Southwest Texas State University

Kelly Wyatt
Umpqua Community College

Elise Price
Tarrant County Junior College

John R. Martin
Tarrant County Junior College

Larry Dilley
Central Missouri State University

Denise Hennicke
Collin County Community College

Nancy Heary
Indiana University at Kokomo

Steven Heath
Southern Utah University

Users' Questionnaire

Thanks also to the people listed below for taking the time to fill out the detailed questionnaire on their experiences with the previous edition of this book.

Ann Smith
Hutchinson Community College

Cheryl Ooten
Rancho Santiago College

Don Skow
UT-Pan American

Daniel S. Chesley
Anne Arundel Community College

Alice Neath
Butte Community College

Trish Cabral
Butte College

Stanley Carter
Central Missouri State University

Louise Rodgers
Central Missouri State University

Sue Richardson
Clatsop Community College

Milford Stevens
Clatsop Community College

Jim Barton
College of the Sequoias

Tom Rodgers
Collin County Community College

Mollie A. Steward
Cumberland County Community College

Gerald Krauss
Gannon University

Diana Fernandez
Hillsborough Community College

S. Marie Kunkle
Indiana University at Kokomo

Carol Huffman
Iowa Western Community College

Dean Strenger
Irvine Valley College

Masato Hayachi
Irvine Valley College

Lyle E. Gleesing
Ithaca Community College

Susan Brockman
Kodiak College

Deanna Li
North Seattle Community College

Melba Morgan
Northeast Mississippi Community College

Kenneth Kochey
Northampton County Area Community College

Michael Pittner
Ohlone Community College

Ron Staszkow
Ohlone Community College

Bob Bradshaw
Ohlone Community College

Virginia Tebelskis
Ohlone Community College

Joyce Roling
Oklahoma Junior College

Kathy Mowers
Owensboro Community College

Tommie Ann Hill
Prairie View A&M University

Joan Capps
Raritan Valley Community College

Kim Castagna
West Hills College

Ali Moshgi
Richland Community College

Harry Baldwin
San Diego City College

E.J. Brown
St. Clair Community College

Ed Zimmerman
Tacoma Community College

Wayne James
University of South Dakota

Curtis Olson
University of South Dakota

B. Jan Davis
University of Southern Mississippi

Dean Christianson
University of the Pacific

Jack Heller
West Los Angeles College

Thomas Neils
Wisconsin Lutheran College

Manuel Zax
Worcester State College

Preface to the Student

I I I I I I I I I I

Trigonometry can be a very enjoyable subject to study. You will find that there are many interesting and useful problems that trigonometry can be used to solve. Many of my trigonometry students, however, are apprehensive at first because they are worried that they will not understand the topics we cover. When I present a new topic that they do not grasp completely, they think something is wrong with them for not understanding it.

On the other hand, some students are excited about the course from the beginning. They are not worried about understanding trigonometry and, in fact, expect to find some topics difficult.

What is the difference between these two types of students?

Those who are excited about the course know from experience (as you do) that a certain amount of confusion is associated with most new topics in mathematics. They don't worry about it because they also know that the confusion gives way to understanding in the process of reading the textbook, working the problems, and getting questions answered. If they find that they are having difficulty with a topic, they work as many problems as necessary to grasp the subject. They don't wait for the understanding to come to them; they go out and get it by working lots of problems. In contrast, the students who lack confidence tend to give up when they become confused. Instead of working more problems, they sometimes stop working problems altogether—and that of course guarantees that they will remain confused.

If you are worried about this course because you lack confidence in your ability to understand trigonometry, and you want to change the way you feel about mathematics, then look forward to the first topic that causes you some confusion. As soon as that topic comes along, make it your goal to master it, in spite of your apprehension. You will see that each and every topic covered in this course is one you can eventually master, even if your initial introduction to it is accompanied by some confusion. As long as you have passed a college-level intermediate algebra course (or its equivalent), you are ready to take this course.

If you have decided to do well in trigonometry, the following list will be important to you:

1. **Attend all class sessions on time.** You cannot know exactly what goes on in class unless you are there. Missing class and then expecting to find out what went on from someone else is not the same as being there yourself.

2. **Read the book.** It is best to read the section that will be covered in class beforehand. Reading in advance, even if you do not understand everything you read, is still better than going to class with no idea of what will be discussed.

3. **Work problems every day and check your answers.** The key to success in mathematics is working problems. The more problems you work, the better you will become at working them. The answers to the odd-numbered problems are given in the back of the book. When you have finished an assignment, be sure to compare your answers with the ones in the book. If you have made a mistake, find out what it was.

4. **Do it on your own.** Don't be misled into thinking someone else's work is your own. Having someone else show you how to work a problem is not the same as working the problem yourself. It is okay to get help when you are stuck. As a matter of fact, it is a good idea. Just be sure you do the work yourself.

5. **Review every day.** After you have finished the problems your instructor has assigned, take another fifteen minutes and review a section you have already completed. The more you review, the longer you will retain the material you have learned.

6. **Don't expect to understand every new topic the first time you see it.** Sometimes you will understand everything you are doing, and sometimes you won't. That's just the way things are in mathematics. Expecting to understand each new topic the first time you see it can lead to disappointment and frustration. The process of understanding trigonometry takes time. It requires that you read the book, work problems, and get your questions answered.

7. **Spend as much time as it takes for you to master the material.** No set formula exists for the exact amount of time you need to spend on trigonometry to master it. You will find out as you go along what is or isn't enough time for you. If you end up spending two or more hours on each section in order to master the material there, then that's how much time it takes; trying to get by with less will not work.

Keep This Book for Reference

I hope that you will find this book to be one of the more readable mathematics books you have used. I also hope you will consider keeping it as a reference book after you have finished your trigonometry course. It is likely that you will have occasion to refer back to trigonometry from time to time during your college career. If so, it will be much easier for you to find and recall topics if you look them up in the book you used for trigonometry.

1

The Six Trigonometric Functions

▶ **To the Student**

The material in Chapter 1 is some of the most important material in the book. We begin Chapter 1 with a review of some material from geometry (Section 1.1) and algebra (Section 1.2). Section 1.3 contains the definition for the six trigonometric functions. As you will see, this definition is used again and again throughout the book. It is very important that you understand and memorize the definition.

Once we have been introduced to the definition of the six trigonometric functions in Section 1.3, we then move on to study some of the more important consequences of the definition. These consequences take the form of trigonometric identities, the study of which is also important in trigonometry. Our work with trigonometric identities will take up most of Sections 1.4 and 1.5.

You can get a good start in trigonometry by mastering the material in Chapter 1. Any extra time you spend with the definition of the six trigonometric functions and the identities that are derived from it will be well worth it when you proceed to Chapters 2 and 3.

SECTION 1.1   # Angles, Degrees, and Special Triangles

The diagram shown in Figure 1 is called the Spiral of Roots. It is constructed using the Pythagorean Theorem, a theorem we will introduce in this section and then use many times throughout the course. The Spiral of Roots gives us a way to visualize positive square roots: The length of each diagonal line segment corresponds to the positive square root of one of the positive integers. Later in the book we will use trigonometry to find the measure of each of the angles formed at the center of the spiral.

Much of what we will do in this course will have a visual component to it because trigonometry is a very visual subject. Before we begin our study of trigonometry, there are some topics from geometry and algebra to review. Let's begin by looking at some of the terminology associated with angles.

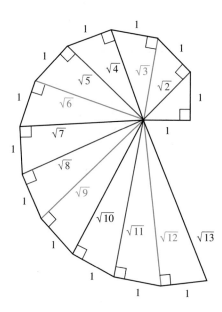

Figure 1

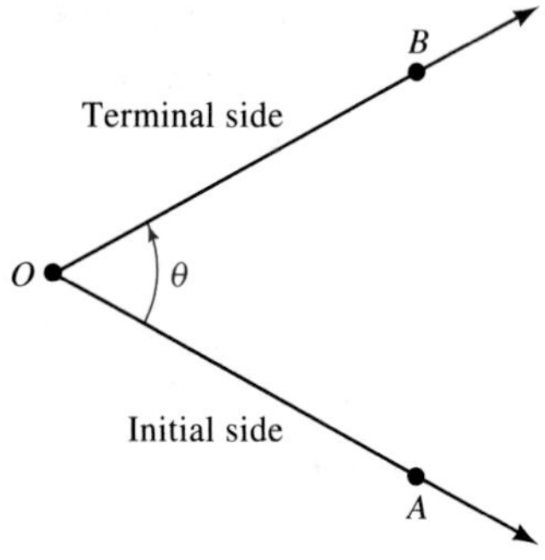

Angles in General

An angle is formed by two rays with the same end point. The common end point is called the *vertex* of the angle, and the rays are called the *sides* of the angle.

In Figure 2, the vertex of angle θ (theta) is labeled O, and A and B are points on each side of θ. Angle θ can also be denoted by AOB, where the letter associated with the vertex is written between the letters associated with the points on each side.

We can think of θ as having been formed by rotating side OA about the vertex to side OB. In this case, we call side OA the *initial side* of θ and side OB the *terminal side* of θ.

Figure 2

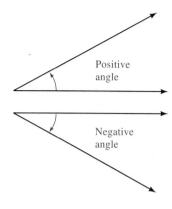

Positive angle

Negative angle

Figure 3

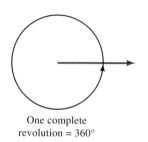

One complete revolution = 360°

Figure 4

When the rotation from the initial side to the terminal side takes place in a counterclockwise direction, the angle formed is considered a *positive angle*. If the rotation is in a clockwise direction, the angle formed is a *negative angle* (Figure 3).

Degree Measure

One way to measure the size of an angle is with degree measure. The angle formed by rotating a ray through one complete revolution has a measure of 360 degrees, written 360° (Figure 4).

One degree (1°), then, is 1/360 of a full rotation. Likewise, 180° is one-half of a full rotation, and 90° is half of that (or a quarter of a rotation). Angles that measure 90° are called *right angles,* while angles that measure 180° are called *straight angles.* Angles that measure between 0° and 90° are called *acute angles,* while angles that measure between 90° and 180° are called *obtuse angles.*

If two angles have a sum of 90°, then they are called *complementary angles,* and we say each is the *complement* of the other. Two angles with a sum of 180° are called *supplementary angles.*

Note   To be precise, we should say "two angles, the sum of the measures of which is 180°, are called supplementary angles" because there is a difference between an angle and its measure. However, in this book, we will not always draw the distinction between an angle and its measure. Many times we will refer to "angle θ" when we actually mean "the measure of angle θ."

Note   The little square by the vertex of the right angle in Figure 5 is used to indicate that the angle is a right angle. You will see this symbol often in the book.

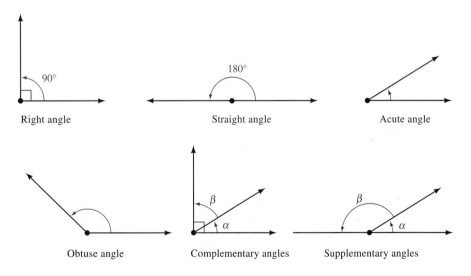

Right angle   Straight angle   Acute angle

Obtuse angle   Complementary angles   Supplementary angles

Figure 5

▶ **EXAMPLE 1**   Give the complement and the supplement of each angle.

a. 40°   **b.** 110°   **c.** θ

Solution

a. The complement of 40° is 50° since 40° + 50° = 90°.
   The supplement of 40° is 140° since 40° + 140° = 180°.
b. The complement of 110° is −20° since 110° + (−20°) = 90°.
   The supplement of 110° is 70° since 110° + 70° = 180°.
c. The complement of θ is 90° − θ since θ + (90° − θ) = 90°.
   The supplement of θ is 180° − θ since θ + (180° − θ) = 180°.   ◀

Special Triangles

A *right triangle* is a triangle in which one of the angles is a right angle. In every right triangle, the longest side is called the *hypotenuse,* and it is always opposite the right angle. The other two sides are called the *legs* of the right triangle. Since the sum of the angles in any triangle is 180°, the other two angles in a right triangle must be complementary, acute angles. The Pythagorean Theorem that we mentioned in the introduction to this section gives us the relationship that exists among the sides of a right triangle. First we state the theorem, then we will prove it.

Pythagorean Theorem

In any right triangle, the square of the length of the longest side (called the hypotenuse) is equal to the sum of the squares of the lengths of the other two sides (called legs).

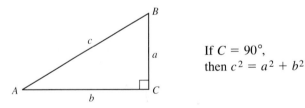

If $C = 90°$,
then $c^2 = a^2 + b^2$

Figure 6

We denote the lengths of the sides of triangle *ABC* in Figure 6 with lowercase letters and the angles or vertices with uppercase letters. It is standard practice in mathematics to label the sides and angles so that *a* is opposite *A*, *b* is opposite *B*, and *c* is opposite *C*.

Next we will prove the Pythagorean Theorem. Part of the proof involves finding the area of a triangle. In any triangle, the area is given by the formula

$$\text{Area} = \frac{1}{2}\,(\text{base})(\text{height})$$

For the right triangle shown in Figure 6, the base is b, and the height is a. Therefore the area is $A = \dfrac{1}{2}\,ab$.

A Proof of the Pythagorean Theorem

There are many ways in which to prove the Pythagorean Theorem. The one that we are offering here is based on the diagram below and the formula for the area of a triangle.

Figure 7 is constructed by taking the right triangle in the lower right corner and repeating it three times so that the final diagram is a square in which each side has a length $a + b$.

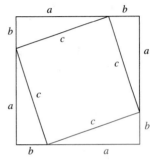

Figure 7

To derive the relationship between a, b, and c, we simply notice that the area of the large square is equal to the sum of the areas of the four triangles and the inner square. In symbols we have

Area of large square		Area of four triangles		Area of inner square
$(a + b)^2$	$=$	$4\left(\dfrac{1}{2}\,ab\right)$	$+$	c^2

We expand the left side using the formula for the square of a binomial, from algebra. We simplify the right side by multiplying 4 with $\dfrac{1}{2}$.

$$a^2 + 2ab + b^2 = 2ab + c^2$$

Adding $-2ab$ to each side, we have the relationship we are after:

$$a^2 + b^2 = c^2$$

▶ **EXAMPLE 2** Solve for x in the right triangle in Figure 8.

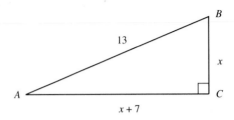

Figure 8

Solution Applying the Pythagorean Theorem gives us a quadratic equation to solve.

$$(x + 7)^2 + x^2 = 13^2$$

$$x^2 + 14x + 49 + x^2 = 169 \qquad \text{Expand } (x + 7)^2 \text{ and } 13^2$$

$$2x^2 + 14x + 49 = 169 \qquad \text{Combine similar terms}$$

$$2x^2 + 14x - 120 = 0 \qquad \text{Add } -169 \text{ to both sides}$$

$$x^2 + 7x - 60 = 0 \qquad \text{Divide both sides by 2}$$

$$(x - 5)(x + 12) = 0 \qquad \text{Factor the left side}$$

$$x - 5 = 0 \quad \text{or} \quad x + 12 = 0 \qquad \text{Set each factor to 0}$$

$$x = 5 \quad \text{or} \quad x = -12$$

Our only solution is $x = 5$. We cannot use $x = -12$ since x is the length of a side of triangle ABC and therefore cannot be negative. ◀

Before leaving the Pythagorean Theorem we should mention something about Pythagoras and his followers, the Pythagoreans. They established themselves as a secret society around the year 540 B.C. The Pythagoreans kept no written record of their work; everything was handed down by spoken word. Their influence was not only in mathematics, but also in religion, science, medicine, and music. Among other things, they discovered the correlation between musical notes and the reciprocals of counting numbers, $\frac{1}{2}, \frac{1}{3}, \frac{1}{4}$, and so on. In their daily lives they followed strict dietary and moral rules to achieve a higher rank in future lives. The British philosopher Bertrand Russell has referred to Pythagoras as "intellectually one of the most important men that ever lived."

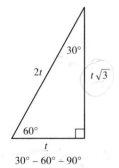

$30° - 60° - 90°$

Figure 9

The 30°–60°–90° Triangle

In any right triangle in which the two acute angles are 30° and 60°, the longest side (the hypotenuse) is always twice the shortest side (the side opposite the 30° angle), and the side of medium length (the side opposite the 60° angle) is always $\sqrt{3}$ times the shortest side (Figure 9).

Note that the shortest side t is opposite the smallest angle 30°. The longest side $2t$ is opposite the largest angle 90°. To verify the relationship between the sides in this triangle, we draw an equilateral triangle (one in which all three sides are equal) and label half the base with t (Figure 10).

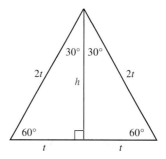

Figure 10

The altitude h (the colored line) bisects the base. We have two $30°–60°–90°$ triangles. The longest side in each is $2t$. We find that h is $t\sqrt{3}$ by applying the Pythagorean Theorem.

$$t^2 + h^2 = (2t)^2$$
$$h = \sqrt{4t^2 - t^2}$$
$$= \sqrt{3t^2}$$
$$= t\sqrt{3}$$

▶ **EXAMPLE 3** If the shortest side of a $30°–60°–90°$ triangle is 5, find the other two sides.

Solution The longest side is 10 (twice the shortest side), and the side opposite the $60°$ angle is $5\sqrt{3}$.

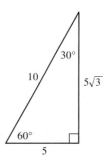

Figure 11 ◀

▶ **EXAMPLE 4** A ladder is leaning against a wall. The top of the ladder is 4 ft above the ground and the bottom of the ladder makes an angle of $60°$ with the ground. How long is the ladder, and how far from the wall is the bottom of the ladder?

Solution The triangle formed by the ladder, the wall, and the ground is a 30°–60°–90° triangle. If we let x represent the distance from the bottom of the ladder to the wall, then the length of the ladder can be represented by $2x$. The distance from the top of the ladder to the ground is $x\sqrt{3}$, since it is opposite the 60° angle. Therefore,

$$x\sqrt{3} = 4$$

$$x = \frac{4}{\sqrt{3}}$$

$$= \frac{4\sqrt{3}}{3} \quad \begin{array}{l}\text{Rationalize the denominator}\\ \text{by multiplying the numerator}\\ \text{and denominator by } \sqrt{3}.\end{array}$$

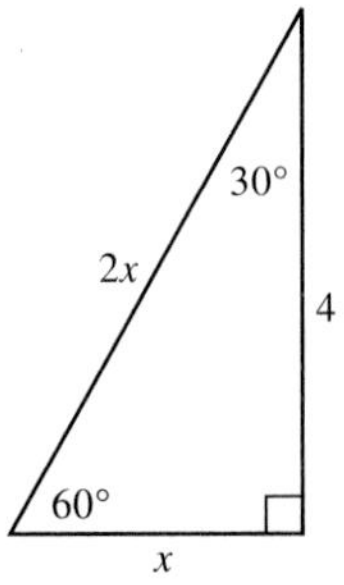

Figure 12

The distance from the bottom of the ladder to the wall, x, is $4\sqrt{3}/3$ feet, so the length of the ladder, $2x$, must be $8\sqrt{3}/3$ feet. Note that these lengths are given in exact values. If we want a decimal approximation for them, we can replace $\sqrt{3}$ with 1.732 to obtain

$$\frac{4\sqrt{3}}{3} \approx \frac{4(1.732)}{3} = 2.309 \text{ feet}$$

$$\frac{8\sqrt{3}}{3} \approx \frac{8(1.732)}{3} = 4.619 \text{ feet}$$

Calculator Note On a calculator with algebraic logic, this last calculation could be done as follows:

8 $\boxed{\times}$ 3 $\boxed{\sqrt{}}$ $\boxed{\div}$ 3 $\boxed{=}$

with the answer rounded to three decimal places (that is, three places past the decimal point). ◀

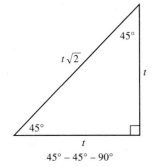

45° – 45° – 90°

Figure 13

The 45°–45°–90° Triangle

If the two acute angles in a right triangle are both 45°, then the two shorter sides (the legs) are equal and the longest side (the hypotenuse) is $\sqrt{2}$ times as long as the shorter sides. That is, if the shorter sides are of length t, then the longest side has length $t\sqrt{2}$ (Figure 13).

To verify this relationship, we simply note that if the two acute angles are equal, then the sides opposite them are also equal. We apply the Pythagorean Theorem to find the length of the hypotenuse.

$$\text{hypotenuse} = \sqrt{t^2 + t^2}$$
$$= \sqrt{2t^2}$$
$$= t\sqrt{2}$$

▶ **EXAMPLE 5**   A 10-foot rope connects the top of a tent pole to the ground. If the rope makes an angle of 45° with the ground, find the length of the tent pole.

Solution   Assuming that the tent pole forms an angle of 90° with the ground, the triangle formed by the rope, tent pole, and the ground is a 45°–45°–90° triangle.

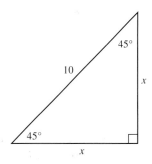

Figure 14

If we let x represent the length of the tent pole, then the length of the rope, in terms of x, is $x\sqrt{2}$. It is also given as 10 feet. Therefore

$$x\sqrt{2} = 10$$
$$x = \frac{10}{\sqrt{2}}$$
$$= 5\sqrt{2}$$

The length of the tent pole is $5\sqrt{2}$ feet. Again, $5\sqrt{2}$ is the exact value of the length of the tent pole. To find a decimal approximation, we replace $\sqrt{2}$ with 1.414 to obtain

$$5\sqrt{2} \approx 5(1.414) = 7.07 \text{ feet}$$ ◀

PROBLEM SET 1.1

Indicate which of the angles below are acute angles and which are obtuse angles. Then give the complement and the supplement of each angle.

1. 10°   2. 50°   3. 45°   4. 90°
5. 120°   6. 160°   7. x 8. y

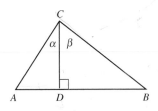

Figure 15

Problems 9 through 14 refer to Figure 15. (*Remember:* The sum of the three angles in any triangle is always 180°.)

9. Find α if $A = 30°$.

10. Find B if $\beta = 45°$.

11. Find α if $A = \alpha$.

12. Find α if $A = 2\alpha$.

13. Find A if $B = 30°$ and $\alpha + \beta = 100°$.

14. Find B if $\alpha + \beta = 80°$ and $A = 80°$.

Figure 16 shows a walkway with a handrail. Angle α is the angle between the walkway and the horizontal, while angle β is the angle between the vertical posts of the handrail and the walkway. Use Figure 16 to work Problems 15 through 18. (Assume that the vertical posts are perpendicular to the horizontal.)

15. Are angles α and β complementary or supplementary angles?

16. If we did not know that the vertical posts were perpendicular to the horizontal, could we answer Problem 15?

17. Find β if $\alpha = 25°$.

18. Find α if $\beta = 52°$.

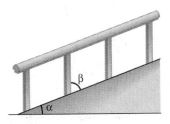

Figure 16

19. The red light on the top of a police car rotates through one complete revolution every 2 seconds. Through how many degrees does it rotate in 1 second?

20. A searchlight rotates through one complete revolution every 4 seconds. How long does it take the light to rotate through 90°?

21. Through how many degrees does the hour hand of a clock move in 4 hours?

22. It takes the earth 24 hours to make one complete revolution on its axis. Through how many degrees does the earth turn in 12 hours?

23. An equilateral triangle is a triangle in which all three sides are equal. What is the measure of each angle in an equilateral triangle?

24. An isosceles triangle is a triangle in which two sides are equal in length. The angle between the two equal sides is called the vertex angle, while the other two angles are called the base angles. If the vertex angle is 40°, what is the measure of the base angles?

Problems 25 through 30 refer to right triangle ABC with $C = 90°$.

25. If $a = 4$ and $b = 3$, find c.

26. If $a = 6$ and $b = 8$, find c.

27. If $a = 8$ and $c = 17$, find b.

28. If $a = 2$ and $c = 6$, find b.

29. If $b = 12$ and $c = 13$, find a.

30. If $b = 10$ and $c = 26$, find a.

Solve for x in each of the following right triangles:

31.

32.

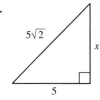

33.

34.

35.

36.

Problems 37 and 38 refer to Figure 17.

37. Find AB if $BC = 4$, $BD = 5$, and $AD = 2$.
38. Find BD if $BC = 5$, $AB = 13$, and $AD = 4$.

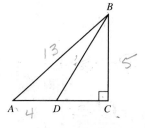

Figure 17

Problems 39 and 40 refer to Figure 18, which shows a circle with center at C and a radius of r, and right triangle ADC.

39. Find r if $AB = 4$ and $AD = 8$.
40. Find r if $AB = 8$ and $AD = 12$.

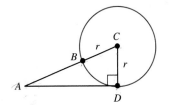

Figure 18

41. The roof of a house is to extend up 13.5 ft above the ceiling, which is 36 ft across. Find the length of one side of the roof.

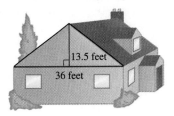

Figure 19

42. A surveyor is attempting to find the distance across a pond. From a point on one side of the pond he walks 25 yd to the end of the pond and then makes a 90° turn and walks another 60 yd before coming to a point directly across the pond from the point at which he started. What is the distance across the pond? (See Figure 20.)

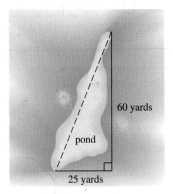

Figure 20

Find the remaining sides of a $30°–60°–90°$ triangle if

43. the shortest side is 1.
44. the shortest side is 3.
45. the longest side is 8.
46. the longest side is 5.
47. the side opposite $60°$ is 6.
48. the side opposite $60°$ is 4.

49. An escalator in a department store is to carry people a vertical distance of 20 ft between floors. How long is the escalator if it makes an angle of $30°$ with the ground?

50. What is the length of the escalator in Problem 49 if it makes an angle of $60°$ with the ground?

51. A two-person tent is to be made so that the height at the center is 4 ft. If the sides of the tent are to meet the ground at an angle of $60°$, and the tent is to be 6 ft in length, how many square feet of material will be needed to make the tent? (Assume that the tent has a floor and is closed at both ends, and give your answer to the nearest tenth of a square foot.)

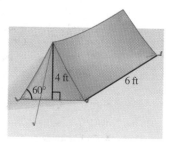

Figure 21

52. If the height at the center of the tent in Problem 51 is to be 3 ft, how many square feet of material will be needed to make the tent?

Find the remaining sides of a 45°–45°–90° triangle if

53. the shorter sides are each $\frac{4}{5}$.

54. the shorter sides are each $\frac{1}{2}$.

55. the longest side is $8\sqrt{2}$.

56. the longest side is $5\sqrt{2}$.

57. the longest side is 4.

58. the longest side is 12.

59. A bullet is fired into the air at an angle of 45°. How far does it travel before it is 1,000 ft above the ground? (Assume that the bullet travels in a straight line, neglect the forces of gravity, and give your answer to the nearest foot.)

60. If the bullet in Problem 59 is traveling at 2,828 ft/sec, how long does it take for the bullet to reach a height of 1,000 ft?

The object shown in Figure 22 is a cube (all edges are equal in length).

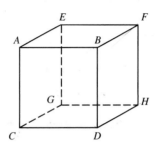

Figure 22

61. If the length of each edge of the cube shown in Figure 22 is 1 inch, find
a. the length of diagonal *CH*.
b. the length of diagonal *CF*.

62. If the length of each edge of the cube shown in Figure 22 is 5 cm, find
a. the length of diagonal *GD*.
b. the length of diagonal *GB*.

63. If the length of each edge of the cube shown in Figure 22 is unknown, we can represent it with the variable *x*. Then we can write formulas for the lengths of any of the diagonals. Finish each of the following statements:
a. If the length of each edge of a cube is *x*, then the length of the diagonal of any face of the cube will be ___.
b. If the length of each edge of a cube is *x*, then the length of any diagonal that passes through the center of the cube will be ___.

64. What is the measure of ∠*GDH* in Figure 22?

65. Suppose the length of diagonal *CF* in Figure 22 is 3 ft. What is the length of edge *CD*?

66. Suppose the length of diagonal *CF* in Figure 22 is $5\sqrt{3}$ ft. What is the length of edge *CD*?

Research Projects

67. Although Pythagoras preceded William Shakespeare by 2,000 years, the philosophy of the Pythagoreans is mentioned in Shakespeare's *The Merchant of*

Venice. Here is a quote from that play:

Thou almost mak'st me waver in my faith,
To hold opinion with Pythagoras,
That souls of animals infuse themselves
Into the trunks of men.

What part of the philosophy of the Pythagoreans was Shakespeare referring to with this quote? What present-day religions share a similar belief?

68. Copy Figure 23 below and use it to derive the Pythagorean Theorem. The right triangle with sides *a*, *b*, and *c* has been repeated four times. The area of the large square is equal to the sum of the areas of the four triangles and the area of the small square in the center. The key to this derivation is in finding the length of a side in the small square.

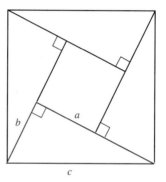

Figure 23

The Rectangular Coordinate System

The book *The Closing of the American Mind* by Allan Bloom was published in 1987 and spent many weeks on the bestseller list. In the book, Mr. Bloom recalls being in a restaurant in France and overhearing a waiter call another waiter a "Cartesian." He goes on to say that French people today define themselves in terms of the philosophy of either René Descartes (1595–1650) or Blaise Pascal (1623–1662). Followers of Descartes are sometimes referred to as *Cartesians.* As a philosopher, Descartes is responsible for the statement "I think, therefore I am." In mathematics, Descartes is credited with, among other things, the invention of the rectangular coordinate system, which we sometimes call the *Cartesian coordinate system.* Until Descartes invented his coordinate system in 1637, algebra and geometry were treated as separate subjects. The rectangular coordinate system allows us to connect algebra and geometry by associating geometric shapes with algebraic equations. For example, every nonvertical straight line (a geometric concept) can be paired with an equation of the form $y = mx + b$ (an algebraic concept), where m and b are real numbers, and x and y are variables that we associate with the axes of a coordinate system. In this section we will review some of the concepts developed around the rectangular coordinate system and graphing in two dimensions.

The rectangular (or Cartesian) coordinate system is constructed by drawing two number lines perpendicular to each other.

The horizontal number line is called the *x-axis,* and the vertical number line is called the *y-axis.* Their point of intersection is called the *origin.* The axes divide the plane into four *quadrants* that are numbered I through IV in a counterclockwise direction (Figure 1).

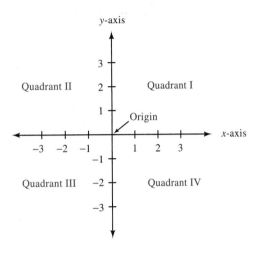

Figure 1

Among other things, the rectangular coordinate system is used to graph ordered pairs (a, b). For the ordered pair (a, b), a is called the *x-coordinate* (or *x*-component), and b is called the *y-coordinate* (or *y*-component). To graph the ordered pair (a, b) on a rectangular coordinate system, we start at the origin and move a units to the right or left (right if a is positive and left if a is negative). We then move b units up or down (up if b is positive and down if b is negative). The point where we end up after these two moves is the graph of the ordered pair (a, b).

▶ **EXAMPLE 1**   Graph the ordered pairs $(1, 5)$, $(-2, 4)$, $(-3, -2)$, $(\frac{1}{2}, -4)$, $(3, 0)$, and $(0, -2)$.

Solution   To graph the ordered pair $(1, 5)$, we start at the origin and move 1 unit to the right and then 5 units up. We are now at the point whose coordinates are $(1, 5)$. The other ordered pairs are graphed in a similar manner, as in Figure 2.

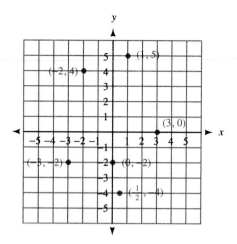

Figure 2

Looking at Figure 2, we see that any point in quadrant I will have both coordinates positive; that is, $(+, +)$. In quadrant II, the form is $(-, +)$. In quadrant III, the form is $(-, -)$, and in quadrant IV it is $(+, -)$. Also, any point on the x-axis will have a y-coordinate of 0 (it has no vertical displacement), and any point on the y-axis will have an x-coordinate of 0 (no horizontal displacement). A summary of this is given in Figure 3.

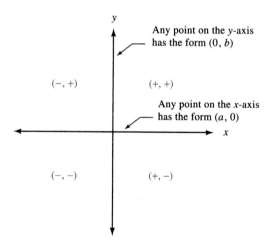

Figure 3

▶ **EXAMPLE 2**   Graph the line $2x - 3y = 6$.

Solution   Since a line is determined by two points, we can graph the line $2x - 3y = 6$ by first finding two ordered pairs that satisfy its equation. To find ordered pairs that satisfy the equation $2x - 3y = 6$, we substitute any convenient number for either variable and then solve the equation that results for the corresponding value of the other variable.

If we let	$x = 0$	If	$y = 0$

If we let $\qquad x = 0$	If $\qquad y = 0$
the equation $\quad 2x - 3y = 6$	the equation $\quad 2x - 3y = 6$
becomes $\quad 2(0) - 3y = 6$	becomes $\quad 2x - 3(0) = 6$
$-3y = 6$	$2x = 6$
$y = -2$	$x = 3$
This gives us $(0, -2)$ as one solution to $2x - 3y = 6$.	This gives us $(3, 0)$ as a second solution.

Graphing the points $(0, -2)$ and $(3, 0)$ and then drawing a line through them, we have the graph of $2x - 3y = 6$ (Figure 4).

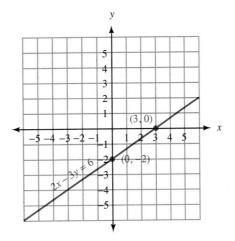

Figure 4

The x-coordinate of the point where our graph crosses the x-axis is called the x-intercept; in this case, it is 3. The y-coordinate of the point where the graph crosses the y-axis is the y-intercept; in this case, -2.   ◀

Example 2 illustrates the connection between algebra and geometry that we mentioned in the introduction to this section. The rectangular coordinate system allows us to associate the equation $2x - 3y = 6$ (an algebraic concept) with a specific straight line (a geometric concept). The study of the relationship between equations in algebra and their associated geometric figures is called *analytic geometry* and is based on the coordinate system credited to Descartes. As you know, not all equations have straight lines for graphs. For example, you may recall from algebra that the graph of an equation of the form

$$x^2 + y^2 = r^2$$

will be a circle with its center at the origin and a radius of r.

▶ **EXAMPLE 3**   Graph each of the following circles:

a. $x^2 + y^2 = 16$ **b.** $x^2 + y^2 = 1$

Solution   The graph of each equation will be a circle with its center at the origin. The first circle will have a radius of 4, and the second one will have a radius of 1. The graphs are shown in Figures 5 and 6 below.

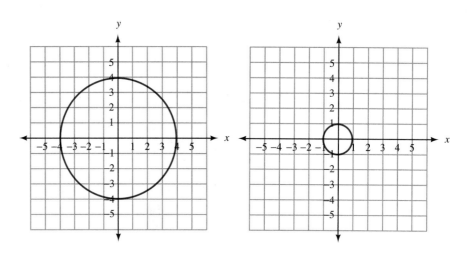

Figure 5   **Figure 6**

The circle shown in Figure 6 is called the *unit circle* because its radius is 1. As you will see, it will be an important part of one of the definitions we will give in Chapter 3.   ◀

Our next definition gives us a formula for finding the distance between any two points on the coordinate system.

The Distance Formula

The distance between any two points (x_1, y_1) and (x_2, y_2) in a rectangular coordinate system is given by the formula

$$r = \sqrt{(x_2 - x_1)^2 + (y_2 - y_1)^2}$$

The distance formula can be derived by applying the Pythagorean Theorem to the right triangle in Figure 7.

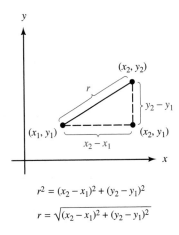

$$r^2 = (x_2 - x_1)^2 + (y_2 - y_1)^2$$

$$r = \sqrt{(x_2 - x_1)^2 + (y_2 - y_1)^2}$$

Figure 7

▶ **EXAMPLE 4**   Find the distance between the points $(2, 1)$ and $(-1, 5)$.

Solution   It makes no difference which of the points we call (x_1, y_1) and which we call (x_2, y_2) because this distance will be the same between the two points regardless.

$$r = \sqrt{(2 + 1)^2 + (1 - 5)^2}$$
$$= \sqrt{3^2 + (-4)^2}$$
$$= \sqrt{9 + 16}$$
$$= \sqrt{25}$$
$$= 5$$

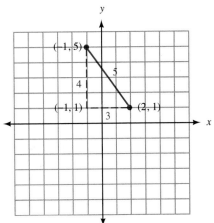

Figure 8   ◀

▶ **EXAMPLE 5**   Find the distance from the origin to the point (x, y).

Solution   The coordinates of the origin are $(0, 0)$. Applying the distance formula, we have

$$r = \sqrt{(x - 0)^2 + (y - 0)^2}$$
$$= \sqrt{x^2 + y^2}$$

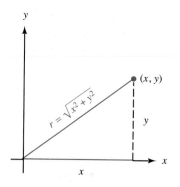

Figure 9

Angles in Standard Position

DEFINITION   An angle is said to be in *standard position* if its initial side is along the positive *x*-axis and its vertex is at the origin.

▶ **EXAMPLE 6**   Draw an angle of 45° in standard position and find a point on the terminal side.

Solution   If we draw 45° in standard position, we see that the terminal side is along the line $y = x$ in quadrant I (Figure 10).

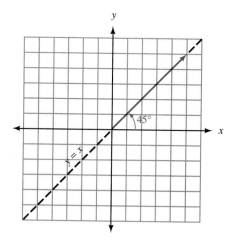

Figure 10

Since the terminal side of $45°$ lies along the line $y = x$ in the first quadrant, any point on the terminal side will have positive coordinates that satisfy the equation $y = x$. Here are some of the points that do just that.

$$(1, 1)\quad (2, 2)\quad (3, 3)\quad (\sqrt{2}, \sqrt{2})\quad \left(\frac{1}{2}, \frac{1}{2}\right)\quad \left(\frac{7}{8}, \frac{7}{8}\right)$$ ◀

Vocabulary

If angle θ is in standard position and the terminal side of θ lies in quadrant I, then we say θ lies in quadrant I and we abbreviate it like this:

$$\theta \in \text{QI}$$

Likewise, $\theta \in \text{QII}$ means θ is in standard position with its terminal side in quadrant II.

If the terminal side of an angle in standard position lies along one of the axes, then that angle is called a *quadrantal angle*. For example, an angle of $90°$ drawn in standard position would be a quadrantal angle, since the terminal side would lie along the positive y-axis. Likewise, $270°$ in standard position is a quadrantal angle because the terminal side would lie along the negative y-axis (Figure 11).

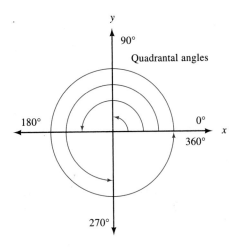

Figure 11

Two angles in standard position with the same terminal side are called *coterminal angles.* Figure 12 shows that 60° and −300° are coterminal angles when they are in standard position.

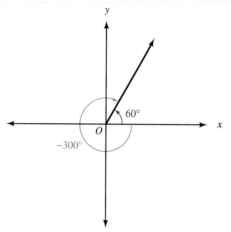

Figure 12

▶ **EXAMPLE 7**   Draw −90° in standard position, name some of the points on the terminal side, and find an angle between 0° and 360° that is coterminal with −90°.

Solution   Figure 13 shows −90° in standard position. Since the terminal side is along the negative y-axis, the points $(0, -1)$, $(0, -2)$, $(0, -\frac{5}{2})$, and, in general, $(0, b)$ where b is a negative number, are all on the terminal side. The angle between 0° and 360° that is coterminal with −90° is 270°.

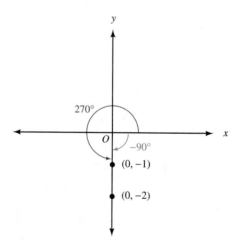

Figure 13

PROBLEM SET 1.2

Graph each of the following ordered pairs on a rectangular coordinate system:

1. $(2, 4)$ **2.** $(2, -4)$ **3.** $(-2, 4)$ **4.** $(-2, -4)$

5. $(-3, -4)$ **6.** $(-4, -2)$ **7.** $(5, 0)$ **8.** $(-3, 0)$

9. $(0, -3)$ **10.** $(0, 2)$ **11.** $\left(-5, \dfrac{1}{2}\right)$ **12.** $\left(5, \dfrac{1}{2}\right)$

Graph each of the following lines:

13. $3x + 2y = 6$ **14.** $3x - 2y = 6$

15. $y = 2x - 1$ **16.** $y = 2x + 3$

17. $y = \dfrac{1}{2}x$ **18.** $y = -\dfrac{1}{2}x$

19. $x = 3$ **20.** $y = -2$

Graph each of the following circles:

21. $x^2 + y^2 = 25$
22. $x^2 + y^2 = 36$
23. $x^2 + y^2 = 5$
24. $x^2 + y^2 = 6$

25. Use the graph of Problem 21 to name the points at which the line $x + y = 5$ will intersect the circle $x^2 + y^2 = 25$.

26. Use the graph of Problem 22 to name the points at which the line $x - y = 6$ will intersect the circle $x^2 + y^2 = 36$.

27. At what points will the line $y = x$ intersect the unit circle $x^2 + y^2 = 1$?

28. At what points will the line $y = -x$ intersect the unit circle $x^2 + y^2 = 1$?

Find the distance between the following points:

29. $(3, 7), (6, 3)$ **30.** $(4, 7), (8, 1)$

31. $(0, 12), (5, 0)$ **32.** $(-3, 0), (0, 4)$

33. $(3, -5), (-2, 1)$ **34.** $(-8, 9), (-3, -2)$

35. $(-1, -2), (-10, 5)$ **36.** $(-3, 8), (-1, 6)$

37. Find the distance from the origin out to the point $(3, -4)$.

38. Find the distance from the origin out to the point $(12, -5)$.

39. Find x so the distance between $(x, 2)$ and $(1, 5)$ is $\sqrt{13}$.

40. Find y so the distance between $(7, y)$ and $(3, 3)$ is 5.

41. An airplane is approaching Los Angeles International Airport at an altitude of 2,640 ft. If the horizontal distance from the plane to the runway is 1.2 mi, use the

Pythagorean Theorem to find the diagonal distance from the plane to the runway. (5,280 ft equals 1 mi.)

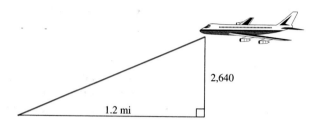

Figure 14

42. In softball, the distance from home plate to first base is 60 ft, as is the distance from first base to second base. If the lines joining home plate to first base and first base to second base form a right angle, how far does a catcher standing on home plate have to throw the ball so that it reaches the shortstop standing on second base?

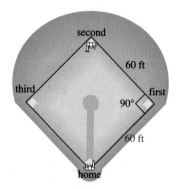

Figure 15

43. If a coordinate system is superimposed on the softball diamond in Problem 42 with the x-axis along the line from home plate to first base and the y-axis on the line from home plate to third base, what would be the coordinates of home plate, first base, second base, and third base?
44. If a coordinate system is superimposed on the softball diamond in Problem 42 with the origin on home plate and the positive x-axis along the line joining home plate to second base, what would be the coordinates of first base and third base?
45. In what two quadrants do all the points have negative x-coordinates?
46. In what two quadrants do all the points have negative y-coordinates?
47. In what two quadrants do all the points have positive y-coordinates?
48. In what two quadrants do all the points have positive x-coordinates?

49. For points (x, y) in quadrant I, the ratio x/y is always positive because x and y are always positive. In what other quadrant is the ratio x/y always positive?

50. For points (x, y) in quadrant II, the ratio x/y is always negative because x is negative and y is positive in quadrant II. In what other quadrant is the ratio x/y always negative?

Draw each of the following angles in standard position and then do the following:
a. Name a point on the terminal side of the angle.
b. Find the distance from the origin to that point.
c. Name another angle that is coterminal with the angle you have drawn.

51. $135°$	**52.** $45°$	**53.** $225°$	**54.** $315°$
55. $90°$	**56.** $360°$	**57.** $-45°$	**58.** $-90°$

59. Draw an angle of $30°$ in standard position. Then find a if the point $(a, 1)$ is on the terminal side of $30°$.

60. Draw $60°$ in standard position. Then find b if the point $(2, b)$ is on the terminal side of $60°$.

61. Draw an angle in standard position whose terminal side contains the point $(3, -2)$. Find the distance from the origin to this point.

62. Draw an angle in standard position whose terminal side contains the point $(2, -3)$. Find the distance from the origin to this point.

63. Plot the points $(0, 0)$, $(5, 0)$, and $(5, 12)$ and show that, when connected, they are the vertices of a right triangle.

64. Plot the points $(0, 2)$, $(-3, 2)$, and $(-3, -2)$ and show that they form the vertices of a right triangle.

Research Projects

65. In the introduction to this section we mentioned two French philosophers, Descartes and Pascal. Many people see the philosophies of the two men as being opposites. Why is this?

66. Pascal has a triangular array of numbers named after him, Pascal's triangle. What part does Pascal's triangle play in the expansion of $(a + b)^n$, where n is a positive integer?

SECTION 1.3 # Definition I: Trigonometric Functions

In this section we begin our work with trigonometry. The formal study of trigonometry dates back to the Greeks, when it was used mainly in the design of clocks and calendars and in navigation. The trigonometry of that period was spherical in nature, as it was based on measurement of arcs and chords associated with spheres. Unlike the trigonometry of the Greeks, our introduction to trigonometry takes place on a rectangular coordinate system. It concerns itself with angles, line segments, and points in the plane.

The definition of the trigonometric functions that begins this section is one of three definitions we will use. For us, it is the most important definition in the book. What should you do with it? Memorize it. Remember, in mathematics, definitions are simply accepted. That is, unlike theorems, there is no proof associated with a definition; we simply accept them exactly as they are written, memorize them, and then use them. When you are finished with this section, be sure you have memorized this first definition. It is the most valuable thing you can do for yourself at this point in your study of trigonometry.

DEFINITION I   If θ is an angle in standard position, and the point (x, y) is any point on the terminal side of θ other than the origin, then the six trigonometric functions of angle θ are defined as follows:

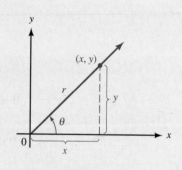

Figure 1

Function		Abbreviation		Definition
The sine of θ	$=$	$\sin \theta$	$=$	$\dfrac{y}{r}$
The cosine of θ	$=$	$\cos \theta$	$=$	$\dfrac{x}{r}$
The tangent of θ	$=$	$\tan \theta$	$=$	$\dfrac{y}{x}$
The cotangent of θ	$=$	$\cot \theta$	$=$	$\dfrac{x}{y}$
The secant of θ	$=$	$\sec \theta$	$=$	$\dfrac{r}{x}$
The cosecant of θ	$=$	$\csc \theta$	$=$	$\dfrac{r}{y}$

Where $x^2 + y^2 = r^2$, or $r = \sqrt{x^2 + y^2}$. That is, r is the distance from the origin to (x, y).

As you can see, the six trigonometric functions are simply names given to the six possible ratios that can be made from the numbers x, y, and r as shown in Figure 1.

▶ **EXAMPLE 1**   Find the six trigonometric functions of θ if θ is in standard position and the point $(-2, 3)$ is on the terminal side of θ.

Solution   We begin by making a diagram showing θ, $(-2, 3)$, and the distance r from the origin to $(-2, 3)$, as shown in Figure 2.

Applying the definition for the six trigonometric functions using the values $x = -2$, $y = 3$, and $r = \sqrt{13}$, we have

$$\sin \theta = \frac{y}{r} = \frac{3}{\sqrt{13}} \qquad \csc \theta = \frac{r}{y} = \frac{\sqrt{13}}{3}$$

$$\cos \theta = \frac{x}{r} = -\frac{2}{\sqrt{13}} \qquad \sec \theta = \frac{r}{x} = -\frac{\sqrt{13}}{2}$$

$$\tan \theta = \frac{y}{x} = -\frac{3}{2} \qquad \cot \theta = \frac{x}{y} = -\frac{2}{3}$$

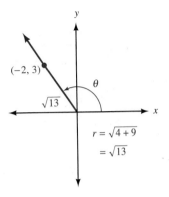

Figure 2   ◀

Note   In algebra, when we encounter expressions like $3/\sqrt{13}$ that contain a radical in the denominator, we usually rationalize the denominator; in this case, by multiplying the numerator and the denominator by $\sqrt{13}$.

$$\frac{3}{\sqrt{13}} = \frac{3}{\sqrt{13}} \cdot \frac{\sqrt{13}}{\sqrt{13}} = \frac{3\sqrt{13}}{13}$$

In trigonometry, it is sometimes convenient to use $3\sqrt{13}/13$, and at other times it is easier to use $3/\sqrt{13}$. For now, let's agree not to rationalize any denominators unless we are told to do so.

Calculator Note   You can use a calculator to see the equivalence of $3\sqrt{13}/13$ and $3/\sqrt{13}$. Here is the sequence of keys to press on a basic scientific calculator to obtain decimal approximations for each expression.

For $\dfrac{3}{\sqrt{13}}$:   3 ÷ 13 √ =

For $\dfrac{3\sqrt{13}}{13}$:   3 × 13 √ ÷ 13 =

In each case, the result should be 0.832050294. Note that the sequence of keys you would press on a more advanced graphing calculator may differ from the ones shown here. The calculator notes in this book are for basic scientific calculators using algebraic logic. If your calculator has a key labeled $\boxed{=}$ then these calculator notes apply to you.

▶ **EXAMPLE 2**   Find the sine and cosine of 45°.

Solution   According to the definition given earlier, we can find sin 45° and cos 45° if we know a point (x, y) on the terminal side of 45°, when 45° is in standard position. Figure 3 is a diagram of 45° in standard position. It shows that a convenient point on the terminal side of 45° is the point (1, 1). (We say it is a convenient point because the coordinates are easy to work with.)

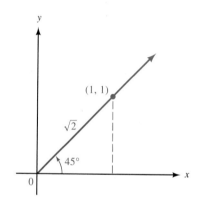

Figure 3

Since $x = 1$ and $y = 1$ and $r = \sqrt{x^2 + y^2}$, we have
$$r = \sqrt{1^2 + 1^2} = \sqrt{2}$$

Substituting these values for x, y, and r into our definition for sine and cosine, we have

$$\sin 45° = \frac{y}{r} = \frac{1}{\sqrt{2}} \qquad \text{and} \qquad \cos 45° = \frac{x}{r} = \frac{1}{\sqrt{2}}$$ ◀

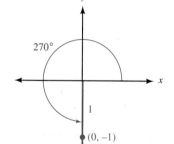

Figure 4

▶ **EXAMPLE 3**   Find the six trigonometric functions of 270°.

Solution   Again, we need to find a point on the terminal side of 270°. From Figure 4, we see that the terminal side of 270° lies along the negative y-axis.

A convenient point on the terminal side of 270° is $(0, -1)$. Therefore,
$$r = \sqrt{x^2 + y^2} = \sqrt{0^2 + (-1)^2} = \sqrt{1} = 1$$

We have $x = 0$, $y = -1$, and $r = 1$. Here are the six trigonometric ratios for $\theta = 270°$.

$$\sin 270° = \frac{y}{r} = \frac{-1}{1} = -1 \qquad\qquad \csc 270° = \frac{r}{y} = \frac{1}{-1} = -1$$

$$\cos 270° = \frac{x}{r} = \frac{0}{1} = 0 \qquad\qquad \sec 270° = \frac{r}{x} = \frac{1}{0} = \text{undefined}$$

$$\tan 270° = \frac{y}{x} = \frac{-1}{0} = \text{undefined} \qquad \cot 270° = \frac{x}{y} = \frac{0}{-1} = 0$$

Note that $\tan 270°$ and $\sec 270°$ are undefined since division by 0 is undefined.

◀

Algebraic Signs of Trigonometric Functions

The algebraic sign, $+$ or $-$, of each of the six trigonometric functions will depend on the quadrant in which θ terminates. For example, in quadrant I all six trigonometric functions are positive since x, y, and r are all positive. In quadrant II, only $\sin \theta$ and $\csc \theta$ are positive since y and r are positive and x is negative. Table 1 shows the signs of all the ratios in each of the four quadrants.

Table 1

For θ in	QI	QII	QIII	QIV
$\sin \theta = \dfrac{y}{r}$ and $\csc \theta = \dfrac{r}{y}$	+	+	−	−
$\cos \theta = \dfrac{x}{r}$ and $\sec \theta = \dfrac{r}{x}$	+	−	−	+
$\tan \theta = \dfrac{y}{x}$ and $\cot \theta = \dfrac{x}{y}$	+	−	+	−

▶ **EXAMPLE 4** If $\sin \theta = -5/13$, and θ terminates in quadrant III, find $\cos \theta$ and $\tan \theta$.

Solution Since $\sin \theta = -5/13$, we know the ratio of y to r, or y/r, is $-5/13$. We can let y be -5 and r be 13 and use these values of y and r to find x.

Note We are not saying that if $y/r = -5/13$, then y *must* be -5 and r *must* be 13. We know from algebra that there are many pairs of numbers whose ratio is $-5/13$, not just -5 and 13. Our definition for sine and cosine, however, indicates we can choose *any* point on the terminal side of θ to find $\sin \theta$ and $\cos \theta$.

To find x, we use the fact that $x^2 + y^2 = r^2$.

$$x^2 + y^2 = r^2$$
$$x^2 + (-5)^2 = 13^2$$
$$x^2 + 25 = 169$$
$$x^2 = 144$$
$$x = \pm 12$$

Is x the number 12 or -12?

Since θ terminates in quadrant III, we know any point on its terminal side will have a negative x-coordinate; therefore,

$$x = -12$$

Using $x = -12$, $y = -5$, and $r = 13$ in our original definition, we have

$$\cos \theta = \frac{x}{r} = \frac{-12}{13} = -\frac{12}{13}$$

and

$$\tan \theta = \frac{y}{x} = \frac{-5}{-12} = \frac{5}{12}$$

◀

As a final note, we should mention that the trigonometric functions of an angle are independent of the choice of the point (x, y) on the terminal side of the angle. Figure 5 shows an angle θ in standard position.

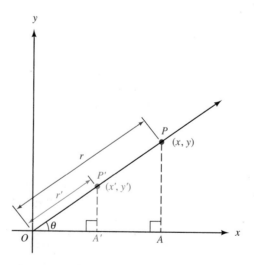

Figure 5

Points $P(x, y)$ and $P'(x', y')$ are both points on the terminal side of θ. Since triangles $P'OA'$ and POA are similar triangles, their corresponding sides are proportional. That is,

$$\sin \theta = \frac{y'}{r'} = \frac{y}{r} \qquad \cos \theta = \frac{x'}{r'} = \frac{x}{r} \qquad \tan \theta = \frac{y'}{x'} = \frac{y}{x}$$

PROBLEM SET 1.3

Find all six trigonometric functions of θ if the given point is on the terminal side of θ.

1. $(3, 4)$	**2.** $(-3, -4)$	**3.** $(-3, 4)$	**4.** $(3, -4)$
5. $(-5, 12)$	**6.** $(-12, 5)$	**7.** $(-1, -2)$	**8.** $(1, 2)$
9. (a, b)	**10.** (m, n)	**11.** $(-3, 0)$	**12.** $(0, -5)$
13. $(\sqrt{3}, -1)$	**14.** $(\sqrt{2}, \sqrt{2})$	**15.** $(-\sqrt{5}, 2)$	**16.** $(-3, \sqrt{7})$
17. $(60, 80)$	**18.** $(-80, 60)$	**19.** $(5a, -12a)$	**20.** $(-9a, 12a)$

21. Use your calculator to find $\sin \theta$ and $\cos \theta$ if the point $(9.36, 7.02)$ is on the terminal side of θ.

22. Use your calculator to find $\sin \theta$ and $\cos \theta$ if the point $(6.36, 2.65)$ is on the terminal side of θ.

Draw each of the following angles in standard position, find a point on the terminal side, and then find the sine, cosine, and tangent of each angle:

23. $135°$	**24.** $225°$	**25.** $90°$	**26.** $180°$
27. $-45°$	**28.** $-90°$	**29.** $0°$	**30.** $-135°$

Indicate the quadrants in which the terminal side of θ must lie in order that

31. $\cos \theta$ is positive	**32.** $\cos \theta$ is negative
33. $\sin \theta$ is negative	**34.** $\sin \theta$ is positive
35. $\tan \theta$ is positive	**36.** $\tan \theta$ is negative
37. $\sin \theta$ is negative and $\tan \theta$ is positive	
38. $\sin \theta$ is positive and $\cos \theta$ is negative	

39. In which quadrant must the terminal side of θ lie if $\sin \theta$ and $\tan \theta$ are to have the same sign?

40. In which quadrant must the terminal side of θ lie if $\cos \theta$ and $\cot \theta$ are to have the same sign?

Find the remaining trigonometric functions of θ if

41. $\sin \theta = 12/13$ and θ terminates in QI

42. $\sin \theta = 12/13$ and θ terminates in QII

43. $\sin \theta = 8/17$ and θ terminates in QI

44. $\sin \theta = 15/17$ and θ terminates in QI

45. $\cos \theta = 24/25$ and θ terminates in QIV

46. $\cos \theta = 7/25$ and θ terminates in QIV

47. $\tan \theta = 3/4$ and θ terminates in QI

48. $\tan \theta = 12/5$ and θ terminates in QI

49. $\sin \theta = -20/29$ and θ terminates in QIII
50. $\cos \theta = -20/29$ and θ terminates in QII
51. $\sec \theta = 13/5$ and $\sin \theta < 0$
52. $\csc \theta = 13/5$ and $\cos \theta < 0$
53. $\cot \theta = 1/2$ and $\cos \theta > 0$
54. $\tan \theta = -1/2$ and $\sin \theta > 0$
55. $\cos \theta = \dfrac{\sqrt{3}}{2}$ and θ terminates in QIV
56. $\cos \theta = \dfrac{\sqrt{2}}{2}$ and θ terminates in QI
57. $\tan \theta = 2$ and θ terminates in QI
58. $\cot \theta = -2$ and θ terminates in QII
59. $\tan \theta = \dfrac{a}{b}$ where a and b are both positive
60. $\cot \theta = \dfrac{m}{n}$ where m and n are both positive
61. Find an angle θ for which $\sin \theta = 1$. (Look for an angle between $0°$ and $180°$.)
62. Find an angle θ in the first quadrant for which $\tan \theta = 1$.
63. Find an angle θ in the third quadrant for which $\tan \theta = 1$.
64. Find an angle θ between $0°$ and $360°$ for which $\cos \theta = -1$.
65. Find $\sin \theta$ and $\cos \theta$ if the terminal side of θ lies along the line $y = 2x$ in quadrant I.
66. Find $\sin \theta$ and $\cos \theta$ if the terminal side of θ lies along the line $y = 2x$ in quadrant III.
67. Find $\sin \theta$ and $\tan \theta$ if the terminal side of θ lies along the line $y = -3x$ in quadrant II.
68. Find $\sin \theta$ and $\tan \theta$ if the terminal side of θ lies along the line $y = -3x$ in quadrant IV.
69. Draw $45°$ and $-45°$ in standard position and then show that $\cos(-45°) = \cos 45°$.
70. Draw $45°$ and $-45°$ in standard position and then show that $\sin(-45°) = -\sin 45°$.
71. Find x if the point $(x, -3)$ is on the terminal side of θ and $\sin \theta = -3/5$.
72. Find y if the point $(5, y)$ is on the terminal side of θ and $\cos \theta = 5/13$.

SECTION 1.4   # Introduction to Identities

You may recall from the work you have done in algebra that an expression such as $\sqrt{x^2 + 9}$ cannot be simplified further because the square root of a sum is not equal to the sum of the square roots. (In other words, it would be a mistake to write $\sqrt{x^2 + 9}$ as $x + 3$.) However, expressions such as $\sqrt{x^2 + 9}$ occur frequently enough in mathematics that we would like to find expressions equivalent to them that do not contain square roots. As it turns out, the relationships that we develop in this section and the next are the key to rewriting expressions such as $\sqrt{x^2 + 9}$ in a more convenient form, which we will show in Section 1.5.

Before we begin our introduction to identities, we need to review some concepts from arithmetic and algebra.

In algebra, statements such as $2x = x + x$, $x^3 = x \cdot x \cdot x$, and $x/4x = \frac{1}{4}$ are called identities. They are identities because they are true for all replacements of the variable for which they are defined.

Note $x/4x$ is not equal to $\frac{1}{4}$ when x is 0. The statement $x/4x = \frac{1}{4}$ is still an identity, however, since it is true for all values of x for which $x/4x$ is defined.

The eight basic trigonometric identities we will work with in this section are all derived from our definition of the trigonometric functions. Since many trigonometric identities have more than one form, we will list the basic identity first and then give the most common equivalent forms of that identity.

Reciprocal Identities

Our definition for the sine and cosecant functions indicates that they are reciprocals; that is,

$$\csc \theta = \frac{1}{\sin \theta} \quad \text{because} \quad \frac{1}{\sin \theta} = \frac{1}{y/r} = \frac{r}{y} = \csc \theta$$

Note that we can also write this same relationship between $\sin \theta$ and $\csc \theta$ in another form as

$$\sin \theta = \frac{1}{\csc \theta} \quad \text{because} \quad \frac{1}{\csc \theta} = \frac{1}{r/y} = \frac{y}{r} = \sin \theta$$

The first identity we wrote, $\csc \theta = 1/\sin \theta$, is the basic identity. The second one, $\sin \theta = 1/\csc \theta$, is an equivalent form of the first.

From the discussion above and from the definition of $\cos \theta$, $\sec \theta$, $\tan \theta$, and $\cot \theta$, it is apparent that $\sec \theta$ is the reciprocal of $\cos \theta$, and $\cot \theta$ is the reciprocal of $\tan \theta$. Table 1 lists three basic reciprocal identities and their common equivalent forms.

Table 1

Reciprocal identities	Equivalent forms
$\csc \theta = \dfrac{1}{\sin \theta}$	$\sin \theta = \dfrac{1}{\csc \theta}$
$\sec \theta = \dfrac{1}{\cos \theta}$	$\cos \theta = \dfrac{1}{\sec \theta}$
$\cot \theta = \dfrac{1}{\tan \theta}$	$\tan \theta = \dfrac{1}{\cot \theta}$

The examples that follow show some of the ways in which we use these reciprocal identities.

▶ Examples

1. If $\sin \theta = \dfrac{3}{5}$, then $\csc \theta = \dfrac{5}{3}$, because

$$\csc \theta = \frac{1}{\sin \theta} = \frac{1}{3/5} = \frac{5}{3}$$

2. If $\cos \theta = -\dfrac{\sqrt{3}}{2}$, then $\sec \theta = -\dfrac{2}{\sqrt{3}}$.

(*Remember:* Reciprocals always have the same algebraic sign.)

3. If $\tan \theta = 2$, then $\cot \theta = \dfrac{1}{2}$.

4. If $\csc \theta = a$, then $\sin \theta = \dfrac{1}{a}$.

5. If $\sec \theta = 1$, then $\cos \theta = 1$ (1 is its own reciprocal).

6. If $\cot \theta = -1$, then $\tan \theta = -1$.   ◀

Ratio Identities

There are two ratio identities, one for $\tan \theta$ and one for $\cot \theta$.

Table 2

Ratio identities		
$\tan \theta = \dfrac{\sin \theta}{\cos \theta}$	because	$\dfrac{\sin \theta}{\cos \theta} = \dfrac{y/r}{x/r} = \dfrac{y}{x} = \tan \theta$
$\cot \theta = \dfrac{\cos \theta}{\sin \theta}$	because	$\dfrac{\cos \theta}{\sin \theta} = \dfrac{x/r}{y/r} = \dfrac{x}{y} = \cot \theta$

▶ **EXAMPLE 7**   If $\sin \theta = -\frac{3}{5}$ and $\cos \theta = \frac{4}{5}$, find $\tan \theta$ and $\cot \theta$.

Solution   Using the ratio identities, we have

$$\tan \theta = \frac{\sin \theta}{\cos \theta} = \frac{-3/5}{4/5} = -\frac{3}{4}$$

$$\cot \theta = \frac{\cos \theta}{\sin \theta} = \frac{4/5}{-3/5} = -\frac{4}{3}$$

Note that, once we found $\tan \theta$, we could have used a reciprocal identity to find $\cot \theta$.

$$\cot \theta = \frac{1}{\tan \theta} = \frac{1}{-3/4} = -\frac{4}{3}$$ ◀

Notation

The notation $\sin^2 \theta$ is a shorthand notation for $(\sin \theta)^2$. It indicates we are to square the number that is the sine of θ.

▶ **EXAMPLES**

8. If $\sin \theta = \dfrac{3}{5}$, then $\sin^2 \theta = \left(\dfrac{3}{5}\right)^2 = \dfrac{9}{25}$.

9. If $\cos \theta = -\dfrac{1}{2}$, then $\cos^3 \theta = \left(-\dfrac{1}{2}\right)^3 = -\dfrac{1}{8}$. ◀

Pythagorean Identities

To derive our first Pythagorean identity, we start with the relationship between x, y, and r as given in the definition of $\sin \theta$ and $\cos \theta$.

$$x^2 + y^2 = r^2$$

$$\frac{x^2}{r^2} + \frac{y^2}{r^2} = 1 \qquad \text{Divide through by } r^2$$

$$\left(\frac{x}{r}\right)^2 + \left(\frac{y}{r}\right)^2 = 1 \qquad \text{Property of exponents}$$

$$(\cos \theta)^2 + (\sin \theta)^2 = 1 \qquad \text{Definition of } \sin \theta \text{ and } \cos \theta$$

$$\cos^2 \theta + \sin^2 \theta = 1 \qquad \text{Notation}$$

This last line is our first Pythagorean identity. We will use it many times throughout the book. It states that, for any angle θ, the sum of the squares of $\sin \theta$ and $\cos \theta$ is *always* 1.

There are two very useful equivalent forms of the first Pythagorean identity. One form occurs when we solve $\cos^2 \theta + \sin^2 \theta = 1$ for $\cos \theta$, and the other form is the result of solving for $\sin \theta$.

Solving for $\cos \theta$ we have

$$\cos^2 \theta + \sin^2 \theta = 1$$

$$\cos^2 \theta = 1 - \sin^2 \theta \qquad \text{Add } -\sin^2 \theta \text{ to both sides}$$

$$\cos \theta = \pm\sqrt{1 - \sin^2 \theta} \qquad \text{Take the square root of both sides}$$

Similarly, solving for $\sin \theta$ gives us

$$\sin^2 \theta = 1 - \cos^2 \theta \qquad \text{or} \qquad \sin \theta = \pm\sqrt{1 - \cos^2 \theta}$$

Our next Pythagorean identity is derived from the first Pythagorean identity, $\cos^2 \theta + \sin^2 \theta = 1$, by dividing both sides by $\cos^2 \theta$. Here is that derivation:

$$\cos^2 \theta + \sin^2 \theta = 1 \qquad \text{First Pythagorean identity}$$

$$\frac{\cos^2 \theta + \sin^2 \theta}{\cos^2 \theta} = \frac{1}{\cos^2 \theta} \qquad \text{Divide each side by } \cos^2 \theta$$

$$\frac{\cos^2 \theta}{\cos^2 \theta} + \frac{\sin^2 \theta}{\cos^2 \theta} = \frac{1}{\cos^2 \theta} \qquad \text{Write the left side as two fractions}$$

$$\left(\frac{\cos \theta}{\cos \theta}\right)^2 + \left(\frac{\sin \theta}{\cos \theta}\right)^2 = \left(\frac{1}{\cos \theta}\right)^2 \qquad \text{A property of exponents}$$

$$1 + \tan^2 \theta = \sec^2 \theta \qquad \text{Ratio and reciprocal identities}$$

This last expression, $1 + \tan^2 \theta = \sec^2 \theta$, is our second Pythagorean identity. To arrive at our third, and last, Pythagorean identity, we proceed as we have above, but instead of dividing each side by $\cos^2 \theta$, we divide by $\sin^2 \theta$. Without showing the work involved in doing so, the result is

$$1 + \cot^2 \theta = \csc^2 \theta$$

We summarize our derivations in Table 3.

Table 3

Pythagorean identities	Equivalent forms
$\cos^2 \theta + \sin^2 \theta = 1$	$\cos \theta = \pm\sqrt{1 - \sin^2 \theta}$
	$\sin \theta = \pm\sqrt{1 - \cos^2 \theta}$
$1 + \tan^2 \theta = \sec^2 \theta$	
$1 + \cot^2 \theta = \csc^2 \theta$	

Notice the $\pm$ sign in the equivalent forms. It occurs as part of the process of taking the square root of both sides of the preceding equation. (*Remember:* We would obtain a similar result in algebra if we solved the equation $x^2 = 9$ to get $x = \pm 3$.) In Example 10 we will see how to deal with the $\pm$ sign.

▶ **EXAMPLE 10**   If $\sin \theta = 3/5$ and θ terminates in quadrant II, find $\cos \theta$ and $\tan \theta$.

Solution   We begin by using the identity $\cos \theta = \pm\sqrt{1 - \sin^2 \theta}$.

$$\text{If} \qquad \sin \theta = \frac{3}{5}$$

$$\text{the identity} \qquad \cos \theta = \pm\sqrt{1 - \sin^2 \theta}$$

becomes

$$\cos \theta = \pm\sqrt{1 - \left(\frac{3}{5}\right)^2}$$

$$= \pm\sqrt{1 - \frac{9}{25}}$$

$$= \pm\sqrt{\frac{16}{25}}$$

$$= \pm\frac{4}{5}$$

Now we know that $\cos \theta$ is either 4/5 or $-4/5$. Looking back to the original statement of the problem, however, we see that θ terminates in quadrant II; therefore, $\cos \theta$ must be negative.

$$\cos \theta = -\frac{4}{5}$$

To find $\tan \theta$, we use a ratio identity.

$$\tan \theta = \frac{\sin \theta}{\cos \theta} = \frac{3/5}{-4/5} = -\frac{3}{4}$$ ◀

▶ **EXAMPLE 11**   If $\cos \theta = 1/2$ and θ terminates in quadrant IV, find the remaining trigonometric ratios for θ.

Solution   The first, and easiest, ratio to find is $\sec \theta$ because it is the reciprocal of $\cos \theta$.

$$\sec \theta = \frac{1}{\cos \theta} = \frac{1}{1/2} = 2$$

Next we find $\sin \theta$. Since θ terminates in QIV, $\sin \theta$ will be negative. Using one of the equivalent forms of our Pythagorean identity, we have

$$\sin \theta = -\sqrt{1 - \cos^2 \theta}$$ Negative sign because $\theta \in$ QIV

$$= -\sqrt{1 - \left(\frac{1}{2}\right)^2}$$ Substitute $\frac{1}{2}$ for $\cos \theta$

$$= -\sqrt{1 - \frac{1}{4}}$$ Square $\frac{1}{2}$ to get $\frac{1}{4}$

$$= -\sqrt{\frac{3}{4}}$$ Subtract

$$= -\frac{\sqrt{3}}{2}$$ Take the square root of the numerator and denominator separately

Now that we have $\sin \theta$ and $\cos \theta$, we can find $\tan \theta$ by using a ratio identity.

$$\tan \theta = \frac{\sin \theta}{\cos \theta} = \frac{-\sqrt{3}/2}{1/2} = -\sqrt{3}$$

Cot θ and csc θ are the reciprocals of tan θ and sin θ, respectively. Therefore,

$$\cot \theta = \frac{1}{\tan \theta} = -\frac{1}{\sqrt{3}}$$

$$\csc \theta = \frac{1}{\sin \theta} = -\frac{2}{\sqrt{3}}$$

Here are all six ratios together:

$$\sin \theta = -\frac{\sqrt{3}}{2} \qquad \csc \theta = -\frac{2}{\sqrt{3}}$$

$$\cos \theta = \frac{1}{2} \qquad \sec \theta = 2$$

$$\tan \theta = -\sqrt{3} \qquad \cot \theta = -\frac{1}{\sqrt{3}}$$

◀

As a final note, we should mention that the eight basic identities we have derived here, along with their equivalent forms, are very important in the study of trigonometry. It is essential that you memorize them. It may be a good idea to practice writing them from memory until you can write each of the eight, and their equivalent forms, perfectly. As time goes by, we will increase our list of identities, so you will want to keep up with them as we go along.

PROBLEM SET 1.4

Give the reciprocal of each number.

1. 7 **2.** 4 **3.** $-2/3$ **4.** $-5/13$
5. $-1/\sqrt{2}$ **6.** $-\sqrt{3}/2$ **7.** x **8.** $1/a$

Use the reciprocal identities for the following problems.

9. If sin $\theta = 4/5$, find csc θ.
10. If cos $\theta = \sqrt{3}/2$, find sec θ.
11. If sec $\theta = -2$, find cos θ.
12. If csc $\theta = -13/12$, find sin θ.
13. If tan $\theta = a$ $(a \neq 0)$, find cot θ.
14. If cot $\theta = -b$ $(b \neq 0)$, find tan θ.

Use a ratio identity to find tan θ if

15. $\sin \theta = \frac{3}{5}$ and $\cos \theta = -\frac{4}{5}$
16. $\sin \theta = \frac{2}{\sqrt{5}}$ and $\cos \theta = \frac{1}{\sqrt{5}}$

Use a ratio identity to find cot θ if

17. $\sin \theta = -\frac{5}{13}$ and $\cos \theta = -\frac{12}{13}$
18. $\sin \theta = \frac{2}{\sqrt{13}}$ and $\cos \theta = \frac{3}{\sqrt{13}}$

For Problems 19 through 22, recall that $\sin^2 \theta$ means $(\sin \theta)^2$.

19. If sin $\theta = 1/\sqrt{2}$, find $\sin^2 \theta$.
20. If cos $\theta = 1/3$, find $\cos^2 \theta$.

21. If $\tan \theta = 2$, find $\tan^3 \theta$.

22. If $\sec \theta = -3$, find $\sec^3 \theta$.

For Problems 23 through 26, let $\sin \theta = -12/13$, and $\cos \theta = -5/13$, and find

23. $\tan \theta$ **24.** $\cot \theta$ **25.** $\sec \theta$ **26.** $\csc \theta$

Use the equivalent forms of the first Pythagorean identity on Problems 27 through 36.

27. Find $\sin \theta$ if $\cos \theta = \dfrac{3}{5}$ and θ terminates in QI.

28. Find $\sin \theta$ if $\cos \theta = \dfrac{5}{13}$ and θ terminates in QI.

29. Find $\cos \theta$ if $\sin \theta = \dfrac{1}{3}$ and θ terminates in QII.

30. Find $\cos \theta$ if $\sin \theta = \dfrac{\sqrt{3}}{2}$ and θ terminates in QII.

31. If $\sin \theta = -4/5$ and θ terminates in QIII, find $\cos \theta$.
32. If $\sin \theta = -4/5$ and θ terminates in QIV, find $\cos \theta$.
33. If $\cos \theta = \sqrt{3}/2$ and θ terminates in QI, find $\sin \theta$.
34. If $\cos \theta = -1/2$ and θ terminates in QII, find $\sin \theta$.

35. If $\sin \theta = \dfrac{1}{\sqrt{5}}$ and $\theta \in$ QII, find $\cos \theta$.

36. If $\cos \theta = -\dfrac{1}{\sqrt{10}}$ and $\theta \in$ QIII, find $\sin \theta$.

37. Find $\tan \theta$ if $\sin \theta = 1/3$ and θ terminates in QI.
38. Find $\cot \theta$ if $\sin \theta = 2/3$ and θ terminates in QII.
39. Find $\sec \theta$ if $\tan \theta = 8/15$ and θ terminates in QIII.
40. Find $\csc \theta$ if $\cot \theta = -24/7$ and θ terminates in QIV.
41. Find $\csc \theta$ if $\cot \theta = -21/20$ and $\sin \theta > 0$.
42. Find $\sec \theta$ if $\tan \theta = 7/24$ and $\cos \theta < 0$.

Find the remaining trigonometric ratios of θ if

43. $\cos \theta = 12/13$ and θ terminates in QI
44. $\sin \theta = 12/13$ and θ terminates in QI
45. $\sin \theta = -1/2$ and θ is not in QIII
46. $\cos \theta = -1/3$ and θ is not in QII
47. $\sec \theta = 2$ and $\sin \theta$ is positive   **48.** $\csc \theta = 2$ and $\cos \theta$ is negative

49. $\cos \theta = \dfrac{2}{\sqrt{13}}$ and $\theta \in$ QIV   **50.** $\sin \theta = \dfrac{3}{\sqrt{10}}$ and $\theta \in$ QII

51. $\sec \theta = -3$ and $\theta \in$ QIII   **52.** $\sec \theta = -4$ and $\theta \in$ QII
53. $\csc \theta = a$ and θ terminates in QI
54. $\sec \theta = b$ and θ terminates in QI

Using your calculator and rounding your answers to the nearest hundredth, find the remaining trigonometric ratios of θ if

55. $\sin \theta = 0.23$ and $\theta \in$ QI

56. $\cos \theta = 0.51$ and $\theta \in$ QI

57. $\sec \theta = -1.24$ and $\theta \in$ QII

58. $\csc \theta = -2.45$ and $\theta \in$ QIII

Recall from algebra that the slope of the line through (x_1, y_1) and (x_2, y_2) is

$$m = \frac{y_2 - y_1}{x_2 - x_1}$$

It is the change in the y-coordinates divided by the change in the x-coordinates.

59. The line $y = 3x$ passes through the points $(0, 0)$ and $(1, 3)$. Find its slope.

60. Suppose the angle formed by the line $y = 3x$ and the positive x-axis is θ. Find the tangent of θ.

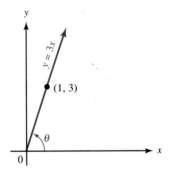

Figure 1

61. Find the slope of the line $y = mx$. [It passes through the origin and the point $(1, m)$.]

62. Find $\tan \theta$ if θ is the angle formed by the line $y = mx$ and the positive x-axis.

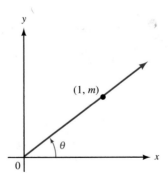

Figure 2

More on Identities

The topics we will cover in this section are an extension of the work we did with identities in Section 1.4. The first topic involves writing any of the six trigonometric functions in terms of any of the others. Let's look at an example.

▶ **EXAMPLE 1**   Write $\tan \theta$ in terms of $\sin \theta$.

Solution   When we say we want $\tan \theta$ written in terms of $\sin \theta$, we mean that we want to write an expression that is equivalent to $\tan \theta$ but involves no trigonometric function other than $\sin \theta$. Let's begin by using a ratio identity to write $\tan \theta$ in terms of $\sin \theta$ and $\cos \theta$.

$$\tan \theta = \frac{\sin \theta}{\cos \theta}$$

Now we need to replace $\cos \theta$ with an expression involving only $\sin \theta$. Since $\cos \theta = \pm\sqrt{1 - \sin^2 \theta}$ we have

$$\tan \theta = \frac{\sin \theta}{\cos \theta}$$

$$= \frac{\sin \theta}{\pm\sqrt{1 - \sin^2 \theta}}$$

$$= \pm\frac{\sin \theta}{\sqrt{1 - \sin^2 \theta}}$$

This last expression is equivalent to $\tan \theta$ and is written in terms of $\sin \theta$ only. (In a problem like this, it is okay to include numbers and algebraic symbols with $\sin \theta$.)   ◀

Here is another example.

▶ **EXAMPLE 2**   Write $\sec \theta \tan \theta$ in terms of $\sin \theta$ and $\cos \theta$, and then simplify.

Solution   Since $\sec \theta = 1/\cos \theta$ and $\tan \theta = \sin \theta/\cos \theta$, we have

$$\sec \theta \tan \theta = \frac{1}{\cos \theta} \cdot \frac{\sin \theta}{\cos \theta}$$

$$= \frac{\sin \theta}{\cos^2 \theta}$$ ◀

The next examples show how we manipulate trigonometric expressions using algebraic techniques.

▶ **EXAMPLE 3** Add $\dfrac{1}{\sin \theta} + \dfrac{1}{\cos \theta}$.

Solution We can add these two expressions in the same way we would add $\frac{1}{3}$ and $\frac{1}{4}$—by first finding a least common denominator (LCD), and then writing each expression again with the LCD for its denominator.

$$\frac{1}{\sin \theta} + \frac{1}{\cos \theta} = \frac{1}{\sin \theta} \cdot \frac{\cos \theta}{\cos \theta} + \frac{1}{\cos \theta} \cdot \frac{\sin \theta}{\sin \theta} \qquad \text{The LCD is } \sin \theta \cos \theta$$

$$= \frac{\cos \theta}{\sin \theta \cos \theta} + \frac{\sin \theta}{\cos \theta \sin \theta}$$

$$= \frac{\cos \theta + \sin \theta}{\sin \theta \cos \theta}$$ ◀

▶ **EXAMPLE 4** Multiply $(\sin \theta + 2)(\sin \theta - 5)$.

Solution We multiply these two expressions in the same way we would multiply $(x + 2)(x - 5)$. (In some algebra books this kind of multiplication is accomplished using the FOIL method.)

$$(\sin \theta + 2)(\sin \theta - 5) = \sin \theta \sin \theta - 5 \sin \theta + 2 \sin \theta - 10$$
$$= \sin^2 \theta - 3 \sin \theta - 10 \qquad ◀$$

In the examples that follow, we want to use the eight basic identities we developed in Section 1.4, along with some techniques from algebra, to show that some more complicated identities are true.

▶ **EXAMPLE 5** Show that the following statement is true by transforming the left side into the right side.

$$\cos \theta \tan \theta = \sin \theta$$

Solution We begin by writing everything on the left side in terms of $\sin \theta$ and $\cos \theta$.

$$\cos \theta \tan \theta = \cos \theta \cdot \frac{\sin \theta}{\cos \theta}$$

$$= \frac{\cos \theta \sin \theta}{\cos \theta}$$

$$= \sin \theta \qquad \qquad \text{Divide out the } \cos \theta \text{ common to the numerator and denominator}$$

Since we have succeeded in transforming the left side into the right side, we have shown that the statement $\cos \theta \tan \theta = \sin \theta$ is an identity. ◀

▶ **EXAMPLE 6**   Prove the identity $(\sin \theta + \cos \theta)^2 = 1 + 2 \sin \theta \cos \theta$

Solution   Let's agree to prove the identities in this section, and the problem set that follows, by transforming the left side into the right side. In this case, we begin by expanding $(\sin \theta + \cos \theta)^2$. (Remember from algebra, $(a + b)^2 = a^2 + 2ab + b^2$.)

$$(\sin \theta + \cos \theta)^2 = \sin^2 \theta + 2 \sin \theta \cos \theta + \cos^2 \theta$$

Now we can rearrange the terms on the right side to get $\sin^2 \theta$ and $\cos^2 \theta$ together.

$$= (\sin^2 \theta + \cos^2 \theta) + 2 \sin \theta \cos \theta$$
$$= 1 + 2 \sin \theta \cos \theta \qquad \blacktriangleleft$$

In the introduction to Section 1.4, we mentioned that trigonometric identities are the key to writing the expression $\sqrt{x^2 + 9}$ without the square root symbol. Our next example shows how we do this.

▶ **EXAMPLE 7**   Simplify the expression $\sqrt{x^2 + 9}$ as much as possible after substituting $3 \tan \theta$ for x.

Solution   Our goal is to write the expression $\sqrt{x^2 + 9}$ without a square root by first making the substitution $x = 3 \tan \theta$.

If $\qquad\qquad\qquad x = 3 \tan \theta$
then the expression $\quad \sqrt{x^2 + 9}$
becomes

$$\sqrt{(3 \tan \theta)^2 + 9} = \sqrt{9 \tan^2 \theta + 9}$$
$$= \sqrt{9(\tan^2 \theta + 1)}$$
$$= \sqrt{9 \sec^2 \theta}$$
$$= 3 \,|\sec \theta| \qquad \blacktriangleleft$$

Note 1   We must use the absolute value symbols unless we know that $3 \sec \theta$ is positive. Remember, in algebra, $\sqrt{a^2} = a$ only when a is positive or 0. If it is possible that a is negative, then $\sqrt{a^2} = |a|$.

Note 2   After reading through Example 7, you may be wondering if it is mathematically correct to make the substitution $x = 3 \tan \theta$. After all, we can assume that x is a real number, but which real number it is we do not know. How do we know that any real number can be written as $3 \tan \theta$? We will take care of this in Section 4.2 by showing that for any real number x, there is a value of θ between $-90°$ and $90°$ for which $x = 3 \tan \theta$.

Finally, we should mention that the ability to prove identities in trigonometry is not always obtained immediately. It usually requires a lot of practice. The more you work at it, the better you will become at it. In the meantime, if you are having trouble, check first to see that you have memorized the eight basic identities —reciprocal, ratio, Pythagorean, and their equivalent forms—as given in Section 1.4.

PROBLEM SET 1.5

Write each of the following in terms of sin θ only:

1. cos θ **2.** csc θ **3.** cot θ **4.** sec θ

Write each of the following in terms of cos θ only:

5. sec θ **6.** sin θ **7.** tan θ **8.** csc θ

Write each of the following in terms of sin θ and cos θ, and then simplify if possible:

9. csc θ cot θ **10.** sec θ cot θ

11. csc θ tan θ **12.** sec θ tan θ csc θ

13. $\dfrac{\sec \theta}{\csc \theta}$ **14.** $\dfrac{\csc \theta}{\sec \theta}$ **15.** $\dfrac{\sec \theta}{\tan \theta}$ **16.** $\dfrac{\csc \theta}{\cot \theta}$

17. $\dfrac{\tan \theta}{\cot \theta}$ **18.** $\dfrac{\cot \theta}{\tan \theta}$ **19.** $\dfrac{\sin \theta}{\csc \theta}$ **20.** $\dfrac{\cos \theta}{\sec \theta}$

21. tan θ + sec θ **22.** cot θ − csc θ

23. sin θ cot θ + cos θ **24.** cos θ tan θ + sin θ

25. sec θ − tan θ sin θ **26.** csc θ − cot θ cos θ

Add and subtract as indicated. Then simplify your answers if possible. Leave all answers in terms of sin θ and/or cos θ.

27. $\dfrac{\sin \theta}{\cos \theta} + \dfrac{1}{\sin \theta}$ **28.** $\dfrac{\cos \theta}{\sin \theta} + \dfrac{\sin \theta}{\cos \theta}$

29. $\dfrac{1}{\sin \theta} - \dfrac{1}{\cos \theta}$ **30.** $\dfrac{1}{\cos \theta} - \dfrac{1}{\sin \theta}$

31. $\sin \theta + \dfrac{1}{\cos \theta}$ **32.** $\cos \theta + \dfrac{1}{\sin \theta}$

33. $\dfrac{1}{\sin \theta} - \sin \theta$ **34.** $\dfrac{1}{\cos \theta} - \cos \theta$

Multiply.

35. (sin θ + 4)(sin θ + 3)   **36.** (cos θ + 2)(cos θ − 5)

37. (2 cos θ + 3)(4 cos θ − 5)   **38.** (3 sin θ − 2)(5 cos θ − 4)

39. (1 − sin θ)(1 + sin θ)   **40.** (1 − cos θ)(1 + cos θ)

41. (1 − tan θ)(1 + tan θ)   **42.** (1 − cot θ)(1 + cot θ)

43. (sin θ − cos θ)^2 **44.** (cos θ + sin θ)^2

45. (sin θ − 4)^2 **46.** (cos θ − 2)^2

Show that each of the following statements is an identity by transforming the left side of each one into the right side.

47. $\cos \theta \tan \theta = \sin \theta$

48. $\sin \theta \cot \theta = \cos \theta$

49. $\sin \theta \sec \theta \cot \theta = 1$

50. $\cos \theta \csc \theta \tan \theta = 1$

51. $\dfrac{\sin \theta}{\csc \theta} = \sin^2 \theta$

52. $\dfrac{\cos \theta}{\sec \theta} = \cos^2 \theta$

53. $\dfrac{\csc \theta}{\cot \theta} = \sec \theta$

54. $\dfrac{\sec \theta}{\tan \theta} = \csc \theta$

55. $\dfrac{\sec \theta}{\csc \theta} = \tan \theta$

56. $\dfrac{\csc \theta}{\sec \theta} = \cot \theta$

57. $\dfrac{\sec \theta \cot \theta}{\csc \theta} = 1$

58. $\dfrac{\csc \theta \tan \theta}{\sec \theta} = 1$

59. $\sin \theta \tan \theta + \cos \theta = \sec \theta$

60. $\cos \theta \cot \theta + \sin \theta = \csc \theta$

61. $\tan \theta + \cot \theta = \sec \theta \csc \theta$

62. $\tan^2 \theta + 1 = \sec^2 \theta$

63. $\csc \theta - \sin \theta = \dfrac{\cos^2 \theta}{\sin \theta}$

64. $\sec \theta - \cos \theta = \dfrac{\sin^2 \theta}{\cos \theta}$

65. $\csc \theta \tan \theta - \cos \theta = \dfrac{\sin^2 \theta}{\cos \theta}$

66. $\sec \theta \cot \theta - \sin \theta = \dfrac{\cos^2 \theta}{\sin \theta}$

67. $(1 - \cos \theta)(1 + \cos \theta) = \sin^2 \theta$

68. $(1 + \sin \theta)(1 - \sin \theta) = \cos^2 \theta$

69. $(\sin \theta + 1)(\sin \theta - 1) = -\cos^2 \theta$

70. $(\cos \theta + 1)(\cos \theta - 1) = -\sin^2 \theta$

71. $\dfrac{\cos \theta}{\sec \theta} + \dfrac{\sin \theta}{\csc \theta} = 1$

72. $1 - \dfrac{\sin \theta}{\csc \theta} = \cos^2 \theta$

73. $(\sin \theta - \cos \theta)^2 - 1 = -2 \sin \theta \cos \theta$

74. $(\cos \theta + \sin \theta)^2 - 1 = 2 \sin \theta \cos \theta$

75. $\sin \theta(\sec \theta + \csc \theta) = \tan \theta + 1$

76. $\sec \theta(\sin \theta + \cos \theta) = \tan \theta + 1$

77. $\sin \theta(\sec \theta + \cot \theta) = \tan \theta + \cos \theta$

78. $\cos \theta(\csc \theta + \tan \theta) = \cot \theta + \sin \theta$

79. $\sin \theta(\csc \theta - \sin \theta) = \cos^2 \theta$

80. $\cos \theta(\sec \theta - \cos \theta) = \sin^2 \theta$

81. Simplify the expression $\sqrt{x^2 + 4}$ as much as possible after substituting $2 \tan \theta$ for x.

82. Simplify the expression $\sqrt{x^2 + 1}$ as much as possible after substituting $\tan \theta$ for x.

83. Simplify the expression $\sqrt{9 - x^2}$ as much as possible after substituting $3 \sin \theta$ for x.

84. Simplify the expression $\sqrt{25 - x^2}$ as much as possible after substituting $5 \sin \theta$ for x.

85. Simplify the expression $\sqrt{4x^2 + 16}$ as much as possible after substituting $2 \tan \theta$ for x.

86. Simplify the expression $\sqrt{4x^2 + 100}$ as much as possible after substituting $5 \tan \theta$ for x.

CHAPTER 1   SUMMARY

Examples

We will use the margin to give examples of the topics being reviewed, whenever it is appropriate.

The number in brackets next to each heading indicates the section in which that topic is discussed.

1.

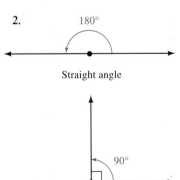

Positive angle

Initial side

Negative angle

Angles [1.1]

An angle is formed by two rays with a common end point. The common end point is called the *vertex*, and the rays are called *sides*, of the angle. If we think of an angle as being formed by rotating the initial side about the vertex to the terminal side, then a counterclockwise rotation gives a *positive angle*, and a clockwise rotation gives a *negative angle*.

2.

180°

Straight angle

90°

Right angle

Degree Measure [1.1]

There are 360° in a full rotation. This means that 1° is $\frac{1}{360}$ of a full rotation.

An angle that measures 90° is a *right angle*. An angle that measures 180° is a *straight angle*. Angles that measure between 0° and 90° are called *acute angles*, and angles that measure between 90° and 180° are called *obtuse angles*.

3. If ABC is a right triangle with $C = 90°$, and if $a = 4$ and $c = 5$, then

$$4^2 + b^2 = 5^2$$
$$16 + b^2 = 25$$
$$b^2 = 9$$
$$b = 3$$

Pythagorean Theorem [1.1]

In any right triangle, the square of the length of the longest side (the *hypotenuse*) is equal to the sum of the squares of the lengths of the other two sides *(legs)*.

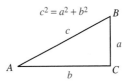

4. The distance between $(2, 7)$ and $(-1, 3)$ is

$$r = \sqrt{(2 + 1)^2 + (7 - 3)^2}$$
$$= \sqrt{9 + 16}$$
$$= \sqrt{25}$$
$$= 5$$

Distance Formula [1.2]

The distance r between the points (x_1, y_1) and (x_2, y_2) is given by the formula

$$r = \sqrt{(x_2 - x_1)^2 + (y_2 - y_1)^2}$$

5. $135°$ in standard position is

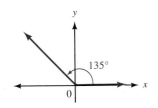

Standard Position for Angles [1.2]

An angle is said to be in standard position if its vertex is at the origin and its initial side is along the positive x-axis.

6. If $(-3, 4)$ is on the terminal side of θ, then

$$r = \sqrt{9 + 16} = 5$$

and

$$\sin \theta = \frac{4}{5} \qquad \csc \theta = \frac{5}{4}$$
$$\cos \theta = -\frac{3}{5} \qquad \sec \theta = -\frac{5}{3}$$
$$\tan \theta = -\frac{4}{3} \qquad \cot \theta = -\frac{3}{4}$$

Trigonometric Functions [1.3]

If θ is an angle in standard position and (x, y) is any point on the terminal side of θ (other than the origin), then

$$\sin \theta = \frac{y}{r} \qquad \csc \theta = \frac{r}{y}$$
$$\cos \theta = \frac{x}{r} \qquad \sec \theta = \frac{r}{x}$$
$$\tan \theta = \frac{y}{x} \qquad \cot \theta = \frac{x}{y}$$

Where $x^2 + y^2 = r^2$, or $r = \sqrt{x^2 + y^2}$. That is, r is the distance from the origin to (x, y).

7. If $\sin \theta > 0$, and $\cos \theta > 0$, then θ must terminate in QI.

If $\sin \theta > 0$, and $\cos \theta < 0$, then θ must terminate in QII.

If $\sin \theta < 0$, and $\cos \theta < 0$, then θ must terminate in QIII.

If $\sin \theta < 0$, and $\cos \theta > 0$, then θ must terminate in QIV.

Signs of the Trigonometric Functions [1.3]

The algebraic signs, $+$ or $-$, of the six trigonometric functions depend on the quadrant in which θ terminates.

For θ in	QI	QII	QIII	QIV
$\sin \theta$ and $\csc \theta$	$+$	$+$	$-$	$-$
$\cos \theta$ and $\sec \theta$	$+$	$-$	$-$	$+$
$\tan \theta$ and $\cot \theta$	$+$	$-$	$+$	$-$

8. If $\sin \theta = 1/2$ with θ in QI, then

$$\cos \theta = \sqrt{1 - \sin^2 \theta}$$
$$= \sqrt{1 - 1/4} = \frac{\sqrt{3}}{2}$$

$$\tan \theta = \frac{\sin \theta}{\cos \theta} = \frac{1/2}{\sqrt{3}/2} = \frac{1}{\sqrt{3}}$$

$$\cot \theta = \frac{1}{\tan \theta} = \sqrt{3}$$

$$\sec \theta = \frac{1}{\cos \theta} = \frac{2}{\sqrt{3}}$$

$$\csc \theta = \frac{1}{\sin \theta} = 2$$

Basic Identities [1.4]

Reciprocal

$$\csc \theta = \frac{1}{\sin \theta} \qquad \sec \theta = \frac{1}{\cos \theta} \qquad \cot \theta = \frac{1}{\tan \theta}$$

Ratio

$$\tan \theta = \frac{\sin \theta}{\cos \theta} \qquad \cot \theta = \frac{\cos \theta}{\sin \theta}$$

Pythagorean

$$\cos^2 \theta + \sin^2 \theta = 1$$
$$1 + \tan^2 \theta = \sec^2 \theta$$
$$1 + \cot^2 \theta = \csc^2 \theta$$

CHAPTER 1 TEST

Solve for x in each of the following right triangles:

1.

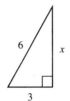

2.

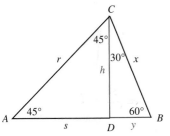

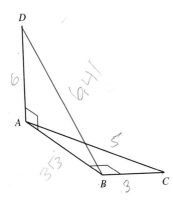

Figure 1

Figure 2

Problems 3 and 4 refer to Figure 1. (*Note: s* is the distance from *A* to *D*, and *y* is the distance from *D* to *B*.)

3. In order, find *h*, *r*, *y*, and *x* if $s = 5\sqrt{3}$.

4. In order, find *x*, *h*, *s*, and *r* if *y* = 3.

5. Figure 2 shows two right triangles drawn at 90° to one another. Find the length of *DB* if *DA* = 6, *AC* = 5, and *BC* = 3.

6. Through how many degrees does the hour hand of a clock move in 3 hours?

7. Find the remaining sides of a 30°–60°–90° triangle if the longest side is 5.

8. Graph the line $2x + 3y = 6$.

9. Find the distance between the points $(4, -2)$ and $(-1, 10)$.

10. Find the distance from the origin to the point (a, b).

11. Find *x* so that the distance between $(-2, 3)$ and $(x, 1)$ is $\sqrt{13}$.

Find sin θ, cos θ, and tan θ for each of the following values of θ:

12. 90°   **13.** −45°

In which quadrant will θ lie if

14. sin θ < 0 and cos θ > 0   **15.** csc θ > 0 and cos θ < 0

Find all six trigonometric functions for θ if the given point lies on the terminal side of θ in standard position.

16. $(-6, 8)$ **17.** $(-3, -1)$

Find the remaining trigonometric functions of θ if

18. sin θ = 1/2 and θ terminates in QII

19. tan θ = 12/5 and θ terminates in QIII

20. Find sin θ and cos θ if the terminal side of θ lies along the line $y = -2x$ in quadrant IV.

21. If sin θ = −3/4, find csc θ.

22. If sec θ = −2, find cos θ.

23. If sin θ = 1/3, find $\sin^3 \theta$.

24. If sec θ = 3 with θ in QIV, find cos θ, sin θ, and tan θ.

25. If $\sin \theta = \dfrac{1}{a}$ with θ in QI, find cos θ, csc θ, and cot θ.

26. Multiply $(\sin \theta + 3)(\sin \theta - 7)$.

27. Expand and simplify $(\cos \theta - \sin \theta)^2$.

28. Subtract $\dfrac{1}{\sin \theta} - \sin \theta$.

Show that each of the following statements is an identity by transforming the left side of each one into the right side.

29. $\dfrac{\cot \theta}{\csc \theta} = \cos \theta$

30. $\cot \theta + \tan \theta = \csc \theta \sec \theta$

31. $(1 - \sin \theta)(1 + \sin \theta) = \cos^2 \theta$

32. $\sin \theta(\csc \theta + \cot \theta) = 1 + \cos \theta$

2

Right Triangle Trigonometry

ı ı ı ı ı ı ı ı ı ı

▶ **To the Student**

We begin this chapter with a second definition for the six trigonometric functions. This second definition will be given in terms of right triangles. As you will see, it will not conflict with our original definition. Since this new definition defines the six trigonometric functions in terms of the sides of a right triangle, there are a number of interesting applications. One of the first applications will be to find the six trigonometric functions of 30°, 45°, and 60°, since these angles can be found in the special triangles we introduced in Chapter 1.

At the end of each Problem Set in this chapter, you will find a set of review problems. These review problems are intended to refresh your memory on topics we have covered previously and, at times, to get you ready for the section that follows. We will continue to place review problems at the end of the Problem Sets throughout the remainder of the book.

Definition II: Right Triangle Trigonometry

The word *trigonometry* is derived from two Greek words: *tri'gonon,* which translates as *triangle*, and *met'ron,* which means *measure*. Trigonometry, then, is triangle measure. In this section, we will give a second definition for the trigonometric functions that is, in fact, based on "triangle measure." We will define the trigonometric functions as the ratios of sides in right triangles. As you will see, this new definition does not conflict with the definition from Chapter 1 for the trigonometric functions.

DEFINITION II   If triangle ABC is a right triangle with $C = 90°$, then the six trigonometric functions for A are defined as follows:

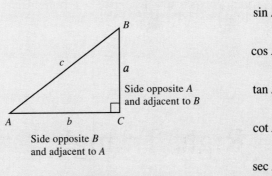

$$\sin A = \frac{\text{side opposite } A}{\text{hypotenuse}} = \frac{a}{c}$$

$$\cos A = \frac{\text{side adjacent } A}{\text{hypotenuse}} = \frac{b}{c}$$

$$\tan A = \frac{\text{side opposite } A}{\text{side adjacent } A} = \frac{a}{b}$$

$$\cot A = \frac{\text{side adjacent } A}{\text{side opposite } A} = \frac{b}{a}$$

$$\sec A = \frac{\text{hypotenuse}}{\text{side adjacent } A} = \frac{c}{b}$$

$$\csc A = \frac{\text{hypotenuse}}{\text{side opposite } A} = \frac{c}{a}$$

Figure 1

▶ **EXAMPLE 1**   Triangle ABC is a right triangle with $C = 90°$. If $a = 6$ and $c = 10$, find the six trigonometric functions of A.

Solution   We begin by making a diagram of ABC, and then use the given information and the Pythagorean Theorem to solve for b.

$$b = \sqrt{c^2 - a^2}$$
$$= \sqrt{100 - 36}$$
$$= \sqrt{64}$$
$$= 8$$

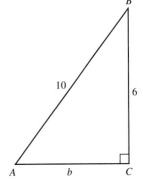

Figure 2

Now we write the six trigonometric functions of A using $a = 6$, $b = 8$, and $c = 10$.

$$\sin A = \frac{a}{c} = \frac{6}{10} = \frac{3}{5} \qquad \csc A = \frac{c}{a} = \frac{5}{3}$$

$$\cos A = \frac{b}{c} = \frac{8}{10} = \frac{4}{5} \qquad \sec A = \frac{c}{b} = \frac{5}{4}$$

$$\tan A = \frac{a}{b} = \frac{6}{8} = \frac{3}{4} \qquad \cot A = \frac{b}{a} = \frac{4}{3}$$

◀

Now that we have done an example using our new definition, let's see how our new definition compares with Definition I from the previous chapter. We can place right triangle ABC on a rectangular coordinate system so that A is in standard position. We then note that a point on the terminal side of A is (b, a).

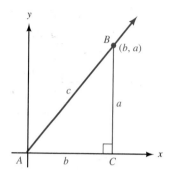

Figure 3

From Definition I in Chapter 1, we have	From Definition II in this chapter, we have
$\sin A = \dfrac{a}{c}$	$\sin A = \dfrac{a}{c}$
$\cos A = \dfrac{b}{c}$	$\cos A = \dfrac{b}{c}$
$\tan A = \dfrac{a}{b}$	$\tan A = \dfrac{a}{b}$

The two definitions agree as long as A is an acute angle. If A is not an acute angle, then Definition II does not apply, since in right triangle ABC, A must be an acute angle.

Here is another definition that we will need before we can take Definition II any further.

DEFINITION   Sine and *co*sine are *co*functions, as are tangent and *co*tangent, and secant and *co*secant. We say sine is the cofunction of cosine, and cosine is the cofunction of sine.

Now let's see what happens when we apply Definition II to B in right triangle ABC.

$$\sin B = \frac{\text{side opposite } B}{\text{hypotenuse}} = \frac{b}{c} = \cos A$$

$$\cos B = \frac{\text{side adjacent } B}{\text{hypotenuse}} = \frac{a}{c} = \sin A$$

$$\tan B = \frac{\text{side opposite } B}{\text{side adjacent } B} = \frac{b}{a} = \cot A$$

$$\cot B = \frac{\text{side adjacent } B}{\text{side opposite } B} = \frac{a}{b} = \tan A$$

$$\sec B = \frac{\text{hypotenuse}}{\text{side adjacent } B} = \frac{c}{a} = \csc A$$

$$\csc B = \frac{\text{hypotenuse}}{\text{side opposite } B} = \frac{c}{b} = \sec A$$

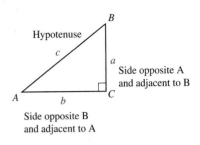

Figure 4

As you can see, every trigonometric function of A is equal to the cofunction of B. That is, $\sin A = \cos B$, $\sec A = \csc B$, and $\tan A = \cot B$, to name a few. Since A and B are the acute angles in a right triangle, they are always complementary angles; that is, their sum is always $90°$. What we actually have here is another property of trigonometric functions: The sine of an angle is the cosine of its complement, the secant of an angle is the cosecant of its complement, and the tangent of an angle is the cotangent of its complement. Or, in symbols,

$$\text{if } A + B = 90°, \text{ then} \quad \begin{aligned} \sin A &= \cos B \\ \sec A &= \csc B \\ \tan A &= \cot B \end{aligned}$$

and so on.

We generalize this discussion with the following theorem.

Cofunction Theorem

A trigonometric function of an angle is always equal to the cofunction of the complement of the angle.

To clarify this further, if two angles are complementary, such as $40°$ and $50°$, then a trigonometric function of one is equal to the cofunction of the other. That is, $\sin 40° = \cos 50°$, $\sec 40° = \csc 50°$, and $\tan 40° = \cot 50°$.

▶ **EXAMPLE 2** Fill in the blanks so that each expression becomes a true statement.

    a. $\sin \underline{\quad} = \cos 30°$ **b.** $\tan y = \cot \underline{\quad}$ **c.** $\sec 75° = \csc \underline{\quad}$

Solution Using the theorem on cofunctions of complementary angles, we fill in the blanks as follows:

a. $\sin \underline{60°} = \cos 30°$ since sine and cosine are cofunctions
                                           and $60° + 30° = 90°$

b. $\tan y = \cot \underline{(90° - y)}$ since tangent and cotangent are
cofunctions and $y + (90° - y) = 90°$

c. $\sec 75° = \csc \underline{15°}$ since secant and cosecant are cofunctions
and $75° + 15° = 90°$ ◀

For our next application of Definition II, we need to recall the two special triangles we introduced in Chapter 1. They are the 30°–60°–90° triangle and the 45°–45°–90° triangle.

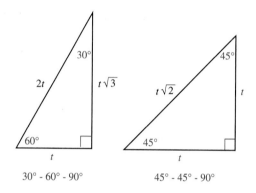

Figure 5

We can use these two special triangles to find the trigonometric functions of 30°, 45°, and 60°. For example,

$$\sin 60° = \frac{\text{side opposite } 60°}{\text{hypotenuse}} = \frac{t\sqrt{3}}{2t} = \frac{\sqrt{3}}{2} \qquad \text{Dividing out the}$$
$$\tan 30° = \frac{\text{side opposite } 30°}{\text{side adjacent } 30°} = \frac{t}{t\sqrt{3}} = \frac{1}{\sqrt{3}} \qquad \text{common factor } t$$

We could go on in this manner—using Definition II and the two special triangles—to find the six trigonometric ratios for 30°, 45°, and 60°. Instead, let us vary things a little and use the information just obtained for sin 60° and tan 30° and the theorem on cofunctions of complementary angles to find cos 30° and cot 60°.

$$\cos 30° = \sin 60° = \frac{\sqrt{3}}{2} \qquad \text{Cofunction Theorem}$$
$$\cot 60° = \tan 30° = \frac{1}{\sqrt{3}}$$

To vary things even more, we can use some reciprocal identities to find csc 60° and cot 30°.

$$\csc 60° = \frac{1}{\sin 60°} = \frac{2}{\sqrt{3}} \quad \text{or} \quad \frac{2\sqrt{3}}{3} \quad \text{If denominator is rationalized}$$

$$\cot 30° = \frac{1}{\tan 30°} = \sqrt{3}$$

The idea behind using the different methods listed here is not to make things confusing. We have a number of tools at hand, and it does not hurt to show the different ways they can be used.

If we were to continue finding sine, cosine, and tangent for these special angles, we would obtain the results summarized in Table 1.

Table 1 Exact Values

θ	30°	45°	60°
$\sin \theta$	$\dfrac{1}{2}$	$\dfrac{1}{\sqrt{2}}$ or $\dfrac{\sqrt{2}}{2}$	$\dfrac{\sqrt{3}}{2}$
$\cos \theta$	$\dfrac{\sqrt{3}}{2}$	$\dfrac{1}{\sqrt{2}}$ or $\dfrac{\sqrt{2}}{2}$	$\dfrac{1}{2}$
$\tan \theta$	$\dfrac{1}{\sqrt{3}}$ or $\dfrac{\sqrt{3}}{3}$	1	$\sqrt{3}$

Table 1 is called a table of exact values to distinguish it from a table of approximate values. Later in this chapter we will work with tables of approximate values.

Calculator Note   To use a calculator to find the sine, cosine, or tangent of an angle, you first enter the angle (with the calculator in the degree mode) and then press the sin, cos, or tan key.* For example, to find sin 30°, you would use the following sequence

$$30 \quad \boxed{\sin}$$

The calculator will then display 0.5. The values the calculator gives for the trigonometric functions of an angle are approximations except for a few cases such as sin 30° = 0.5. If you use a calculator to find sin 60°, the calculator will display 0.866025403, which is a nine-digit decimal approximation of $\dfrac{\sqrt{3}}{2}$. You can check this by using your calculator to find an approximation to $\dfrac{\sqrt{3}}{2}$ to see that it agrees with

---

* Remember, the calculator notes in this book are for basic scientific calculators that use algebraic logic (usually identified by the presence of the $\boxed{=}$ key).

the calculator value of sin 60°. In the meantime, remember that if you are asked to find the exact value of a trigonometric function, you must use the values given in the table of exact values.

▶ **EXAMPLE 3** Use the exact values from the table to show that the following are true.

$$\textbf{a. } \cos^2 30° + \sin^2 30° = 1 \qquad \textbf{b. } \cos^2 45° + \sin^2 45° = 1$$

Solution

a. $\cos^2 30° + \sin^2 30° = \left(\dfrac{\sqrt{3}}{2}\right)^2 + \left(\dfrac{1}{2}\right)^2 = \dfrac{3}{4} + \dfrac{1}{4} = 1$

b. $\cos^2 45° + \sin^2 45° = \left(\dfrac{1}{\sqrt{2}}\right)^2 + \left(\dfrac{1}{\sqrt{2}}\right)^2 = \dfrac{1}{2} + \dfrac{1}{2} = 1$ ◀

▶ **EXAMPLE 4** Let $x = 30°$ and $y = 45°$ in each of the expressions that follow, and then simplify each expression as much as possible.

$$\textbf{a. } 2 \sin x \qquad \textbf{b. } \sin 2y \qquad \textbf{c. } 4 \sin (3x - 90°)$$

Solution

a. $2 \sin x = 2 \sin 30° = 2 \left(\dfrac{1}{2}\right) = 1$

b. $\sin 2y = \sin 2 (45°) = \sin 90° = 1$

c. $4 \sin (3x - 90°) = 4 \sin [3(30°) - 90°] = 4 \sin 0° = 4(0) = 0$ ◀

As you will see as we progress through the book, there are times when it is appropriate to use Definition II for the trigonometric functions and times when it is more appropriate to use Definition I. To illustrate, suppose we wanted to find sin θ, cos θ, and tan θ for θ = 90°. We would not be able to use Definition II because we can't draw a right triangle in which one of the *acute* angles is 90°. Instead, we would draw 90° in standard position, locate a point on the terminal side, and use Definition I to find sin 90°, cos 90°, and tan 90°. Figure 6 illustrates this.

A point on the terminal side of 90° is (0, 1). The distance from the origin to (0, 1) is 1. Therefore, $x = 0$, $y = 1$, and $r = 1$. From Definition I we have:

$$\sin 90° = \dfrac{1}{1} = 1$$

$$\cos 90° = \dfrac{0}{1} = 0$$

$$\tan 90° = \dfrac{1}{0}, \text{ which is undefined}$$

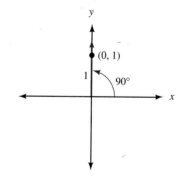

Figure 6

To show how the properties, definitions, and theorems you are learning can be used together, let's use the above information, along with the Cofunction Theorem and a ratio identity, to find sin θ, cos θ, and tan θ for $\theta = 0°$.

First, since $0°$ and $90°$ are complementary angles, we can use the Cofunction Theorem to write

A trigonometric function of an angle		is equal to the cofunction of its complement
sin 0°	=	cos 90° = 0
cos 0°	=	sin 90° = 1

Now that we have sin $0°$ and cos $0°$, we can find tan $0°$ by using a ratio identity.

$$\tan 0° = \frac{\sin 0°}{\cos 0°} = \frac{0}{1} = 0$$

To conclude this section, we take the information just obtained for $0°$ and $90°$, along with the exact values in Table 1, and summarize them in Table 2. To make the information in Table 2 a little easier to memorize, we have written some of the exact values differently than we usually do. For example, in Table 2 we have written 2 as $\sqrt{4}$, 0 as $\sqrt{0}$, and 1 as $\sqrt{1}$.

Table 2

θ	0°	30°	45°	60°	90°
sin θ	$\dfrac{\sqrt{0}}{2}$	$\dfrac{\sqrt{1}}{2}$	$\dfrac{\sqrt{2}}{2}$	$\dfrac{\sqrt{3}}{2}$	$\dfrac{\sqrt{4}}{2}$
cos θ	$\dfrac{\sqrt{4}}{2}$	$\dfrac{\sqrt{3}}{2}$	$\dfrac{\sqrt{2}}{2}$	$\dfrac{\sqrt{1}}{2}$	$\dfrac{\sqrt{0}}{2}$

PROBLEM SET 2.1

Problems 1 through 6 refer to right triangle ABC with $C = 90°$. In each case, use the given information to find the six trigonometric functions of A.

1. $b = 3, c = 5$
2. $b = 5, c = 13$
3. $a = 2, b = 1$
4. $a = 3, b = 2$
5. $a = 2, b = \sqrt{5}$
6. $a = 3, b = \sqrt{7}$

In each right triangle below, find sin A, cos A, tan A, and sin B, cos B, tan B.

7.

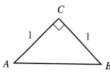

8.

9.

10.

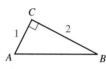

11.

12.

13.

14.

Use the Cofunction Theorem to fill in the blanks so that each expression becomes a true statement.

15. sin 10° = cos _____
16. cos 40° = sin _____
17. tan 8° = cot _____
18. cot 12° = tan _____
19. sec 73° = csc _____
20. csc 63° = sec _____
21. sin x = cos _____
22. sin y = cos _____
23. tan $(90° - x)$ = cot _____
24. tan $(90° - y)$ = cot _____

Simplify each expression by first substituting values from the table of exact values and then simplifying the resulting expression.

25. 4 sin 30°
26. 5 sin² 30°
27. (2 cos 30°)²
28. sin³ 30°
29. sin 30° cos 45°
30. sin 30° + cos 45°
31. (sin 60° + cos 60°)²
32. sin² 60° + cos² 60°
33. sin² 45° − 2 sin 45° cos 45° + cos² 45°
34. (sin 45° − cos 45°)²
35. (tan 45° + tan 60°)²
36. tan² 45° + tan² 60°

For each expression that follows, replace x with $30°$, y with $45°$, and z with $60°$, and then simplify as much as possible.

37. $2 \sin x$

38. $4 \cos y$

39. $4 \cos (z - 30°)$

40. $-2 \sin (y + 45°)$

41. $-3 \sin 2x$

42. $3 \sin 2y$

43. $2 \cos (3x - 45°)$

44. $2 \sin (90° - z)$

Find exact values for each of the following:

45. $\sec 30°$ **46.** $\csc 30°$ **47.** $\csc 60°$ **48.** $\sec 60°$

49. $\cot 45°$ **50.** $\cot 30°$ **51.** $\sec 45°$ **52.** $\csc 45°$

Use a calculator to find an approximation to each of the following expressions. Round your answers to the nearest ten-thousandth.

53. $\sin 60°$ **54.** $\cos 30°$ **55.** $\sqrt{3}/2$ **56.** $\sqrt{2}/2$

57. $\sin 45°$ **58.** $\cos 45°$ **59.** $1/\sqrt{2}$ **60.** $1/\sqrt{3}$

Problems 61 through 64 refer to right triangle ABC with $C = 90°$. In each case, use a calculator to find $\sin A$, $\cos A$, $\sin B$, and $\cos B$. Round your answers to the nearest hundredth.

61. $a = 3.42, c = 5.70$

62. $b = 8.88, c = 9.62$

63. $a = 19.44, b = 5.67$

64. $a = 11.28, b = 8.46$

65. Suppose each edge of the cube shown in Figure 7 is 5 inches long. Find the sine and cosine of the angle formed by diagonals CF and CH.

66. Suppose each edge of the cube shown in Figure 7 is 3 inches long. Find the sine and cosine of the angle formed by diagonals DE and DG.

67. Find the sine and cosine of the angle formed by diagonals CF and CH in Figure 7.

68. Find the sine and cosine of the angle formed by diagonals DE and DG in Figure 7.

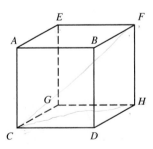

Figure 7

Review Problems

From here on, each Problem Set will end with a series of review problems. In mathematics, it is very important to review. The more you review, the better you will understand the topics we cover and the longer you will remember them. Also, there will be times when material that seemed confusing earlier will be less confusing the second time around.

The problems that follow review material we covered in Section 1.2.

Find the distance between each pair of points.

69. (5, 1)   (2, 5)   **70.** (3, −2)   (−1, −4)

71. Find x so that the distance between $(x, 2)$ and $(1, 5)$ is $\sqrt{13}$.
72. Graph the line $2x − 3y = 6$.   **73.** Graph the line $y = 2x − 1$.

Draw each angle in standard position and name a point on the terminal side.

74. 135°   **75.** 45°

For each angle below, name a coterminal angle between 0° and 360°.

76. −90°   **77.** −135°   **78.** −210°   **79.** −300°

Calculators and Trigonometric Functions of an Acute Angle

In this section, we will see how a basic scientific calculator can be used to find approximations for trigonometric functions of angles between 0° and 90°. We will also practice using a calculator to find an angle between 0° and 90° given a value for one of its trigonometric functions. Before we begin our work with calculators, we need to look at degree measure in more detail.

We previously defined 1 degree (1°) to be $\frac{1}{360}$ of a full rotation. A degree itself can be broken down further. If we divide 1° into 60 equal parts, each one of the parts is called 1 minute, denoted 1′. One minute is $\frac{1}{60}$ of a degree; in other words, there are 60 minutes in every degree. The next smaller unit of angle measure is a second. One second, 1″, is $\frac{1}{60}$ of a minute. There are 60 seconds in every minute.

$$1° = 60' \qquad \text{or} \qquad 1' = \left(\frac{1}{60}\right)^{\circ}$$

$$1' = 60'' \qquad \text{or} \qquad 1'' = \left(\frac{1}{60}\right)'$$

Table 1 shows how to read angles written in degree measure.

Table 1

The expression	Is read
52° 10′	52 degrees, 10 minutes
5° 27′ 30″	5 degrees, 27 minutes, 30 seconds
13° 24′ 15″	13 degrees, 24 minutes, 15 seconds

▶ **EXAMPLE 1**   Add 48° 49′ and 72° 26′.

Solution   We can add in columns with degrees in the first column and minutes in the second column.

$$48° \ 49′$$
$$+ \ 72° \ 26′$$
$$120° \ 75′$$

Since 60 minutes is equal to 1 degree, we can carry 1 degree from the minutes column to the degrees column.

$$120° \ 75′ = 121° \ 15′$$ ◀

▶ **EXAMPLE 2**   Subtract 24° 14′ from 90°.

Solution   To subtract 24° 14′ from 90°, we will have to "borrow" 1° and write that 1° as 60′.

$$90° \quad = \quad 89° \ 60′ \quad \text{(Still 90°)}$$
$$\underline{-24° \ 14′} \quad \underline{-24° \ 14′}$$
$$65° \ 46′$$ ◀

Decimal Degrees

An alternative to using minutes and seconds to break down degrees into smaller units is decimal degrees. For example, 30.5°, 101.75°, and 62.831° are measures of angles written in decimal degrees.

To convert from decimal degrees to degrees and minutes, we simply multiply the fractional part of the angle (the part to the right of the decimal point) by 60 to convert it to minutes.

▶ **EXAMPLE 3**   Change 27.25° to degrees and minutes.

Solution   Multiplying 0.25 by 60, we have the number of minutes equivalent to 0.25°.

$$27.25° = 27° + 0.25°$$
$$= 27° + 0.25(60′)$$
$$= 27° + 15′$$
$$= 27° \ 15′$$

Of course in actual practice, we would not show all of these steps. They are shown here simply to indicate why we multiply only the decimal part of the decimal degree by 60 to change to degrees and minutes.   ◀

Calculator Note   Some scientific calculators have a key that automatically converts angles given in decimal degrees to degrees and minutes. Consult the manual that came with your calculator to see if yours has this key.

▶ **EXAMPLE 4**   Change $10° \, 45'$ to decimal degrees.

Solution   We have to reverse the process we used in Example 3. To change $45'$ to a decimal, we must divide by 60.

$$10° \, 45' = 10° + 45'$$
$$= 10° + \left(\frac{45}{60}\right)°$$
$$= 10° + 0.75°$$
$$= 10.75°$$ ◀

Calculator Note   On a calculator, the result given in Example 4 is accomplished as follows:

$$45 \; \boxed{\div} \; 60 \; \boxed{+} \; 10 \; \boxed{=}$$

The process of converting back and forth between decimal degrees and degrees and minutes can become more complicated when we use decimal numbers with more digits or when we convert to degrees, minutes, and seconds. In this book, most of the angles written in decimal degrees will be written to the nearest tenth or, at most, the nearest hundredth. The angles written in degrees, minutes, and seconds will rarely go beyond the minutes column.

Table 2 lists the most common conversions between decimal degrees and minutes.

Table 2

Decimal degree	Minutes
$0.1°$	$6'$
$0.2°$	$12'$
$0.3°$	$18'$
$0.4°$	$24'$
$0.5°$	$30'$
$0.6°$	$36'$
$0.7°$	$42'$
$0.8°$	$48'$
$0.9°$	$54'$
$1.0°$	$60'$

Trigonometric Functions and Acute Angles

Up until now, the only angles we have been able to determine trigonometric functions for have been angles for which we could find a point on the terminal side or angles that were part of special triangles. We can find decimal approximations for trigonometric functions of any acute angle by using a calculator with keys for sine, cosine, and tangent.

▶ **EXAMPLE 5**   Use a calculator to find cos 37.8°.

Solution   First, be sure your calculator is in degree mode. Then enter 37.8 and press the key labeled cos.

$$37.8 \quad \boxed{\cos}$$

Your calculator will display a number that rounds to 0.7902. The number 0.7902 is just an approximation of cos 37.8°. Cos 37.8° is actually an irrational number, as are the trigonometric functions of most angles.   ◀

▶ **EXAMPLE 6**   Find sin 58° 45′.

Solution   We must first change 58° 45′ to decimal degrees. Doing so yields 58.75°. When 58.75 is displayed on the calculator, we use the $\boxed{\sin}$ key to find its sine. If we were to work the complete problem on a calculator, from beginning to end, we would have

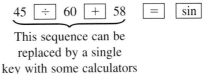

$$45 \quad \boxed{\div} \quad 60 \quad \boxed{+} \quad 58 \qquad \boxed{=} \quad \boxed{\sin}$$

This sequence can be
replaced by a single
key with some calculators

Rounding to four places past the decimal point we have

$$\sin 58° \, 45′ = 0.8549$$ ◀

▶ **EXAMPLE 7**   Find sec 78°.

Solution   Standard calculators rarely have a key for the secant function, so we use a reciprocal identity and the cosine key.

Since $\sec 78° = \dfrac{1}{\cos 78°}$, the calculator sequence is

$$78 \quad \boxed{\cos} \quad \boxed{1/x}$$

Rounding to four digits past the decimal point we have

$$\sec 78° = 4.8097$$ ◀

▶ **EXAMPLE 8**   To further justify the Cofunction Theorem introduced in the previous section, use a calculator to find sin 24.3° and cos 65.7°.

Solution   Note that the sum of 24.3° and 65.7° is 24.3° + 65.7° = 90°; the two angles are complementary. Using a calculator, and rounding our answers as we have above, we find that the sine of 24.3° is the cosine of its complement 65.7°.

$$\sin 24.3° = 0.4115 \qquad \text{and} \qquad \cos 65.7° = 0.4115$$ ◀

▶ **EXAMPLE 9**   Use a calculator to find $\cos^2 35° + \sin^2 35°$.

Solution   We know this expression should simplify to 1 by the Pythagorean identity $\cos^2 \theta + \sin^2 \theta = 1$. On a calculator we press the keys in the following order:

$$35 \boxed{\cos} \boxed{x^2} \boxed{+} 35 \boxed{\sin} \boxed{x^2} \boxed{=}$$

Your calculator should display the number 1. However, some calculators may display 0.9999999 or some other number that is close to 1.   ◀

Next we want to use our calculators to find an angle given the value of one of the trigonometric functions of the angle. It is somewhat like going in the opposite direction from that shown in the previous examples.

▶ **EXAMPLE 10**   Find the acute angle θ for which $\tan \theta = 3.152$. Round your answer to the nearest tenth of a degree.

Solution   We are looking for the angle whose tangent is 3.152. To find this angle on a calculator, we must use the $\tan^{-1}$ or arctan function. This is usually accomplished by pressing the key labeled $\boxed{\text{inv}}$, $\boxed{\text{arc}}$, or $\boxed{\text{2nd}}$, and then the $\boxed{\tan}$ key. Check your manual to see how your calculator does this. In the index, look under inverse trigonometric functions. In this book, we use the following sequence to represent the sequence of keys to press to solve this problem. (Your calculator may require a different sequence.)

$$3.152 \boxed{\text{inv}} \boxed{\tan}$$

To the nearest tenth of a degree, the answer is 72.4°. That is,

If $\tan \theta = 3.152$, then $\theta = 72.4°$.   ◀

▶ **EXAMPLE 11**   To the nearest minute, find the acute angle A for which $\sin A = 0.3733$.

Solution   First we find the angle in decimal degrees, then we convert to degrees and minutes. The sequence

$$0.3733 \boxed{\text{inv}} \boxed{\sin}$$

gives $A = 21.919°$. To convert the decimal part of the angle to minutes, we multiply 60 by 0.919:

$$60(0.919) = 55.14$$

To the nearest minute we have $A = 21° 55'$.   ◀

▶ **EXAMPLE 12**   To the nearest hundredth of a degree, find the acute angle B for which $\sec B = 1.0768$.

Solution   Since we do not have a secant key on our calculator, we must use a reciprocal identity first to see how we can convert this problem into a problem involving $\cos B$.

If sec $B = 1.0768$,

$$\text{then} \quad \frac{1}{\sec B} = \frac{1}{1.0768} \qquad \text{Take the reciprocal of each side}$$

$$\cos B = \frac{1}{1.0768} \qquad \text{Reciprocal identity}$$

From this last line we see that the sequence of keys to press is

$$1.0768 \quad \boxed{1/x} \quad \boxed{\text{inv}} \quad \boxed{\text{cos}}$$

To the nearest hundredth of a degree, our answer is $B = 21.77°$.   ◄

▶ **EXAMPLE 13**   Find the acute angle C for which $\cot C = 0.0975$. Round to the nearest degree.

Solution   First, we rewrite the problem in terms of $\tan C$.
If $\cot C = 0.0975$,

$$\text{then} \quad \frac{1}{\cot C} = \frac{1}{0.0975} \qquad \text{Take the reciprocal of each side}$$

$$\tan C = \frac{1}{0.0975} \qquad \text{Reciprocal identity}$$

From this last line we see that the sequence of keys to press is

$$0.0975 \quad \boxed{1/x} \quad \boxed{\text{inv}} \quad \boxed{\text{tan}}$$

To the nearest degree, our answer is $C = 84°$.   ◄

PROBLEM SET 2.2

Add and subtract as indicated.

1. $(37° \, 45') + (26° \, 24')$
2. $(41° \, 20') + (32° \, 16')$
3. $(51° \, 55') + (37° \, 45')$
4. $(63° \, 38') + (24° \, 52')$
5. $(61° \, 33') + (45° \, 16')$
6. $(77° \, 21') + (23° \, 16')$
7. $90° - (34° \, 12')$
8. $90° - (62° \, 25')$
9. $180° - (120° \, 17')$
10. $180° - (112° \, 19')$
11. $(76° \, 24') - (22° \, 34')$
12. $(89° \, 38') - (28° \, 58')$
13. $(70° \, 40') - (30° \, 50')$
14. $(80° \, 50') - (50° \, 56')$

Convert each of the following to degrees and minutes. Round to the nearest minute.

15. $35.4°$
16. $63.2°$
17. $16.25°$
18. $18.75°$
19. $92.55°$
20. $34.45°$
21. $19.9°$
22. $18.8°$

Change each of the following to decimal degrees. If rounding is necessary, round to the nearest hundredth of a degree.

23. $45° \, 12'$
24. $74° \, 18'$
25. $62° \, 36'$
26. $21° \, 48'$
27. $17° \, 20'$
28. $29° \, 40'$
29. $48° \, 27'$
30. $78° \, 21'$

Use a calculator to find each of the following. Round all answers to four places past the decimal point.

31. sin 27.2°   **32.** cos 82.9°   **33.** cos 18°   **34.** sin 42°
35. tan 87.32°   **36.** tan 81.43°   **37.** cot 31°   **38.** cot 24°
39. sec 48.2°   **40.** sec 71.8°   **41.** csc 14.15°   **42.** csc 12.21°

Use a calculator to find each of the following. (*Remember:* You must convert to decimal degrees before you use the sin, cos, or tan key.) Round all answers to four places past the decimal point.

43. cos 24° 30'   **44.** sin 35° 10'
45. tan 42° 15'   **46.** tan 19° 45'
47. sin 56° 40'   **48.** cos 66° 40'
49. cot 88° 18'   **50.** cot 50° 30'
51. csc 36° 36'   **52.** csc 24° 24'
53. sec 45° 54'   **54.** sec 84° 48'

Find θ if θ is between 0° and 90°. Round your answers to the nearest tenth of a degree.

55. cos θ = 0.9770   **56.** sin θ = 0.3971
57. tan θ = 0.6873   **58.** cos θ = 0.5490
59. sin θ = 0.9813   **60.** tan θ = 0.6273
61. sec θ = 1.0191   **62.** sec θ = 1.0801
63. csc θ = 1.8214   **64.** csc θ = 1.4293
65. cot θ = 0.6873   **66.** cot θ = 0.4327

Use a calculator to find a value of θ between 0° and 90° that satisfies each statement below. Write your answer in degrees and minutes rounded to the nearest minute.

67. sin θ = 0.7038   **68.** cos θ = 0.9153
69. cos θ = 0.4112   **70.** sin θ = 0.9954
71. tan θ = 1.1953   **72.** tan θ = 1.7391
73. csc θ = 3.9451   **74.** sec θ = 2.1609
75. cot θ = 5.5764   **76.** cot θ = 4.6252
77. sec θ = 1.0129   **78.** csc θ = 7.0683

To further justify the Cofunction Theorem, use your calculator to find a value for each pair of trigonometric functions below. In each case, the trigonometric functions are cofunctions of one another, and the angles are complementary angles. Round your answers to four places past the decimal point.

79. sin 23°, cos 67°   **80.** sin 33°, cos 57°
81. sec 34.5°, csc 55.5°   **82.** sec 56.7°, csc 33.3°
83. tan 4° 30', cot 85° 30'   **84.** tan 10° 30', cot 79° 30'

Work each of the following problems on your calculator. Do not write down or round off any intermediate answers. Each answer should be 1.

85. $\cos^2 37° + \sin^2 37°$ **86.** $\cos^2 85° + \sin^2 85°$
87. $\sin^2 10° + \cos^2 10°$ **88.** $\sin^2 8° + \cos^2 8°$

89. What happens when you try to find A for $\sin A = 1.234$ on your calculator? Why does this happen?

90. What happens when you try to find B for $\sin B = 4.321$ on your calculator? Why does this happen?

91. What happens when you try to find $\tan 90°$ on your calculator? Why does this happen? (After you have answered this question, try Problem 93.)

92. What happens when you try to find $\cot 0°$ on your calculator? Why does this happen? (After you have answered this question, try Problem 94.)

93. Find each of the following. Round your answers to the nearest whole number.

$$\tan 86°, \tan 87°, \tan 88°, \tan 89°, \tan 89.9°$$

94. Find each of the following. Round your answers to the nearest whole number.

$$\cot 4°, \cot 3°, \cot 2°, \cot 1°, \cot 0.1°$$

Review Problems

The problems that follow review material we covered in Section 1.3.

Find $\sin \theta$, $\cos \theta$, and $\tan \theta$ if the given point is on the terminal side of θ.

95. $(3, -2)$ **96.** $(-\sqrt{3}, 1)$

Find $\sin \theta$, $\cos \theta$, and $\tan \theta$ for each value of θ. (Do not use calculators.)

97. $90°$ **98.** $135°$

Find the remaining trigonometric functions of θ if

99. $\cos \theta = -5/13$ and θ terminates in QIII
100. $\tan \theta = -3/4$ and θ terminates in QII

In which quadrant must the terminal side of θ lie if

101. $\sin \theta > 0$ and $\cos \theta < 0$ **102.** $\tan \theta > 0$ and $\sec \theta < 0$

SECTION 2.3   # Solving Right Triangles

The first Ferris wheel was designed and built by American engineer George W. G. Ferris in 1893. The diameter of this wheel was 250 feet. It had 36 cars, each of which held 40 passengers. The top of the wheel was 264 feet above the ground. It took 20 minutes to complete one revolution. As you will see as we progress through the book, trigonometric functions can be used to model the motion of a rider on a Ferris wheel. The model can be used to give information about the position of the rider at any time during a ride. For instance, in the last example in this section, we will use Definition II for the trigonometric functions to find the height a rider is above the ground at certain positions on a Ferris wheel.

In this section, we will use Definition II for trigonometric functions of an acute angle, along with our calculators, to find the missing parts to some right triangles. Before we begin, however, we need to talk about significant digits.

DEFINITION   The number of *significant digits* (or figures) in a number is found by counting all the digits from left to right beginning with the first nonzero digit on the left.

According to this definition,

0.042 has two significant digits

0.005 has one significant digit

20.5   has three significant digits

6.000 has four significant digits

9,200 has four significant digits

700   has three significant digits

Note   In actual practice it is not always possible to tell how many significant digits an integer like 700 has. For instance, if exactly 700 people signed up to take history at your school, then 700 has three significant digits. On the other hand, if 700 is the result of a calculation in which the answer, 700, has been rounded to the nearest ten, then it has two significant digits. There are ways to write integers like 700 so that the number of significant digits can be determined exactly. One way is with scientific notation. However, to simplify things, in this book we will assume that integers have the greatest possible number of significant digits. In the case of 700, that number is three.

The relationship between the accuracy of the sides of a triangle and the accuracy of the angles in the same triangle is shown in the following table.

Accuracy of sides	Accuracy of angles
Two significant digits	Nearest degree
Three significant digits	Nearest 10 minutes or tenth of a degree
Four significant digits	Nearest minute or hundredth of a degree

We are now ready to use Definition II to solve right triangles. We solve a right triangle by using the information given about it to find all of the missing sides and angles. In all of the examples and in the Problem Set that follows, we will assume *C* is the right angle in all of our right triangles, unless otherwise noted.

Unless stated otherwise, we round our answers so that the number of significant digits in our answers matches the number of significant digits in the least significant number given in the original problem. Also, we round our answers only and not any of the numbers in the intermediate steps. Finally, we are showing the values of the trigonometric functions to four significant digits simply to avoid cluttering the page with long decimal numbers. This does not mean that you should stop halfway through a problem and round the values of trigonometric functions to four significant digits before continuing.

▶ **EXAMPLE 1**   In right triangle ABC, $A = 40°$ and $c = 12$ centimeters. Find a, b, and B.

Solution   We begin by making a diagram of the situation. The diagram is very important because it lets us visualize the relationship between the given information and the information we are asked to find.

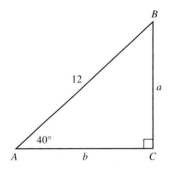

Figure 1

To find B, we use the fact that the sum of the two acute angles in any right triangle is 90°.

$$B = 90° - A$$
$$= 90° - 40°$$
$$B = 50°$$

To find a, we can use the formula for $\sin A$.

$$\sin A = \frac{a}{c}$$

Multiplying both sides of this formula by c and then substituting in our given values of A and c we have

$$a = c \sin A$$
$$= 12 \sin 40°$$
$$= 12(0.6428) \qquad \sin 40° = 0.6428$$
$$a = 7.7 \qquad \text{Answer rounded to two significant digits}$$

Calculator Note   On a calculator, we would use the sin key to find a as follows:

12 $\boxed{\times}$ 40 $\boxed{\sin}$ $\boxed{=}$

There is more than one way to find b.

Using $\cos A = \dfrac{b}{c}$, we have

$$b = c \cos A$$
$$= 12 \cos 40°$$
$$= 12(0.7660)$$
$$b = 9.2$$

Using the Pythagorean Theorem, we have

$$c^2 = a^2 + b^2$$
$$b = \sqrt{c^2 - a^2}$$
$$= \sqrt{12^2 - (7.7)^2}$$
$$= \sqrt{144 - 59.29}$$
$$= \sqrt{84.71}$$
$$b = 9.2$$

Calculator Note To find b using the Pythagorean Theorem and a calculator, we would follow this sequence:

$$12 \;\boxed{x^2}\; \boxed{-}\; 7.7 \;\boxed{x^2}\; \boxed{=}\; \boxed{\sqrt{}} \qquad \blacktriangleleft$$

In Example 2, we are given two sides and asked to find the remaining parts of a right triangle.

▶ **EXAMPLE 2** In right triangle ABC, $a = 2.73$ and $b = 3.41$. Find the remaining side and angles.

Solution Here is a diagram of the triangle.

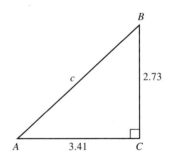

Figure 2

We can find A by using the formula for $\tan A$.

$$\tan A = \frac{a}{b}$$
$$= \frac{2.73}{3.41}$$
$$\tan A = 0.8006$$

Now, to find A, we use a calculator.

$$A = 38.7°$$

Next we find B.

$$B = 90.0° - A$$
$$= 90.0° - 38.7°$$
$$B = 51.3°$$

Notice we are rounding each angle to the nearest tenth of a degree since the sides we were originally given have three significant digits.

We can find c using the Pythagorean Theorem or one of our trigonometric functions. Let's start with a trigonometric function.

If $\sin A = \dfrac{a}{c}$

then $c = \dfrac{a}{\sin A}$ Multiply each side by c, then divide each side by $\sin A$

$$= \frac{2.73}{\sin 38.7°}$$
$$= \frac{2.73}{0.6252}$$
$$= 4.37 \qquad \text{To three significant digits}$$

Using the Pythagorean Theorem, we obtain the same result.

If $c^2 = a^2 + b^2$

then $c = \sqrt{a^2 + b^2}$
$$= \sqrt{(2.73)^2 + (3.41)^2}$$
$$= \sqrt{19.081}$$
$$= 4.37$$
◀

▶ **EXAMPLE 3**   The circle in Figure 3 has its center at C and a radius of 18 inches. If triangle ADC is a right triangle, find x, the distance from A to B.

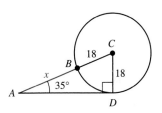

Figure 3

Solution   In triangle ADC, the side opposite A is 18 and the hypotenuse is $x + 18$. We can use $\sin A$ to write an equation that will allow us to solve for x.

$$\sin 35° = \frac{18}{x + 18}$$

$$(x + 18) \sin 35° = 18 \qquad \text{Multiply each side by } x + 18$$

$$x + 18 = \frac{18}{\sin 35°} \qquad \text{Divide each side by } \sin 35°$$

$$x = \frac{18}{\sin 35°} - 18 \qquad \text{Subtract 18 from each side}$$

$$= \frac{18}{0.5736} - 18$$

$$= 13 \text{ inches} \qquad \text{To two significant digits} \qquad \blacktriangleleft$$

▶ **EXAMPLE 4**   In Figure 4, the distance from A to D is 32 feet. Use the information in Figure 4 to solve for x, the distance between D and C.

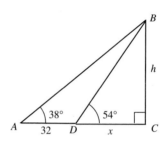

Figure 4

Solution   To find x we write two equations, each of which contains the variables x and h. Then we solve each equation for h and set the two expressions for h equal to each other.

$$\left. \begin{array}{l} \tan 54° = \dfrac{h}{x} \quad \Longrightarrow h = x \tan 54° \\[2mm] \tan 38° = \dfrac{h}{x + 32} \Longrightarrow h = (x + 32) \tan 38° \end{array} \right\}$$

Two equations involving both x and h — Solve each equation for h

Setting the two expressions for h equal to each other gives us an equation that involves only x.

Since $h = h$

we have $x \tan 54° = (x + 32) \tan 38°$

$$x \tan 54° = x \tan 38° + 32 \tan 38° \qquad \text{Distributive property}$$

$$x \tan 54° - x \tan 38° = 32 \tan 38° \qquad \text{Subtract } x \tan 38° \text{ from each side}$$

$$x(\tan 54° - \tan 38°) = 32 \tan 38°$$

Factor x from each term on the left side

$$x = \frac{32 \tan 38°}{\tan 54° - \tan 38°}$$

Divide each side by the coefficient of x

$$= \frac{32(0.7813)}{1.3764 - 0.7813}$$

$$= 42 \text{ feet}$$

To two significant digits ◀

▶ **EXAMPLE 5** In the introduction to this section, we gave some of the facts associated with the first Ferris wheel. Figure 5 is a simplified model of that Ferris wheel. If θ is the central angle formed as a rider moves from position P_0 to position P_1, find the rider's height above the ground h when θ is 45.0°.

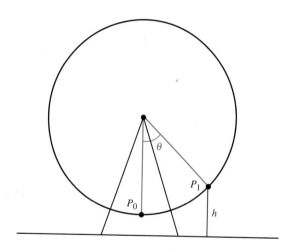

Figure 5

Solution We know from the introduction to this section that the diameter of the first Ferris wheel was 250 feet, which means the radius was 125 feet. Since the top of the wheel was 264 feet above the ground, the distance from the ground to the bottom of the wheel was 14 feet (the distance to the top minus the diameter of the wheel). To form a right triangle, we draw a horizontal line from P_1 to the vertical line connecting the center of the wheel with P_0. This information is shown in Figure 6.

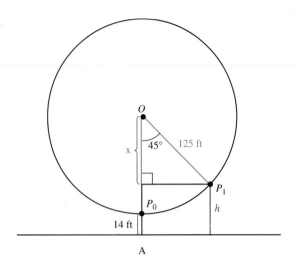

Figure 6

The key to solving this problem is recognizing that x is the difference between OA (the distance from the center of the wheel to the ground) and h. Since OA is 139 feet (the radius of the wheel plus the distance between the bottom of the wheel and the ground: $125 + 14 = 139$), we have

$$x = 139 - h$$

We use a cosine ratio to write an equation that contains h.

$$\cos 45° = \frac{x}{125} = \frac{139 - h}{125}$$

Solving for h we have

$$
\begin{aligned}
h &= 139 - 125 \cos 45.0° \\
&= 139 - 125(0.7071) \\
&= 139 - 88.4 \\
&= 50.6 \text{ ft}
\end{aligned}
$$

If $\theta = 45.0°$, a rider at position P_1 is $\frac{1}{8}$ of the way around the wheel. At that point, the rider is approximately 50.6 feet above the ground. ◀

PROBLEM SET 2.3

Problems 1 through 14 refer to right triangle ABC with $C = 90°$. Begin each problem by drawing a picture of the triangle with both the given and asked for information labeled appropriately. Also, write your answers for angles in decimal degrees.

1. If $A = 42°$ and $c = 15$ ft, find a.
2. If $A = 42°$ and $c = 89$ cm, find b.
3. If $A = 34°$ and $a = 22$ m, find c.
4. If $A = 34°$ and $b = 55$ m, find c.
5. If $B = 24.5°$ and $c = 2.34$ ft, find a.
6. If $B = 16.9°$ and $c = 7.55$ cm, find b.
7. If $B = 55.33°$ and $b = 12.34$ yd, find a.
8. If $B = 77.66°$ and $a = 43.21$ inches, find b.
9. If $a = 16$ cm and $b = 26$ cm, find A.
10. If $a = 42.3$ inches and $b = 32.4$ inches, find B.
11. If $b = 6.7$ m and $c = 7.7$ m, find A.
12. If $b = 9.8$ mm and $c = 12$ mm, find B.
13. If $c = 45.54$ ft and $a = 23.32$ ft, find B.
14. If $c = 5.678$ ft and $a = 4.567$ ft, find A.

Problems 15 through 32 refer to right triangle ABC with $C = 90°$. In each case, solve for all the missing parts using the given information. (In Problems 27 through 32, write your angles in decimal degrees.)

15. $A = 25°$, $c = 24$ m
16. $A = 41°$, $c = 36$ m
17. $A = 32.6°$, $a = 43.4$ inches
18. $A = 48.3°$, $a = 3.48$ inches
19. $A = 10° 42'$, $b = 5.932$ cm
20. $A = 66° 54'$, $b = 28.28$ cm
21. $B = 76°$, $c = 5.8$ ft
22. $B = 21°$, $c = 4.2$ ft
23. $B = 26° 30'$, $b = 324$ mm
24. $B = 53° 30'$, $b = 725$ mm
25. $B = 23.45°$, $a = 5.432$ mi
26. $B = 44.44°$, $a = 5.555$ mi
27. $a = 37$ ft, $b = 87$ ft
28. $a = 91$ ft, $b = 85$ ft
29. $a = 2.75$ cm, $c = 4.05$ cm
30. $a = 62.3$ cm, $c = 73.6$ cm
31. $b = 12.21$ inches, $c = 25.52$ inches
32. $b = 377.3$ inches, $c = 588.5$ inches

In Problems 33 and 34, use the information given in the diagram to find A to the nearest degree.

33.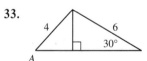

34.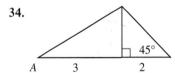

The circle in Figure 7 has a radius of r and center at C. The distance from A to B is x. For Problems 35 through 38, redraw Figure 7, label it as indicated in each problem, and then solve the problem.

35. If $A = 31°$ and $r = 12$, find x.
36. If $C = 26°$ and $r = 20$, find x.
37. If $A = 45°$ and $x = 15$, find r.
38. If $C = 65°$ and $x = 22$, find r.

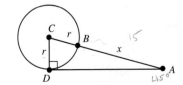

Figure 7

Figure 8 shows two right triangles drawn at 90° to each other. For Problems 39 through 42, redraw Figure 8, label it as the problem indicates, and then solve the problem.

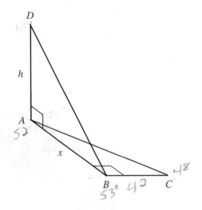

Figure 8

39. If $\angle ABD = 27°$, $C = 62°$, and $BC = 42$, find x and then find h.
40. If $\angle ABD = 53°$, $C = 48°$, and $BC = 42$, find x and then find h.
41. If $AC = 32$, $h = 19$, and $C = 41°$, find $\angle ABD$.
42. If $AC = 19$, $h = 32$, and $C = 49°$, find $\angle ABD$.

In Figure 9, the distance from A to D is y, the distance from D to C is x, and the distance from C to B is h. Use Figure 9 to solve Problems 43 through 48.

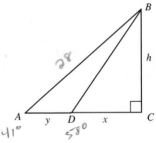

Figure 9

43. If $A = 41°$, $\angle BDC = 58°$, $AB = 18$, and $DB = 14$, find x, then y.
44. If $A = 32°$, $\angle BDC = 48°$, $AB = 17$, and $DB = 12$, find x, then y.
45. If $A = 41°$, $\angle BDC = 58°$, and $AB = 28$, find h, then x.
46. If $A = 32°$, $\angle BDC = 48°$, and $AB = 56$, find h, then x.
47. If $A = 43°$, $\angle BDC = 57°$, and $y = 10$, find x.
48. If $A = 32°$, $\angle BDC = 41°$, and $y = 14$, find x.
49. Suppose each edge of the cube shown in Figure 10 is 5 inches long. Find the measure of the angle formed by diagonals CF and CH. Round your answer to the nearest tenth of a degree.
50. Suppose each edge of the cube shown in Figure 10 is 3 inches long. Find the

measure of the angle formed by diagonals *DE* and *DG*. Round your answer to the nearest tenth of a degree.

51. Find the measure of the angle formed by diagonals *CF* and *CH* in Figure 10. Round your answer to the nearest tenth of a degree.

52. Find the measure of the angle formed by diagonals *DE* and *DG* in Figure 10. Round your answer to the nearest tenth of a degree.

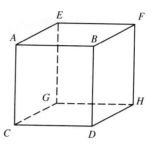

Figure 10

Repeat Example 5 from this section for the following values of θ.

53. $\theta = 30°$

54. $\theta = 60°$

55. $\theta = 120°$

56. $\theta = 135°$

57. In 1897, a Ferris wheel was built in Vienna that still stands today. It is named the Reisenrad, which translates to the *Great Wheel*. The diameter of the Reisenrad is 197 ft. The top of the wheel stands 209 ft above the ground. Figure 11 is a model of the Reisenrad with angle θ the central angle that is formed as a rider moves from the initial position P_0 to position P_1. The rider is h ft above the ground at position P_1. (Round to the nearest tenth.)

    a. Find h if θ is 120°.
    b. Find h if θ is 210°.
    c. Find h if θ is 315°.

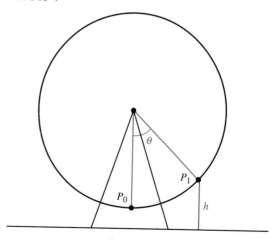

Figure 11

58. A Ferris wheel with a diameter of 165 feet was built in St. Louis in 1986. It is called Colossus. The top of the wheel stands 174 feet above the ground. Use the diagram in Figure 11 as a model of Colossus. (Round to the nearest tenth.)
  a. Find h if θ is 150°.
  b. Find h if θ is 240°.
  c. Find h if θ is 315°.

Review Problems

The following problems review material we covered in Section 1.4.

59. If sec $B = 2$, find cos² B.
60. If csc $B = 3$, find sin² B.
61. If sin $\theta = 1/3$ and θ terminates in QI, find cos θ.
62. If cos $\theta = -2/3$ and θ terminates in QIII, find sin θ.
63. If cos $A = 2/5$ with A in QIV, find sin A.
64. If sin $A = 1/4$ with A in QII, find cos A.

Find the remaining trigonometric ratios for θ, if

65. sin $\theta = \sqrt{3}/2$ with θ in QII
66. cos $\theta = 1/\sqrt{5}$ with θ in QIV
67. sec $\theta = -2$ with θ in QIII
68. csc $\theta = -2$ with θ in QIII

SECTION 2.4 # Applications

The map shown in Figure 1 is called a *topographic map*. This particular map is of Bishop's Peak in San Luis Obispo, California. The curved lines on the map are called

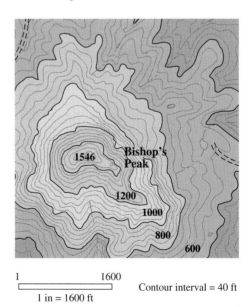

1 in = 1600 ft

Contour interval = 40 ft

Figure 1

contour lines; they are used to show the changes in elevation of the land shown on the map. On this map, the change in elevation between any two contour lines is 40 feet. That is, if you were standing on one contour line and you were to hike to a position two contour lines away from your starting point, your elevation above sea level would change by 80 feet. In general, the closer the contour lines are together, the faster the land rises or falls.

In this section, we will use our knowledge of right triangle trigonometry to solve a variety of problems, some of which involve topographic maps.

▶ **EXAMPLE 1**   The two equal sides of an isosceles triangle are each 24 centimeters. If each of the two equal angles measures 52°, find the length of the base and the altitude.

Solution   An isosceles triangle is any triangle with two equal sides. The angles opposite the two equal sides are called the base angles, and they are always equal. Here is a picture of our isosceles triangle.

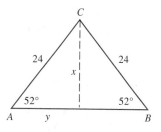

Figure 2

We have labeled the altitude x. We can solve for x using a sine ratio.

$$\text{If } \sin 52° = \frac{x}{24}$$

$$\text{then} \quad x = 24 \sin 52°$$

$$= 24(0.7880)$$

$$= 19 \text{ centimeters} \qquad \text{Rounded to 2 significant digits}$$

We have labeled half the base with y. To solve for y, we can use a cosine ratio.

$$\text{If } \cos 52° = \frac{y}{24}$$

$$\text{then} \quad y = 24 \cos 52°$$

$$= 24(0.6157)$$

$$= 14.8$$

The base is $2y = 2(14.8) = 30$ centimeters to the nearest centimeter.   ◀

For our next applications, we need the following definition.

DEFINITION   An angle measured from the horizontal up is called an *angle of elevation*. An angle measured from the horizontal down is called an *angle of depression.*

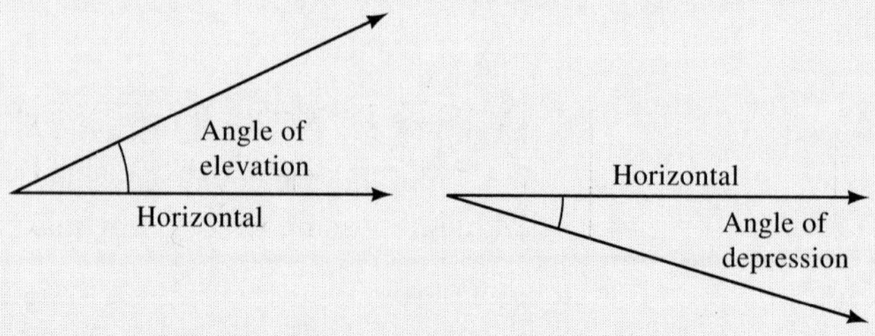

Angle of
elevation

Horizontal

Horizontal

Angle of
depression

Figure 3

These angles of elevation and depression are always considered positive angles. Also, the nonhorizontal side of each angle is sometimes called the *line of sight* of the observer.

▶ **EXAMPLE 2**   If a 75.0-foot flagpole casts a shadow 43.0 feet long, to the nearest 10 minutes what is the angle of elevation of the sun from the tip of the shadow?

Solution   We begin by making a diagram of the situation.

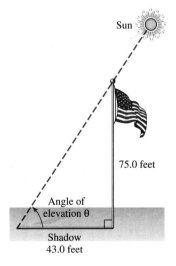

Sun

75.0 feet

Angle of
elevation θ

Shadow
43.0 feet

Figure 4

If we let θ = the angle of elevation of the sun, then

$$\tan \theta = \frac{75.0}{43.0}$$

$$\tan \theta = 1.744$$

which means $\theta = 60° \ 10'$ to the nearest 10 minutes.   ◀

▶ **EXAMPLE 3**   A man climbs 213 meters up the side of a pyramid and finds that the angle of depression to his starting point is 52.6°. How high off the ground is he?

Solution   Again, we begin by making a diagram of the situation.

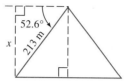

Figure 5

If x is the height above the ground, we can solve for x using a sine ratio.

If $\sin 52.6° = \dfrac{x}{213}$

then $x = 213 \sin 52.6°$

$= 213(0.7944)$

$= 169$ meters to 3 significant digits

The man is 169 meters above the ground.   ◀

▶ **EXAMPLE 4**   Figure 6 shows the topographic map we mentioned in the introduction to this section. Suppose Stacey and Amy are climbing Bishop's Peak. Stacey is at position S, and Amy is at position A. Find the angle of elevation from Amy to Stacey.

Solution   To solve this problem, we have to use two pieces of information from the legend on the map. First, we need to find the horizontal distance between the two people. The legend indicates that 1 inch on the map corresponds to an actual horizontal distance of 1,600 feet. If we measure the distance from Amy to Stacey with a ruler, we find it is $\frac{3}{8}$ inch. Multiplying this by 1,600, we have

$$\frac{3}{8} \cdot 1,600 = 600 \text{ feet}$$

which is the actual horizontal distance from Amy to Stacey.

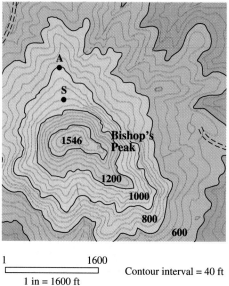

Contour interval = 40 ft

Figure 6

Next, we need the vertical distance Stacey is above Amy. We find this by counting the number of contour intervals between them. There are three. From the legend on the map we know that the elevation changes by 40 feet between any two contour lines. Therefore, Stacey is 120 feet above Amy. Here is a triangle that models the information we have so far:

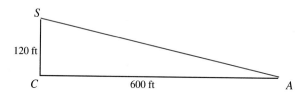

Figure 7

The angle of elevation from Amy to Stacey is angle A. To find A, we use the tangent ratio.

$$\tan A = \frac{120}{600} = 0.2$$

$$A = 11.3° \text{ to the nearest tenth of a degree}$$

Amy must look up at $11.3°$ from straight ahead to see Stacey.

Our next applications are concerned with what is called the *bearing of a line*. It is used in navigation and surveying.

> DEFINITION   The *bearing of a line l* is the acute angle formed by the north-south line and the line *l*. The notation used to designate the bearing of a line begins with N or S (for north or south), followed by the number of degrees in the angle, and ends with E or W (for east or west).

Here are some examples.

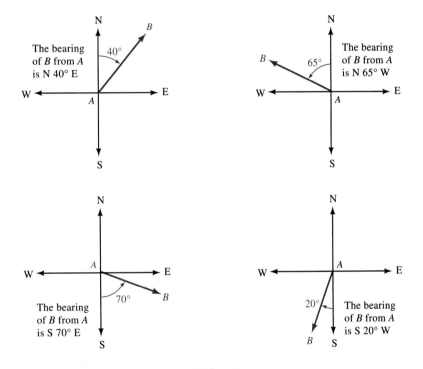

Figure 8

▶ **EXAMPLE 5**   San Luis Obispo, California, is 12 miles due north of Grover Beach. If Arroyo Grande is 4.6 miles due east of Grover Beach, what is the bearing of San Luis Obispo from Arroyo Grande?

Solution   Since we are looking for the bearing of San Luis Obispo *from* Arroyo Grande, we will put our N-S-E-W system on Arroyo Grande.

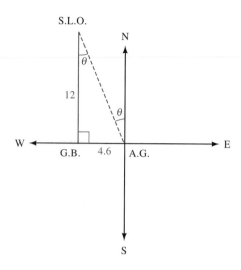

Figure 9

We solve for θ using the tangent ratio.

$$\tan \theta = \frac{4.6}{12}$$

$$\tan \theta = 0.3833$$

which means $\theta = 21°$ to the nearest degree

The bearing of San Luis Obispo from Arroyo Grande is N 21° W.   ◀

▶ **EXAMPLE 6**   A boat travels on a course of bearing N 52° 40′ E for a distance of 238 miles. How many miles north and how many miles east has the boat traveled?

Solution   In the diagram of the situation we put our N-S-E-W system at the boat's starting point (see Figure 10).

Solving for x with a sine ratio and y with a cosine ratio and rounding our answers to three significant digits, we have

If $\sin 52° 40′ = \dfrac{x}{238}$ If $\cos 52° 40′ = \dfrac{y}{238}$

then $x = 238(0.7951)$ then $y = 238(0.6065)$

$= 189$ miles $= 144$ miles

Traveling 238 miles on a line that is N 52° 40′ E will get you to the same place as traveling 144 miles north and then 189 miles east.

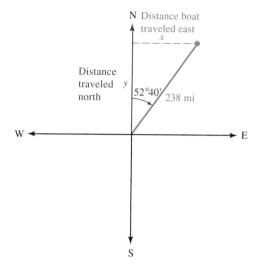

Figure 10

◀

▶ **EXAMPLE 7**   Figure 11 is a diagram that shows how Diane estimates the height of a flagpole. She can't measure the distance between herself and the flagpole directly because there is a fence in the way. So she stands at point *A* facing the pole and finds the angle of elevation from point *A* to the top of the pole to be 61.7°. Then she turns 90° and walks 25.0 feet to point *B*, where she measures the angle between her path and the base of the pole. She finds that angle is 54.5°. Use this information to find the height of the pole.

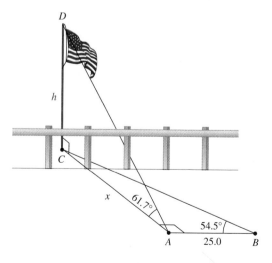

Figure 11

Solution   First we find x in right triangle ABC with a tangent ratio.

$$\tan 54.5° = \frac{x}{25.0}$$

$$x = 25.0 \tan 54.5°$$

$$= 25.0(1.4019)$$

$$= 35.0487$$

Without rounding x, we use it to find h in right triangle ACD using another tangent ratio.

$$\tan 61.7° = \frac{h}{35.0487}$$

$$h = 35.0487(1.8572)$$

$$= 65.1 \text{ feet}   \qquad\qquad \text{To 3 significant digits}$$

Note that if it weren't for the fence, she could measure x directly and use just one triangle to find the height of the flagpole.   ◀

▶ **EXAMPLE 8**   A helicopter is hovering over the desert when it develops mechanical problems and is forced to land. After landing, the pilot radios his position to a pair of radar stations located 25 miles apart along a straight road running north and south. The bearing of the helicopter from one station is N 13° E, and from the other it is S 19° E. After doing a few trigonometric calculations, one of the stations instructs the pilot to walk due west for 3.5 miles to reach the road. Is this information correct?

Solution   Figure 12 is a three-dimensional diagram of the situation. The helicopter is hovering at point D and lands at point C. The radar stations are at A and B, respectively. Since the road runs north and south, the shortest distance from C to the road is due west of C at F. To see if the pilot has the correct information, we must find y, the distance from C to F.

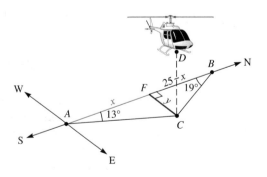

Figure 12

Since the radar stations are 25 miles apart, $AB = 25$. If we let $AF = x$, then $FB = 25 - x$. If we use cotangent ratios in triangles AFC and BFC, we will save ourselves some work.

$$\text{In triangle } AFC, \cot 13° = \frac{x}{y}$$

$$\text{which gives us } x = y \cot 13°$$

$$\text{In triangle } BFC, \cot 19° = \frac{25 - x}{y}$$

$$\text{which gives us } 25 - x = y \cot 19°$$

Solving this equation for x we have

$$-x = -25 + y \cot 19° \qquad \text{Add } -25 \text{ to each side}$$
$$x = 25 - y \cot 19° \qquad \text{Multiply each side by } -1$$

Next, we set our two values of x equal to each other.

$$x = x$$
$$y \cot 13° = 25 - y \cot 19°$$
$$y \cot 13° + y \cot 19° = 25 \qquad\qquad \text{Add } y \cot 19° \text{ to each side}$$
$$y(\cot 13° + \cot 19°) = 25 \qquad\qquad \text{Factor } y \text{ from each term}$$

$$y = \frac{25}{\cot 13° + \cot 19°} \qquad\qquad \text{Divide by the coefficient of } y$$

$$= \frac{25}{4.3315 + 2.9042}$$

$$= \frac{25}{7.2357}$$

$$= 3.5 \text{ miles} \qquad\qquad \text{To 2 significant digits}$$

The information given to the pilot is correct.   ◀

PROBLEM SET 2.4

Figure 13

Solve each of the following problems. In each case, be sure to make a diagram of the situation with all the given information labeled.

1. The two equal sides of an isosceles triangle are each 42 cm. If the base measures 30 cm, find the height and the measure of the two equal angles.

2. An equilateral triangle (one with all sides the same length) has an altitude of 4.3 inches. Find the length of the sides.

3. The height of a right circular cone is 25.3 cm. If the diameter of the base is 10.4 cm, what angle does the side of the cone make with the base? (See Figure 13.)

4. The diagonal of a rectangle is 348 mm, while the longer side is 278 mm. Find the shorter side of the rectangle and the angles the diagonal makes with the sides.

5. How long should an escalator be if it is to make an angle of 33° with the floor and carry people a vertical distance of 21 ft between floors?

6. A road up a hill makes an angle of 5.1° with the horizontal. If the road from the bottom of the hill to the top of the hill is 2.5 mi long, how high is the hill?

7. A 72.5-ft rope from the top of a circus tent pole is anchored to the ground 43.2 ft from the bottom of the pole. What angle does the rope make with the pole? (Assume the pole is perpendicular to the ground.)

8. A ladder is leaning against the top of a 7.0-ft wall. If the bottom of the ladder is 4.5 ft from the wall, what is the angle between the ladder and the wall?

9. If a 73.0-ft flagpole casts a shadow 51.0 ft long, what is the angle of elevation of the sun (to the nearest tenth of a degree)?

10. If the angle of elevation of the sun is 63.4° when a building casts a shadow of 37.5 ft, what is the height of the building?

11. A person standing 150 cm from a mirror notices that the angle of depression from his eyes to the bottom of the mirror is 12°, while the angle of elevation to the top of the mirror is 11°. Find the vertical dimension of the mirror. (See Figure 14.)

12. A person standing on top of a 15-ft high sand pile wishes to estimate the width of the pile. He visually locates two rocks on the ground below at the base of the sand pile. The rocks are on opposite sides of the sand pile, and he and the two rocks are in line with one another. If the angles of depression from the top of the sand pile to each of the rocks are 27° and 19°, how far apart are the rocks?

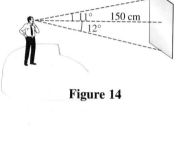

Figure 14

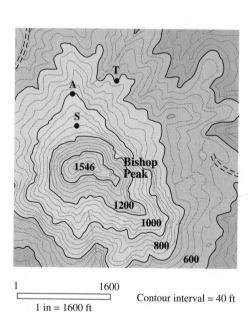

1 in = 1600 ft

Contour interval = 40 ft

Figure 15

Figure 15 shows the topographic map we used in Example 4 of this section. Recall that Stacey is at position *S* and Amy is at position *A*. In Figure 15, Travis, a third hiker, is at position *T*.

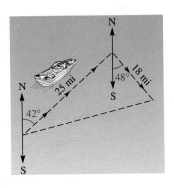

13. If the distance between *A* and *T* on the map in Figure 15 is 0.5 inch, find each of the following:
    a. the horizontal distance between Amy and Travis
    b. the difference in elevation between Amy and Travis
    c. the angle of elevation from Travis to Amy

14. If the distance between *S* and *T* on the map in Figure 15 is $\frac{5}{8}$ inch, find each of the following:
    a. the horizontal distance between Stacey and Travis
    b. the difference in elevation between Stacey and Travis
    c. the angle of elevation from Travis to Stacey

15. A boat leaves the harbor entrance and travels 25 mi in the direction N 42° E. The captain then turns the boat 90° and travels another 18 mi in the direction S 48° E. At that time, how far is the boat from the harbor entrance, and what is the bearing of the boat from the harbor entrance? (See Figure 16.)

Figure 16

16. A man wandering in the desert walks 2.3 mi in the direction S 31° W. He then turns 90° and walks 3.5 mi in the direction N 59° W. At that time, how far is he from his starting point, and what is his bearing from his starting point?

17. Lompoc, California, is 18 mi due south of Nipomo. Buellton, California, is due east of Lompoc and S 65° E from Nipomo. How far is Lompoc from Buellton?

18. A tree on one side of a river is due west of a rock on the other side of the river. From a stake 21 yd north of the rock, the bearing of the tree is S 18.2° W. How far is it from the rock to the tree?

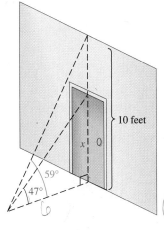

19. A boat travels on a course of bearing N 37° 10′ W for 79.5 mi. How many miles north and how many miles west has the boat traveled?

20. A boat travels on a course of bearing S 63° 50′ E for 100 mi. How many miles south and how many miles east has the boat traveled?

21. From a point on the floor the angle of elevation to the top of a door is 47°, while the angle of elevation to the ceiling above the door is 59°. If the ceiling is 10 ft above the floor, what is the vertical dimension of the door? (See Figure 17.)

22. A man standing on the roof of a building 60.0 ft high looks down to the building next door. He finds the angle of depression to the roof of that building from the roof of his building to be 34.5°, while the angle of depression from the roof of his building to the bottom of the building next door is 63.2°. How tall is the building next door?

Figure 17

23. In Figure 18, a person standing at point *A* notices that the angle of elevation to the top of the antenna is 47° 30′. A second person standing 33.0 ft farther from the antenna than the person at *A* finds the angle of elevation to the top of the antenna to be 42° 10′. How far is the person at *A* from the base of the antenna?

24. Two people decide to find the height of a tree. They position themselves 25 ft apart in line with, and on the same side of, the tree. If they find that the angles of elevation from the ground where they are standing to the top of the tree are 65° and 44°, how tall is the tree?

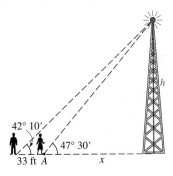

Figure 18

25. An ecologist wishes to find the height of a redwood tree that is on the other side of a creek, as shown in Figure 19. From point *A* he finds that the angle of elevation to the top of the tree is 10.7°. He then walks 24.8 ft at a right angle from point *A* and finds that the angle from the path to the tree at point *B* is 86.6°. What is the height of the tree?

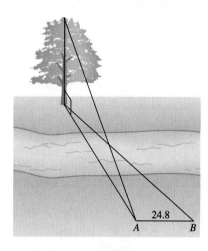

Figure 19

26. A helicopter makes a forced landing at sea. The last radio signal received at station *C* gives the bearing of the helicopter from *C* as N 57.5° E at an altitude of 426 ft. An observer at *C* sights the helicopter and gives ∠*DCB* as 12.3°. How far will a rescue boat at *A* have to travel to reach any survivors at *B*? (See Figure 20.)

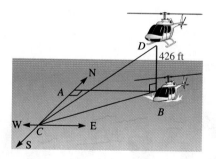

Figure 20

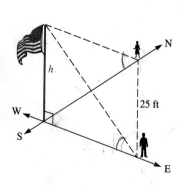

Figure 21

27. Two people decide to estimate the height of a flagpole. One person positions himself due north of the pole and the other person stands due east of the pole. If the two people are the same distance from the pole and 25 ft from each other, find the height of the pole if the angle of elevation from the ground at each person's position to the top of the pole is 56°. (See Figure 21.)

28. To estimate the height of a tree, one person positions himself due south of the tree, while another person stands due east of the tree. If the two people are the same distance from the tree and 35 ft from each other, what is the height of the tree if the angle of elevation from the ground at each person's position to the top of the tree is 48°?

29. A satellite is circling 112 mi above the earth, as shown in Figure 22. When the satellite is directly above point B, angle A is found to be 76.6°. Use this information to find the radius of the earth.

30. Suppose Figure 22 is an exaggerated diagram of a plane flying above the earth. If the plane is 4.55 mi above the earth and the radius of the earth is 4,000 mi, how far is it from the plane to the horizon? What is the measure of angle A?

31. A ship is anchored off a long straight shoreline that runs north and south. From two observation points 15 mi apart on shore, the bearings of the ship are N 31° E and S 53° E. What is the shortest distance from the ship to the shore?

32. Pat and Tim position themselves 2.5 mi apart to watch a missile launch from Vandenberg Air Force Base. When the missile is launched, Pat estimates its bearing from him to be S 75° W, while Tim estimates the bearing of the missile from his position to be N 65° W. If Tim is due south of Pat, how far is Tim from the missile when it is launched?

Figure 23 shows the Spiral of Roots we mentioned in the previous chapter. Notice that we have labeled the angles at the center of the spiral with θ_1, θ_2, θ_3, and so on.

33. Find the values of θ_1, θ_2, and θ_3, accurate to the nearest hundredth of a degree.

34. If θ_n stands for the nth angle formed at the center of the Spiral of Roots, find a formula for $\sin \theta_n$.

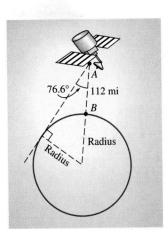

Figure 22

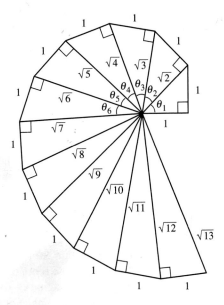

Figure 23

Review Problems

The following problems review material we covered in Section 1.5.

35. Expand and simplify: $(\sin \theta - \cos \theta)^2$

36. Subtract: $\dfrac{1}{\cos \theta} - \cos \theta$

Show that each of the following statements is true by transforming the left side of each one into the right side.

37. $\sin \theta \cot \theta = \cos \theta$

38. $\cos \theta \csc \theta \tan \theta = 1$

39. $\dfrac{\sec \theta}{\tan \theta} = \csc \theta$

40. $(1 - \cos \theta)(1 + \cos \theta) = \sin^2 \theta$

41. $\sec \theta - \cos \theta = \dfrac{\sin^2 \theta}{\cos \theta}$

42. $1 - \dfrac{\cos \theta}{\sec \theta} = \sin^2 \theta$

Research Project

43. One of the items we discussed in this section was topographic maps. The process of making one of these maps is an interesting one. It involves aerial photography and different colored projections of the resulting photographs. Research the process used to draw the contour lines on a topographic map, and then give a detailed explanation of that process.

SECTION 2.5

Vectors: A Geometric Approach

By most accounts, the study of vectors is based on the work of Irish mathematician Sir William Hamilton (1805–1865). Hamilton was actually studying complex numbers (a topic we will cover in Chapter 8) when he made the discoveries that led to what we now call vectors.

Today, vectors are treated both algebraically and geometrically. In this section, we will focus our attention on the geometric representation of vectors. We will cover the algebraic approach later in the book in Section 7.5. We begin with a discussion of vector quantities and scalar quantities.

Many of the quantities that describe the world around us have both magnitude and direction, while others have only magnitude. Quantities that have magnitude and direction are called *vector quantities,* while quantities with magnitude only are called *scalars.* Some examples of vector quantities are force, velocity, and acceleration. For example, a car traveling 50 miles per hour due south has a different velocity from another car traveling due north at 50 miles per hour, while a third car traveling at 25 miles per hour due north has a velocity that is different from both of the first two.

One way to represent vector quantities is with arrows. The direction of the arrow represents the direction of the vector quantity, and the length of the arrow corresponds to the magnitude. For example, the velocities of the three cars we mentioned above could be represented like this:

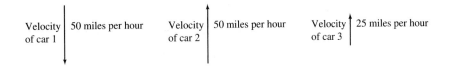

Figure 1

Notation

To distinguish between vectors and scalars, we will write the letters used to represent vectors with boldface type, such as **U** or **V**. (When you write them on paper, put an arrow above them like this: $\vec{U}$ or $\vec{V}$.) The magnitude of a vector is represented with absolute value symbols. For example, the magnitude of **V** is written $|\mathbf{V}|$. Table 1 illustrates further.

Table 1

Notation	The quantity is
V	a vector
$\vec{V}$	a vector
$\overrightarrow{AB}$	a vector
x	a scalar
$|\mathbf{V}|$	the magnitude of vector **V**, a scalar

Equality for Vectors

The position of a vector in space is unimportant. Two vectors are equivalent if they have the same magnitude and direction.

In Figure 2, $\mathbf{V}_1 = \mathbf{V}_2 \neq \mathbf{V}_3$. The vectors $\mathbf{V}_1$ and $\mathbf{V}_2$ are equivalent because they have the same magnitude and the same direction. Notice also that $\mathbf{V}_3$ and $\mathbf{V}_4$ have the same magnitude but opposite directions. This means that $\mathbf{V}_4$ is the opposite of $\mathbf{V}_3$, or $\mathbf{V}_4 = -\mathbf{V}_3$.

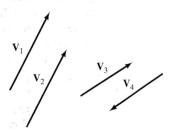

Figure 2

Addition and Subtraction of Vectors

The sum of the vectors **U** and **V**, written **U** + **V**, is called the *resultant vector*. It is the vector that extends from the tail of **U** to the tip of **V** when the tail of **V** is placed at the tip of **U**, as illustrated in Figure 3. Note that this diagram shows the resultant vector to be a diagonal in the parallelogram that has **U** and **V** as adjacent sides. This being the case, we could also add the vectors by putting the tails of **U** and **V** together to form the adjacent sides of that same parallelogram, as shown in Figure 4. In either case, the resultant vector is the diagonal that starts at the tail of **U**.

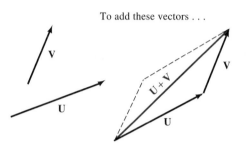

To add these vectors . . .

. . . place the tail of one on the tip of the other . . .   **Figure 3**

. . .or place their tails together.

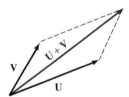

Figure 4

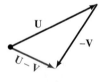

Figure 5

To subtract one vector from another, we can add its opposite. That is,

$$\mathbf{U} - \mathbf{V} = \mathbf{U} + (-\mathbf{V})$$

If **U** and **V** are the vectors shown in Figure 3, then their difference, **U** − **V**, is shown in Figure 5.

▶ **EXAMPLE 1**   A boat is crossing a river that runs due north. The boat is pointed due east and is moving through the water at 12 miles per hour. If the current of the river is a constant 5.1 miles per hour, find the true course of the boat through the water to two significant digits.

Solution   Problems like this are a little difficult to read the first time they are encountered. Even though the boat is ''headed'' due east, the current is pushing it a little toward the north, so it is actually on a course that will take it east and a little north. By representing the heading of the boat and the current of the water with vectors, we can find the true course of the boat.

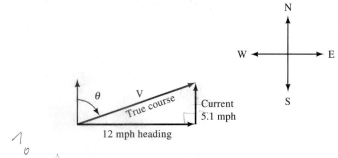

Figure 6

We find θ using a tangent ratio.

$$\tan \theta = \frac{12}{5.1} = 2.3529$$

so

$$\theta = 67° \text{ to the nearest degree}$$

If we let **V** represent the true course of the boat, then we can find the magnitude of **V** using the Pythagorean Theorem or a trigonometric ratio. Using the sine ratio, we have

$$\sin \theta = \frac{12}{|\mathbf{V}|}$$

$$|\mathbf{V}| = \frac{12}{\sin \theta}$$

$$= \frac{12}{\sin 67°}$$

$$= 13$$

The true course of the boat is 13 miles per hour at N 67° E. That is, the vector **V**, which represents the motion of the boat with respect to the banks of the river, has a magnitude of 13 miles per hour and a direction of N 67° E. ◀

Navigation

In Section 2.4, we used bearing to give the position of one point from another and to give the direction of a moving object. Another way to specify the direction of a moving object is with an angle between the north-south line and the vector representing the path of the object. By agreement, the angle is always measured clockwise from due north.

Figure 7 shows two vectors that represent the velocities of two ships—one traveling at 14 miles per hour in the direction 120° from due north, and the other traveling at 12 miles per hour at 240° from due north. Note that this method of giving the path of a moving object is a little simpler than the bearing method used in Section 2.4. To give the path of the ship traveling at 14 miles per hour using the bearing method, we would say it is on a course with bearing S 60° E.

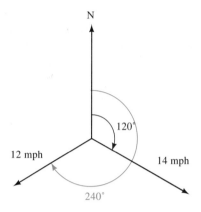

Figure 7

In Example 1 we found that although a boat is headed in one direction, its actual course may be in a different direction because of the current of the river. The same type of thing can happen with a plane if the air currents are in a different direction from the direction in which the plane is headed.

To generalize the vocabulary associated with these situations, we say the direction in which a plane or boat is headed is called its *heading,* while the direction in which the plane or boat is moving with respect to the ground is called its *true course.* Both heading and true course can be given, as above, by angles measured clockwise from due north. We further note that the airspeed of a plane is the speed with which it is moving through the air. The ground speed is its speed with respect to the ground below.

As illustrated in Figure 8, the vector representing ground speed and true course is the resultant vector found by adding the vector representing air speed and heading and the vector representing wind speed and direction.

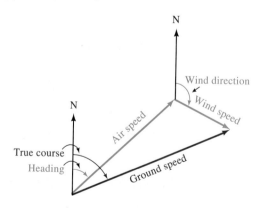

Figure 8

▶ **EXAMPLE 2**   A plane is flying at 170 miles per hour with heading 52.5°. The wind currents are a constant 35.0 miles per hour at 142.5° from due north. Find the ground speed and true course of the plane.

Solution   We represent the velocity of the plane and the velocity of the wind with vectors, the resultant sum of which will give the ground speed and true course of the plane. Figure 9 illustrates. (Note that when we say the wind currents are at 142.5° from due north, we are assuming the 142.5° is measured clockwise from due north.)

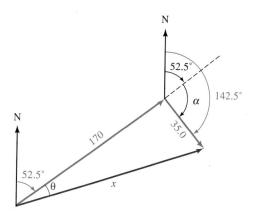

Figure 9

As you can see, we have extended a dotted line in the direction of our heading vector. Note that α is a right angle, since the difference of 142.5 and 52.5 is 90. (If the triangle formed were not a right triangle, we would use the methods in Chapter 7 to solve this problem.)

The ground speed of the plane can be found using the Pythagorean Theorem.

$$x = \sqrt{170^2 + 35.0^2}$$
$$= \sqrt{30,125}$$
$$= 174 \text{ miles per hour} \quad \text{To 3 significant digits}$$

To find the true course of the plane, we first find θ using a tangent ratio.

$$\tan \theta = \frac{35.0}{170} = 0.2059$$

giving $\theta = 11.6°$ To the nearest tenth of a degree

The true course of the plane is $52.5° + 11.6° = 64.1°$. ◀

Horizontal and Vertical Vectors

Many times it is convenient to write vectors in terms of horizontal and vertical vectors. To do so, we first superimpose a coordinate system on the vector in question

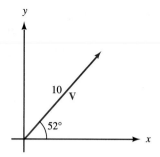

Figure 10

so that the tail of the vector is at the origin. Figure 10 shows a vector with magnitude 10 making an angle of 52° with the horizontal.

Two horizontal and vertical vectors whose sum is **V** are shown in Figure 11.

Note that in Figure 11 we labeled the horizontal vector as $\mathbf{V}_x$ and the vertical as $\mathbf{V}_y$. We can find the magnitudes of these vectors by using sine and cosine ratios.

$$|\mathbf{V}_x| = |\mathbf{V}| \cos 52° = 10(0.6157) = 6.2 \text{ to the nearest tenth}$$
$$|\mathbf{V}_y| = |\mathbf{V}| \sin 52° = 10(0.7880) = 7.9 \text{ to the nearest tenth}$$

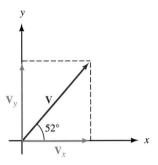

Figure 11

▶ **EXAMPLE 3**   A bullet is fired into the air with an initial velocity of 1,500 feet per second at an angle of 30° from the horizontal. Find the horizontal and vertical vectors of the velocity vector.

Solution   Figure 12 is a diagram of the situation.

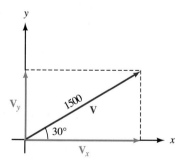

Figure 12

The magnitudes of $\mathbf{V}_x$ and $\mathbf{V}_y$ from Figure 12 to two significant digits are as follows:

$$|\mathbf{V}_x| = 1,500 \cos 30° = 1,300 \text{ feet per second}$$
$$|\mathbf{V}_y| = 1,500 \sin 30° = 750 \text{ feet per second}$$

The bullet has a horizontal velocity of 1,300 feet per second and a vertical velocity of 750 feet per second.   ◀

The magnitude of a vector can be written in terms of the magnitude of its horizontal and vertical vectors.

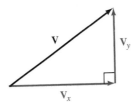

Figure 13

By the Pythagorean Theorem we have

$$|\mathbf{V}| = \sqrt{|\mathbf{V}_x|^2 + |\mathbf{V}_y|^2}$$

▶ **EXAMPLE 4**   An arrow is shot into the air so that its horizontal velocity is 25 feet per second and its vertical velocity is 15 feet per second. Find the velocity of the arrow.

Solution   Figure 14 shows the velocity vector along with the angle of elevation of the velocity vector.

The magnitude of the velocity is given by

$$|\mathbf{V}| = \sqrt{25^2 + 15^2}$$
$$= 29 \text{ feet per second, to the nearest whole number}$$

We can find the angle of elevation using a tangent ratio.

$$\tan \theta = \frac{|\mathbf{V}_y|}{|\mathbf{V}_x|}$$

$$= \frac{15}{25}$$

$$\tan \theta = 0.6$$

so $\theta = 31°$ to the nearest degree

The arrow was shot into the air at 29 feet per second at an angle of elevation of 31°.   ◀

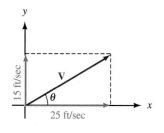

Figure 14

▶ **EXAMPLE 5**   A boat travels 72 miles on a course with direction 27° and then changes its course to travel 37 miles at 55°. How far north and how far east has the boat traveled on this 109-mile trip?

Solution   We can solve this problem by representing each part of the trip with a vector and then writing each vector in terms of its horizontal and vertical vectors. Figure 15 shows the vectors that represent the two parts of the trip. As Figure 15 indicates, the total distance traveled east is given by the sum of the horizontal components, while the total distance traveled north is given by the sum of the vertical components.

$$\text{Total distance traveled east} = |\mathbf{U}_x| + |\mathbf{V}_x|$$
$$= 72 \cos 63° + 37 \cos 35°$$
$$= 63 \text{ miles}$$
$$\text{Total distance traveled north} = |\mathbf{U}_y| + |\mathbf{V}_y|$$
$$= 72 \sin 63° + 37 \sin 35°$$
$$= 85 \text{ miles}$$

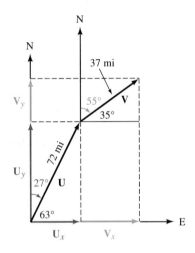

Figure 15

Force

Another important vector quantity is *force*. We can loosely define force as a push or a pull. The most intuitive force in our lives is the force of gravity that pulls us towards the center of the earth. The magnitude of this force is our weight; the direction of this force is always straight down toward the center of the earth.

Imagine a 10-pound bronze sculpture sitting on a coffee table. The force of gravity pulls the sculpture downward with a force of magnitude 10 pounds. At the same time, the table pushes the sculpture upward with a force of magnitude 10 pounds. The net result is that the sculpture remains motionless; the two forces, represented by vectors, add to zero.

Although there may be many forces acting on an object at the same time, if the object is stationary, the sum of the forces must be 0. This leads us to our next definition.

Figure 16

DEFINITION   Static Equilibrium

When an object is stationary (at rest) we say it is in the state of *static equilibrium*. When an object is in this state, the sum of the forces acting on the object must be 0.

Figure 17

▶ **EXAMPLE 6**   Danny is 5 years old and weighs 42.0 pounds. He is sitting on a swing when his sister Stacey pulls him and the swing back horizontally through an angle of 30.0° and then stops. Find the tension in the ropes of the swing. (Figure 17 is a diagram of the situation.)

Solution   As you can see from Figure 17, there are three forces acting on Danny (and the swing), which we have labeled **W, H,** and **T.** The vector **W** is due to the force of gravity pulling him toward the center of the earth. Its magnitude is $|\mathbf{W}| = 42.0$ pounds, and its direction is straight down. The vector **H** represents the force with which Stacey is pulling Danny horizontally, and **T** is the force acting on Danny in the direction of the ropes. We call this force the *tension* in the ropes.

If we rearrange the vectors in the diagram in Figure 17, we can get a better picture of the situation. Since Stacey is holding Danny in the position shown in Figure 17, he is in the state of static equilibrium. Therefore, the sum of the forces acting on him is 0. It is because of this fact that we can redraw our diagram so that the tip of **T** coincides with the tail of **W**, forming a right triangle.

The lengths of the sides in the right triangle shown in Figure 18 are given by the magnitudes of the vectors. We use right triangle trigonometry to find the magnitude of **T.**

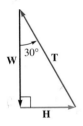

Figure 18

$$\cos 30.0° = \frac{|\mathbf{W}|}{|\mathbf{T}|} \qquad \text{Definition of cosine}$$

$$|\mathbf{T}| = \frac{|\mathbf{W}|}{\cos 30.0°} \qquad \text{Solve for } |\mathbf{T}|$$

$$= \frac{42.0}{0.8660} \qquad \text{The magnitude of } \mathbf{W} \text{ is } 42.0$$

$$= 48.5 \text{ pounds} \quad \text{To the nearest tenth of a pound}$$

Next, let's find the magnitude of the force with which Stacey pulls on Danny to keep him in static equilibrium.

$$\tan 30.0° = \frac{|\mathbf{H}|}{|\mathbf{W}|} \qquad \text{Definition of tangent}$$

$$|\mathbf{H}| = |\mathbf{W}| \tan 30.0° \quad \text{Solve for } |\mathbf{H}|$$

$$= 42.0(0.5774)$$

$$= 24.2 \text{ pounds} \qquad \text{To the nearest tenth of a pound}$$

Stacey must pull horizontally with a force of magnitude 24.2 pounds to hold Danny at an angle of 30.0° from vertical.   ◀

PROBLEM SET 2.5

Draw vectors representing the following velocities:

1. 30 mph due north
2. 30 mph due south
3. 30 mph due east
4. 30 mph due west
5. 50 cm/sec N 30° W
6. 50 cm/sec N 30° E
7. 20 ft/min S 60° E
8. 20 ft/min S 60° W

Draw vectors representing the course of a ship that travels

9. 75 mi on a course with direction 30°
10. 75 mi on a course with direction 330°
11. 25 mi on a course with direction 135°
12. 25 mi on a course with direction 225°

13. A boat is crossing a river that runs due north. The heading of the boat is due east, and it is moving through the water at 12.0 mph. If the current of the river is a constant 3.25 mph, find the true course of the boat.
14. A boat is crossing a river that runs due east. The heading of the boat is due south, and its speed is 11.0 ft/sec. If the current of the river is 2.50 ft/sec, find the true course of the boat.
15. A plane has an airspeed of 195 mph and a heading of 30.0°. The air currents are moving at a constant 32.5 mph at 300.0° from due north. Find the ground speed and true course of the plane.
16. A plane has an airspeed of 140 mph and a heading of 130°. The wind is blowing at a constant 14.0 mph at 40° from due north. Find the ground speed and true course of the plane.
17. A ship headed due east is moving through the water at a constant 12 mph. However, the true course of the ship is 78°. If the current of the water is running due north at a constant rate of speed, find the speed of the current.
18. A plane headed due east is moving through the air at a constant 180 mph. Its true course, however, is 65.0°. If the wind currents are moving due north at a constant rate, find the speed of these currents.
19. A person is riding in a hot air balloon. For the first hour and a half the wind current is a constant 22.0 mph in the direction N 37.5° E. Then the wind current changes to 18.5 mph and heads the balloon in the direction S 52.5° E. If this continues for another 2 hours, how far is the balloon from its starting point? What is the bearing of the balloon from its starting point? (See Figure 19.)
20. Two planes take off at the same time from an airport. The first plane is flying at 255 mph on a course of 135°. The second plane is flying in the direction 225° at 275 mph. If there are no wind currents blowing, how far apart are they after 2 hours? What is the bearing of the second plane from the first after 2 hours?
21. If a navigation error puts a plane 3.0° off course, how far off course is the plane after flying 130 mi?
22. A ship is 2.8° off course. If the ship is traveling at 14 mph, how far off course will it be after 2 hours?

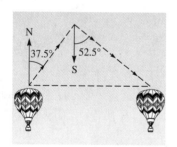

Figure 19

Each problem below refers to a vector **V** with magnitude $|\mathbf{V}|$ that forms an angle θ with the positive x-axis. In each case, give the magnitude of the horizontal and vertical vectors of **V**, namely $\mathbf{V}_x$ and $\mathbf{V}_y$, respectively.

23. $|\mathbf{V}| = 13.8$, $\theta = 24.2°$

24. $|\mathbf{V}| = 17.6$, $\theta = 67.2°$

25. $|\mathbf{V}| = 420$, $\theta = 36° \, 10'$

26. $|\mathbf{V}| = 380$, $\theta = 16° \, 40'$

27. $|\mathbf{V}| = 64$, $\theta = 150°$

28. $|\mathbf{V}| = 48$, $\theta = 120°$

For each problem below, the magnitude of the horizontal and vertical vectors, $\mathbf{V}_x$ and $\mathbf{V}_y$, of vector **V** are given. In each case find the magnitude of **V**.

29. $|\mathbf{V}_x| = 35.0$, $|\mathbf{V}_y| = 26.0$

30. $|\mathbf{V}_x| = 45.0$, $|\mathbf{V}_y| = 15.0$

31. $|\mathbf{V}_x| = 4.5$, $|\mathbf{V}_y| = 3.8$

32. $|\mathbf{V}_x| = 2.2$, $|\mathbf{V}_y| = 5.8$

33. A bullet is fired into the air with an initial velocity of 1,200 ft/sec at an angle of 45° from the horizontal. Find the magnitude of the horizontal and vertical vectors of the velocity vector.

34. A bullet is fired into the air with an initial velocity of 1,800 ft/sec at an angle of 60° from the horizontal. Find the horizontal and vertical vectors of the velocity.

35. Use the results of Problem 33 to find the horizontal distance traveled by the bullet in 3 seconds. (Neglect the resistance of air on the bullet.)

36. Use the results of Problem 34 to find the horizontal distance traveled by the bullet in 2 seconds.

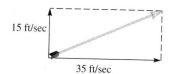

15 ft/sec

35 ft/sec

Figure 20

37. An arrow is shot into the air so that its horizontal velocity is 35.0 ft/sec and its vertical velocity is 15.0 ft/sec. Find the velocity of the arrow. (See Figure 20.)

38. The horizontal and vertical components of the velocity of an arrow shot into the air are 15.0 ft/sec and 25.0 ft/sec, respectively. Find the velocity of the arrow.

39. A ship travels 135 km on a 138° course. How far east and how far south has it traveled?

40. A plane flies for 3 hours at 230 km/hr on a 215° course. How far west and how far south does it travel in the 3 hours?

41. A plane travels 175 mi on a course of 18° and then changes its course to 49° and travels another 120 mi. Find the total distance traveled north and the total distance traveled east.

42. A ship travels on a course of 168° for 68 mi and then changes its course to 120° and travels another 112 mi. Find the total distance south and the total distance east that the ship traveled.

43. Repeat the swing problem shown in Example 6 if Stacey pulls Danny through an angle of 45.0° and then holds him in static equilibrium. Find the magnitude of both **H** and **T**.

44. The diagram in Figure 18 of this section would change if Stacey were to push Danny forward through an angle of 30° and then hold him in that position. Draw the diagram that corresponds to this new situation.

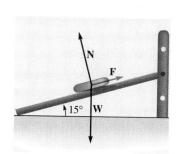

Figure 21

45. A 10-pound weight is lying on a situp bench at the gym. If the bench is inclined at an angle of 15°, there are three forces acting on the weight, as shown in Figure 21. **N** is called the normal force and it acts in the direction perpendicular to the

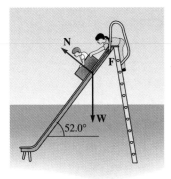

Figure 22

bench. **F** is the force due to friction that holds the weight on the bench. If the weight does not move, then the sum of these three forces is 0. Find the magnitude of **N** and the magnitude of **F**.

46. Repeat Problem 45 for a 25.0-pound weight and a bench inclined at 10.0°.

47. Danny and Stacey have gone from the swing (Example 6) to the slide at the park. The slide is inclined at an angle of 52.0°. Danny weighs 42.0 pounds. He is sitting in a cardboard box with a piece of wax paper on the bottom. Stacey is at the top of the slide holding on to the cardboard box (Figure 22). Find the magnitude of the force Stacey must pull with, in order to keep Danny from sliding down the slide. (We are assuming that the wax paper makes the slide into a frictionless surface, so that the only force keeping Danny from sliding is the force with which Stacey pulls.)

48. Repeat Problem 47 for a slide inclined at an angle of 65.0°.

Review Problems

The problems that follow review material we covered in Section 1.3.

49. Draw 135° in standard position, locate a convenient point on the terminal side, and then find sin 135°, cos 135°, and tan 135°.

50. Draw −270° in standard position, locate a convenient point on the terminal side, and then find sine, cosine, and tangent of −270°.

51. Find $\sin \theta$ and $\cos \theta$ if the terminal side of θ lies along the line $y = 2x$ in quadrant I.

52. Find $\sin \theta$ and $\cos \theta$ if the terminal side of θ lies along the line $y = -x$ in quadrant II.

53. Find x if the point $(x, -8)$ is on the terminal side of θ and $\sin \theta = -\frac{4}{5}$.

54. Find y if the point $(-6, y)$ is on the terminal side of θ and $\cos \theta = -\frac{3}{5}$.

CHAPTER 2 SUMMARY

Examples

1.

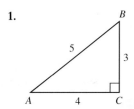

Definition II for Trigonometric Functions [2.1]

If triangle ABC is a right triangle with $C = 90°$, then the six trigonometric functions for angle A are

$$\sin A = \frac{\text{side opposite } A}{\text{hypotenuse}} = \frac{a}{c}$$

$$\sin A = \frac{3}{5} = \cos B$$

$$\cos A = \frac{\text{side adjacent } A}{\text{hypotenuse}} = \frac{b}{c}$$

$$\cos A = \frac{4}{5} = \sin B$$

$$\tan A = \frac{\text{side opposite } A}{\text{side adjacent } A} = \frac{a}{b}$$

$$\tan A = \frac{3}{4} = \cot B$$

$$\cot A = \frac{\text{side adjacent } A}{\text{side opposite } A} = \frac{b}{a}$$

$$\cot A = \frac{4}{3} = \tan B$$

$$\sec A = \frac{\text{hypotenuse}}{\text{side adjacent } A} = \frac{c}{b}$$

$$\sec A = \frac{5}{4} = \csc B$$

$$\csc A = \frac{\text{hypotenuse}}{\text{side opposite } A} = \frac{c}{a}$$

$$\csc A = \frac{5}{3} = \sec B$$

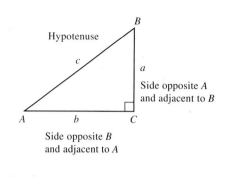

Hypotenuse

Side opposite A and adjacent to B

Side opposite B and adjacent to A

2.
$\sin 3° = \cos 87°$
$\cos 10° = \sin 80°$
$\tan 15° = \cot 75°$
$\cot A = \tan (90° - A)$
$\sec 30° = \csc 60°$
$\csc 45° = \sec 45°$

Cofunction Theorem [2.1]

A trigonometric function of an angle is always equal to the cofunction of its complement. In symbols, since the complement of x is $90° - x$, we have

$$\sin x = \cos (90° - x)$$

$$\cos x = \sin (90° - x)$$

$$\tan x = \cot (90° - x)$$

3. The values given in the table are called *exact values* because they are not decimal approximations as you would find on a calculator.

Trigonometric Functions of Special Angles [2.1]

θ	30°	45°	60°
$\sin \theta$	$\dfrac{1}{2}$	$\dfrac{1}{\sqrt{2}}$ or $\dfrac{\sqrt{2}}{2}$	$\dfrac{\sqrt{3}}{2}$
$\cos \theta$	$\dfrac{\sqrt{3}}{2}$	$\dfrac{1}{\sqrt{2}}$ or $\dfrac{\sqrt{2}}{2}$	$\dfrac{1}{2}$
$\tan \theta$	$\dfrac{1}{\sqrt{3}}$ or $\dfrac{\sqrt{3}}{3}$	1	$\sqrt{3}$

4.
$47° 30'$
$+23° 50'$
$\overline{70° 80'} = 71° 20'$

Degrees, Minutes, and Seconds [2.2]

There are 360° (degrees) in one complete rotation, 60' (minutes) in one degree, and 60″ (seconds) in one minute. This is equivalent to saying 1 minute is 1/60 of a degree, and 1 second is 1/60 of a minute.

5.
$$74.3° = 74° + 0.3°$$
$$= 74° + 0.3(60')$$
$$= 74° + 18'$$
$$= 74° 18'$$
$$42° 48' = 42° + \left(\frac{48}{60}\right)°$$
$$= 42° + 0.8°$$
$$= 42.8°$$

Converting to and from Decimal Degrees [2.2]

To convert from decimal degrees to degrees and minutes, multiply the fractional part of the angle (that which follows the decimal point) by 60 to get minutes.

To convert from degrees and minutes to decimal degrees, divide minutes by 60 to get the fractional part of the angle.

6. These angles and sides correspond in accuracy.

$a = 24$	$A = 39°$
$a = 5.8$	$A = 45°$
$a = 62.3$	$A = 31.3°$
$a = 0.498$	$A = 42.9°$
$a = 2.77$	$A = 37° 10'$
$a = 49.87$	$A = 43° 18'$
$a = 6.932$	$A = 24.81°$

Significant Digits [2.3]

The number of *significant digits* (or figures) in a number is found by counting the number of digits from left to right, beginning with the first nonzero digit on the left and disregarding the decimal point.

The relationship between the accuracy of the sides in a triangle and the accuracy of the angles in the same triangle is given below.

Accuracy of sides	Accuracy of angles
Two significant digits	Nearest degree
Three significant digits	Nearest 10 minutes or tenth of a degree
Four significant digits	Nearest minute or hundredth of a degree

7.

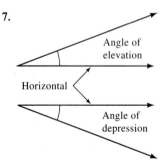

Angle of Elevation and Angle of Depression [2.4]

An angle measured from the horizontal up is called an *angle of elevation*. An angle measured from the horizontal down is called an *angle of depression*.

8.

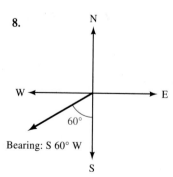

Direction [2.4, 2.5]

There are two ways to specify the direction of a line or vector.

    a. The first method of giving direction is called *bearing*. The notation used to designate bearing begins with N or S, followed by the number of degrees in the angle, and ends with E or W, as in S 60° W.

    b. We can also give the direction of a line or vector by simply stating the angle through which the line or vector has been rotated clockwise from due north, as in a direction of 240°.

9. If car 1 is traveling at 50 mph due south, car 2 at 50 mph due north, and car 3 at 25 mph due north, then the velocities of the three cars can be represented with vectors.

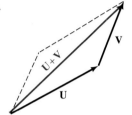

50	50	25
Velocity	Velocity	Velocity
of car 1	of car 2	of car 3

10.

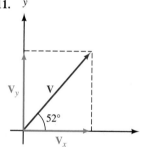

11.

Vectors [2.5]

Quantities that have both magnitude and direction are called *vector quantities,* while quantities that have only magnitude are called *scalar quantities.* We represent vectors graphically by using arrows. The length of the arrow corresponds to the magnitude of the vector, and the direction of the arrow corresponds to the direction of the vector. In symbols, we denote the magnitude of vector **V** with $|\mathbf{V}|$.

Addition of Vectors [2.5]

The sum of the vectors **U** and **V**, written **U** + **V**, is the vector that extends from the tail of **U** to the tip of **V** when the tail of **V** coincides with the tip of **U**.

Horizontal and Vertical Vectors [2.5]

The horizontal and vertical vectors of vector **V** are the horizontal and vertical vectors whose sum is **V**. The horizontal vector is denoted by $\mathbf{V}_x$, and the vertical vector is denoted by $\mathbf{V}_y$.

CHAPTER 2 TEST

Find $\sin A$, $\cos A$, $\tan A$, and $\sin B$, $\cos B$, and $\tan B$ in right triangle ABC, with $C = 90°$, if

1. $a = 1$ and $b = 2$ **2.** $b = 3$ and $c = 6$
3. $a = 3$ and $c = 5$ **4.** $a = 5$ and $b = 12$

Fill in the blanks to make each statement true.

5. $\sin 14° = \cos$ ____ **6.** $\sec$ ____ $= \csc 73°$

Simplify each expression as much as possible.

7. $\sin^2 45° + \cos^2 30°$ **8.** $\tan 45° + \cot 45°$

9. $\sin^2 60° - \cos^2 30°$

10. $\dfrac{1}{\sec 30°}$

11. Add $48° \ 18'$ and $24° \ 52'$.

12. Subtract $15° \ 32'$ from $25° \ 15'$.

Convert to degrees and minutes.

13. $73.2°$

14. $16.45°$

Convert to decimal degrees.

15. $2° \ 48'$

16. $79° \ 30'$

Use a calculator to find the following:

17. $\sin 24° \ 20'$

18. $\cos 37.8°$

19. $\tan 63° \ 50'$

20. $\cot 71° \ 20'$

Use a calculator to find θ if θ is an acute angle and

21. $\tan \theta = 0.0816$

22. $\sec \theta = 1.923$

23. $\sin \theta = 0.9465$

24. $\cos \theta = 0.9730$

Give the number of significant digits in each number.

25. 49.35

26. 0.0028

The following problems refer to right triangle ABC with $C = 90°$. In each case, find all the missing parts.

27. $a = 68.0$ and $b = 104$

28. $a = 24.3$ and $c = 48.1$

29. $b = 305$ and $B = 24.9°$

30. $c = 0.462$ and $A = 35° \ 30'$

31. If the altitude of an isosceles triangle is 25 cm and each of the two equal angles measures $17°$, how long are the two equal sides?

32. If the angle of elevation of the sun is $75° \ 30'$, how tall is a post that casts a shadow 1.5 ft long?

33. Two guy wires from the top of a 35-ft tent pole are anchored to the ground below by two stakes so that the two stakes and the tent pole lie along the same line. If the angles of depression from the top of the pole to each of the stakes are $47°$ and $43°$, how far apart are the stakes? (Assume the tent pole is perpendicular to the ground and between the two stakes.)

34. If vector **V** has magnitude 5.0 and makes an angle of $30°$ with the positive x-axis, find the magnitude of the horizontal and vertical vectors of **V**.

35. Vector **V** has a horizontal vector with magnitude 11 and a vertical vector with magnitude 31. What is the angle formed by **V** and the positive x-axis?

36. A bullet is fired into the air with an initial velocity of 800 feet per second at an angle of $62°$ from the horizontal. Find the magnitude of the horizontal and vertical vectors of the velocity vector.

37. A ship travels 120 mi on a $120°$ course. How far east and how far south has the ship traveled?

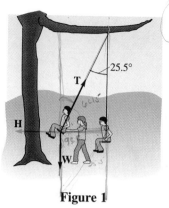

Figure 1

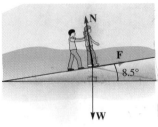

Figure 2

38. A plane has an airspeed of 245 mph and a heading of 128.5°. The wind is blowing at a constant 45.5 mph at 38.5° from due north. Find the ground speed and true course of the plane.

39. Tyler and his cousin Kelly have attached a rope to the branch of a tree and tied a board to the other end to form a swing. Tyler sits on the board while his cousin pushes him through an angle of 25.5° and holds him there. If Tyler weighs 95.5 pounds, find the magnitude of the force Kelly must push with horizontally to keep Tyler in static equilibrium. See Figure 1.

40. After they are done swinging, Tyler and Kelly decide to rollerskate. They come to a hill that is inclined at 8.5°. Tyler pushes Kelly halfway up the hill and then holds her there (Figure 2). If Kelly weighs 58.0 pounds, find the magnitude of the force Tyler must push with to keep Kelly from rolling down the hill. (We are assuming that the rollerskates make the hill into a frictionless surface so that the only force keeping Kelly from rolling backwards down the hill is the force Tyler is pushing with.)

3

Radian Measure

│ │ │ │ │ │ │ │ │ │

▶ **To the Student**

In Chapters 1 and 2, we used degree measure exclusively to give the measure of angles. We will begin this chapter with another kind of angle measure called radian measure. Radian measure gives us a way to measure angles with real numbers instead of degrees. As you will see, there are a number of situations that occur in trigonometry for which real numbers are the more appropriate measure for angles.

The most important material in the chapter, from the standpoint of what is needed to continue on through the book, is the material in the first three sections. These are the sections in which you will be introduced to, and become familiar with, radian measure. In Chapter 4, when we graph the different trigonometric functions, we will work almost exclusively in radian measure.

SECTION 3.1 **Reference Angle**

In the previous chapter we found exact values for trigonometric functions of certain angles between $0°$ and $90°$. By using what are called *reference angles,* we can find exact values for trigonometric functions of angles outside the interval $0°$ to $90°$.

> DEFINITION   The *reference angle* (sometimes called related angle) for any angle θ in standard position is the positive acute angle between the terminal side of θ and the *x*-axis. In this book, we will denote the reference angle for θ by $\hat{\theta}$.

Note that, for this definition, $\hat{\theta}$ is always positive and always between 0° and 90°. That is, a reference angle is always an acute angle.

▶ **EXAMPLE 1**   Name the reference angle for each of the following angles.

a. 30°   **b.** 135°   **c.** 240°   **d.** 330°

Solution   We draw each angle in standard position. The reference angle is the positive acute angle formed by the terminal side of the angle in question and the *x*-axis.

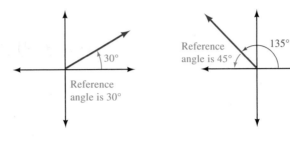

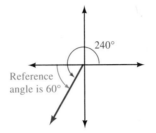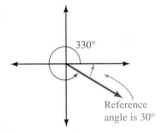

Figure 1 ◀

We can generalize the results of Example 1 as follows: If θ is a positive angle between 0° and 360°, then

$$\text{If } \theta \in \text{QI, then } \hat{\theta} = \theta$$
$$\text{If } \theta \in \text{QII, then } \hat{\theta} = 180° - \theta$$
$$\text{If } \theta \in \text{QIII, then } \hat{\theta} = \theta - 180°$$
$$\text{If } \theta \in \text{QIV, then } \hat{\theta} = 360° - \theta$$

We can use our information on reference angles and the signs of the trigonometric functions to write the following theorem.

Reference Angle Theorem

A trigonometric function of an angle and its reference angle differ at most in sign.

We will not give a detailed proof of this theorem, but rather, justify it by example. Let's look at the sines of all the angles between $0°$ and $360°$ that have a reference angle of $30°$. These angles are $30°$, $150°$, $210°$, and $330°$.

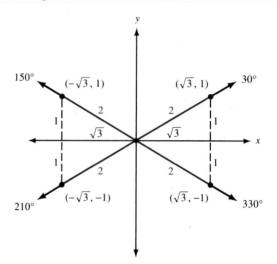

Figure 2

$$\sin 150° = \sin 30° = \frac{1}{2}$$

They differ in sign only.

$$\sin 210° = \sin 330° = -\frac{1}{2}$$

As you can see, any angle with a reference angle of $30°$ will have a sine of $\frac{1}{2}$ or $-\frac{1}{2}$. The sign, $+$ or $-$, will depend on the quadrant in which the angle terminates. Using this discussion as justification, we write the following steps used to find trigonometric functions of angles between $0°$ and $360°$.

Step 1. Find $\hat{\theta}$, the reference angle.

Step 2. Determine the sign of the trigonometric function based on the quadrant in which θ terminates.

Step 3. Write the original trigonometric function of θ in terms of the same trigonometric function of $\hat{\theta}$.

Step 4. Find the trigonometric function of $\hat{\theta}$.

▶ **EXAMPLE 2**   Find the exact value of $\sin 240°$.

Solution   For this first example, we will list the steps just given as we use them. Figure 3 is a diagram of the situation.

Step 1. We find $\hat{\theta}$ by subtracting $180°$ from θ.

$$240° - 180° = 60°$$

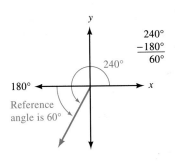

Figure 3

Step 2. Since θ terminates in quadrant III, and the sine function is negative in quadrant III, our answer will be negative. That is, in this case, $\sin\theta = -\sin\hat{\theta}$.

Step 3. Using the results of Steps 1 and 2, we write

$$\sin 240° = -\sin 60°$$

Step 4. We finish by finding $\sin 60°$.

$$\sin 240° = -\sin 60° = -\frac{\sqrt{3}}{2}$$

▶ **EXAMPLE 3**   Find the exact value of tan 315°.

Solution   The reference angle is $360° - 315° = 45°$. Since 315° terminates in quadrant IV, its tangent will be negative.

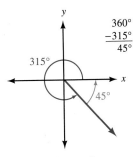

Figure 4

$$\tan 315° = -\tan 45° \qquad \text{Because tangent is negative in QIV}$$
$$= -1$$

▶ **EXAMPLE 4**   Find the exact value of csc 300°.

Solution   The reference angle is $360° - 300° = 60°$. To find the exact value of csc 60°, we use the fact that cosecant and sine are reciprocals.

$$\csc 300° = -\csc 60° \qquad \text{Because cosecant is negative in QIV}$$
$$= -\frac{1}{\sin 60°}$$
$$= -\frac{1}{\sqrt{3}/2}$$
$$= -\frac{2}{\sqrt{3}}$$

Figure 5

Recall from Section 1.2 that coterminal angles always differ from each other by multiples of 360°. For example, −45° and 315° are coterminal, as are 10° and 370°.

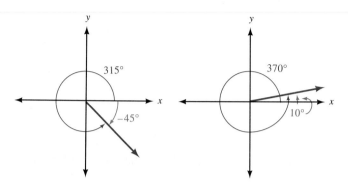

Figure 6

The trigonometric functions of an angle and any angle coterminal to it are always equal. For sine and cosine, we can write this in symbols as follows:

for any integer k,
$$\sin(\theta + 360°k) = \sin\theta \quad \text{and} \quad \cos(\theta + 360°k) = \cos\theta$$

To find values of trigonometric functions for an angle larger than 360° or smaller than 0°, we simply find an angle between 0° and 360° that is coterminal to it and then use the steps outlined in Examples 2 through 4.

▶ **EXAMPLE 5**   Find the exact value of cos 495°.

Solution   By subtracting 360° from 495°, we obtain 135°, which is coterminal to 495°. The reference angle for 135° is 45°. Since 495° terminates in quadrant II, its cosine is negative.

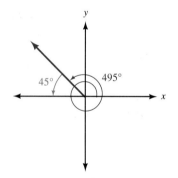

$$
\begin{aligned}
\cos 495° &= \cos 135° && \text{495° and 135° are coterminal} \\
&= -\cos 45° && \text{In QII } \cos\theta = -\cos\hat{\theta} \\
&= -\frac{1}{\sqrt{2}} && \text{Exact value}
\end{aligned}
$$

◀

Figure 7

Approximations

To find trigonometric functions of angles that do not lend themselves to exact values, we use a calculator. To find an approximation for $\sin\theta$, $\cos\theta$, or $\tan\theta$, we simply enter the angle and press the appropriate key on the calculator. Check to see that you can obtain the following values for sine, cosine, and tangent of 250° and −160° on your calculator. (These answers are rounded to the nearest ten-thousandth.)

$$
\begin{array}{ll}
\sin 250° = -0.9397 & \sin(-160°) = -0.3420 \\
\cos 250° = -0.3420 & \cos(-160°) = -0.9397 \\
\tan 250° = 2.7475 & \tan(-160°) = 0.3640
\end{array}
$$

To find csc 250°, sec 250°, and cot 250°, we must also use the reciprocal key, 1/x, on our calculators.

$$\csc 250° = \frac{1}{\sin 250°} = -1.0642 \qquad 250 \boxed{\sin} \boxed{1/x}$$

$$\sec 250° = \frac{1}{\cos 250°} = -2.9238 \qquad 250 \boxed{\cos} \boxed{1/x}$$

$$\cot 250° = \frac{1}{\tan 250°} = 0.3640 \qquad 250 \boxed{\tan} \boxed{1/x}$$

Next we use a calculator to find an approximation for θ—given one of the trigonometric functions of θ and the quadrant in which θ terminates.

▶ **EXAMPLE 6**   Find θ if $\sin \theta = -0.5592$ and θ terminates in QIII with $0° < \theta < 360°$.

Solution   In this example, we must use our calculators in the reverse direction from the way we used them in the previous discussion. Using the $\boxed{\sin^{-1}}$ key, we find the angle whose sine is 0.5592 is 34°. (*Note:* When we refer to the $\boxed{\sin^{-1}}$ key, we are referring to the sequence of calculator keys we press to obtain the inverse sine of a number. In Chapter 2 we showed this sequence as $\boxed{\text{inv}}\ \boxed{\sin}$.)

$$0.5592 \ \boxed{\sin^{-1}} \ \text{gives } \hat{\theta} = 34°$$

This is our reference angle, $\hat{\theta}$. The angle in quadrant III whose reference angle is 34° is $\theta = 180° + 34° = 214°$.

If $\sin \theta = -0.5592$ and θ terminates in QIII,

then $\theta = 180° + 34°$
$= 214°$

If we wanted to list *all* the angles that terminate in quadrant III and have a sine of -0.5592, we would write

$$\theta = 214° + 360°k \qquad \text{where } k = \text{an integer.}$$

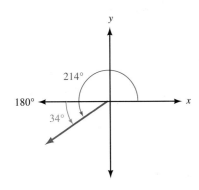

Figure 8

This gives us all angles coterminal with 214°.   ◀

Calculator Note   If you were to try Example 6 on your calculator by simply displaying -0.5592 and then pressing the $\boxed{\sin^{-1}}$ key, you would not obtain 214° for your answer. Instead, you would get approximately $-34°$ for the answer, which is wrong. To see why this happens you will have to wait until we cover inverse trigonometric functions. In the meantime, to use a calculator on this kind of problem, use it to find the reference angle and then proceed as we did in Example 6. That is, you would display 0.5592 and then press $\boxed{\sin^{-1}}$ to obtain approximately 34°, to which you would add 180°, since, in this case, we know the angle terminates in quadrant III.

▶ **EXAMPLE 7** Find θ to the nearest tenth of a degree if $\tan \theta = -0.8541$ and θ terminates in QIV with $0° < \theta < 360°$.

Solution Entering 0.8541 on a calculator and pressing the $\boxed{\tan^{-1}}$ key gives the reference angle as $\hat{\theta} = 40.5°$. The angle in QIV with a reference angle of $40.5°$ is

$$\theta = 360° - 40.5° = 319.5°$$

Again, if we wanted to list *all* angles in quadrant IV with a tangent of -0.8541, we would write

$$\theta = 319.5° + 360°k \qquad k = \text{an integer}$$

to include not only $319.5°$ but all angles coterminal with it.   ◀

▶ **EXAMPLE 8** Find θ if $\sin \theta = -\frac{1}{2}$ and θ terminates in QIII with $0° < \theta < 360°$.

Solution Using our calculators, we find the reference angle to be $30°$. The angle in QIII with a reference angle of $30°$ is $180° + 30° = 210°$.   ◀

▶ **EXAMPLE 9** Find θ to the nearest degree if $\sec \theta = 3.8637$ and θ terminates in QIV with $0° < \theta < 360°$.

Solution To find the reference angle on a calculator, we must use the fact that $\sec \theta$ is the reciprocal of $\cos \theta$. That is,

$$\text{If } \sec \theta = 3.8637, \qquad \text{then } \cos \theta = \frac{1}{3.8637}$$

Therefore, we enter 3.8637 and press the reciprocal key, $\boxed{1/x}$. Then we press the $\boxed{\cos^{-1}}$ key to find our reference angle.

$$3.8637 \quad \boxed{1/x} \quad \boxed{\cos^{-1}}$$

To the nearest degree, the reference angle is $\hat{\theta} = 75°$. Since we want θ to terminate in QIV, we subtract $75°$ from $360°$ to get

$$360° - 75° = 285°$$

We can check our result on a calculator by entering $285°$, finding its cosine, and then finding the reciprocal of the result.

$$285 \quad \boxed{\cos} \quad \boxed{1/x} \quad \text{gives } 3.8637$$ ◀

▶ **EXAMPLE 10** Find θ to the nearest degree if $\cot \theta = -1.6003$ and θ terminates in QII, with $0° < \theta < 360°$.

Solution To find the reference angle on a calculator, we ignore the negative sign in -1.6003 and use the fact that $\cot \theta$ is the reciprocal of $\tan \theta$.

$$\text{If } \cot \theta = 1.6003, \qquad \text{then } \tan \theta = \frac{1}{1.6003}$$

Therefore, we enter 1.6003 and press the reciprocal key, $\boxed{1/x}$. Then we press the $\boxed{\tan^{-1}}$ key to find our reference angle.

$$1.6003 \quad \boxed{1/x} \quad \boxed{\tan^{-1}}$$

To the nearest degree, the reference angle is $\hat{\theta} = 32°$. Since we want θ to terminate in QII, we subtract $32°$ from $180°$ to get $\theta = 148°$.

Again, we can check our result on a calculator by entering $148°$, finding its tangent, and then finding the reciprocal of the result.

$$148 \quad \boxed{\tan} \quad \boxed{1/x} \quad \text{gives} \ -1.6003 \qquad \blacktriangleleft$$

PROBLEM SET 3.1

Draw each of the following angles in standard position and then name the reference angle:

1. 210°	**2.** 150°	**3.** 143.4°	**4.** 253.8°
5. 311.7°	**6.** 93.2°	**7.** 195° 10′	**8.** 171° 40′
9. −300°	**10.** −330°	**11.** −120°	**12.** −150°

Find the exact value of each of the following.

13. cos 225°	**14.** cos 135°	**15.** sin 120°	**16.** sin 210°
17. tan 135°	**18.** tan 315°	**19.** cos 240°	**20.** cos 150°
21. csc 330°	**22.** sec 330°	**23.** sec 300°	**24.** csc 300°
25. sin 390°	**26.** cos 420°	**27.** cot 480°	**28.** cot 510°

Use a calculator to find the following.

29. cos 347°	**30.** cos 238°	**31.** sec 101.8°	**32.** csc 166.7°
33. tan 143.4°	**34.** tan 253.8°	**35.** sec 311.7°	**36.** csc 93.2°
37. cot 390°	**38.** cot 420°	**39.** csc 575.4°	**40.** sec 590.9°
41. cos (−315°)	**42.** sin (−225°)	**43.** tan 195° 10′	**44.** tan 171° 40′
45. sec 314° 40′	**46.** csc 670° 20′	**47.** sin (−120)°	**48.** cos (−150°)

Use a calculator to find θ to the nearest tenth of a degree, if $0° < \theta < 360°$ and

49. sin θ = −0.3090 with θ in QIII	**50.** sin θ = −0.3090 with θ in QIV
51. cos θ = −0.7660 with θ in QII	**52.** cos θ = −0.7660 with θ in QIII
53. tan θ = 0.5890 with θ in QIII	**54.** tan θ = 0.5890 with θ in QI
55. cos θ = 0.2644 with θ in QI	**56.** cos θ = 0.2644 with θ in QIV
57. sin θ = 0.9652 with θ in QII	**58.** sin θ = 0.9652 with θ in QI
59. sec θ = 1.4325 with θ in QIV	**60.** csc θ = 1.4325 with θ in QII
61. csc θ = 2.4957 with θ in QII	**62.** sec θ = −3.4159 with θ in QII
63. cot θ = −0.7366 with θ in QII	**64.** cot θ = −0.1234 with θ in QIV
65. sec θ = −1.7876 with θ in QIII	**66.** csc θ = −1.7876 with θ in QIII

Find θ, if $0° < \theta < 360°$ and

67. $\sin \theta = -\dfrac{\sqrt{3}}{2}$ and θ in QIII

68. $\sin \theta = -\dfrac{1}{\sqrt{2}}$ and θ in QIII

69. $\cos \theta = -\dfrac{1}{\sqrt{2}}$ and θ in QII

70. $\cos \theta = -\dfrac{\sqrt{3}}{2}$ and θ in QIII

71. $\sin \theta = -\dfrac{\sqrt{3}}{2}$ and θ in QIV

72. $\sin \theta = \dfrac{1}{\sqrt{2}}$ and θ in QII

73. $\tan \theta = \sqrt{3}$ and θ in QIII

74. $\tan \theta = \dfrac{1}{\sqrt{3}}$ and θ in QIII

75. $\sec \theta = -2$ with θ in QII

76. $\csc \theta = 2$ with θ in QII

77. $\csc \theta = \sqrt{2}$ with θ in QII

78. $\sec \theta = \sqrt{2}$ with θ in QIV

79. $\cot \theta = -1$ with θ in QIV

80. $\cot \theta = \sqrt{3}$ with θ in QIII

Review Problems

The problems that follow review material we covered in Sections 1.1 and 2.1.

Give the complement and supplement of each angle.

81. $70°$ **82.** $120°$ **83.** x **84.** $90° - y$

85. If the longest side in a $30°$-$60°$-$90°$ triangle is 10, find the length of the other two sides.

86. If the two shorter sides of a $45°$-$45°$-$90°$ triangle are both $\frac{3}{4}$, find the length of the hypotenuse.

Simplify each expression by substituting values from the table of exact values and then simplifying the resulting equation.

87. $\sin 30° \cos 60°$

88. $4 \sin 60° - 2 \cos 30°$

89. $\sin^2 45° + \cos^2 45°$

90. $(\sin 45° + \cos 45°)^2$

SECTION 3.2

Radians and Degrees

If you think back to the work you have done with functions of the form $y = f(x)$ in your algebra class, you will see that the variables x and y were always real numbers. The trigonometric functions we have worked with so far have had the form $y = f(\theta)$, where θ is measured in degrees. In order to apply the knowledge we have about functions from algebra to our trigonometric functions, we need to write our angles as real numbers, not degrees. The key to doing this is called *radian measure*. Radian measure is a relatively new concept in the history of mathematics. It was first introduced by physicist James T. Thomson in a paper published in 1893. The introduction of radian measure will allow us to do a number of useful things. For instance, in Chapter 4 we will graph the function $y = \sin x$ on a rectangular coordinate system, where the units on x and y axes are given by real numbers, just as they would be if we were graphing $y = 2x + 3$ or $y = x^2$. To understand the definition for radian measure, we have to recall from geometry that a central angle in a circle is an angle with its vertex at the center of the circle. Here is the definition for an angle with a measure of 1 radian.

DEFINITION   In a circle, a central angle that cuts off an arc equal in length to the radius of the circle has a measure of 1 *radian*. The following diagram illustrates.

Angle θ has a
measure of 1 radian

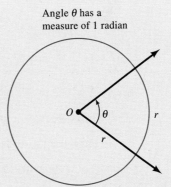

The vertex of θ is at the center of
the circle; the arc cut off by θ is
equal in length to the radius

Figure 1

To find the radian measure of *any* central angle, we must find how many radii are in the arc it cuts off. To do so, we divide the arc length by the radius. If the radius is 2 centimeters and the arc cut off by central angle θ is 6 centimeters, then the radian measure of θ is $\frac{6}{2} = 3$ radians. Here is the formal definition:

DEFINITION   Radian Measure

If a central angle θ, in a circle of radius r, cuts off an arc of length s, then the measure of θ, in radians, is given by s/r.

$\theta \text{ (in radians)} = \frac{s}{r}$

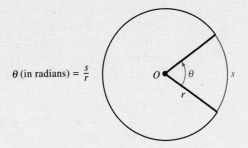

Figure 2

▶ **EXAMPLE 1**   A central angle θ in a circle of radius 3 centimeters cuts off an arc of length 6 centimeters. What is the radian measure of θ?

Solution   We have $r = 3$ centimeters and $s = 6$ centimeters; therefore,

$$\theta \text{ (in radians)} = \frac{s}{r}$$

$$= \frac{6 \text{ centimeters}}{3 \text{ centimeters}}$$

$$= 2$$

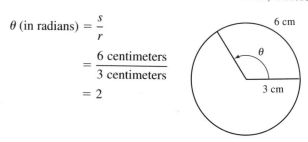

Figure 3

We say the radian measure of θ is 2, or $\theta = 2$ radians.   ◀

Note   It is common practice to omit the word radian when using radian measure. If no units are showing, an angle is understood to be measured in radians; with degree measure, the degree symbol ° must be written.

$$\theta = 2 \text{ means the measure of } \theta \text{ is 2 radians}$$

$$\theta = 2° \text{ means the measure of } \theta \text{ is 2 degrees}$$

To see the relationship between degrees and radians, we can compare the number of degrees and the number of radians in one full rotation.

The angle formed by one full rotation about the center of a circle of radius r will cut off an arc equal to the circumference of the circle. Since the circumference of a circle of radius r is $2\pi r$, we have

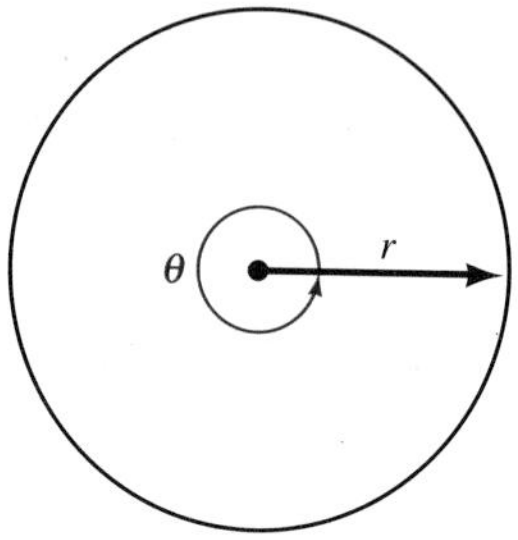

$$\begin{array}{ccc} \theta \text{ measures one} & \theta = \dfrac{2\pi r}{r} = 2\pi & \text{The measure of } \theta \\ \text{full rotation} & & \text{in radians is } 2\pi \end{array}$$

Since one full rotation in degrees is 360°, we have the following relationship between radians and degrees.

$$360° = 2\pi \text{ radians}$$

Dividing both sides by 2 we have

$$180° = \pi \text{ radians}$$

To obtain conversion factors that will allow us to change back and forth between degrees and radians, we divide both sides of this last equation alternately by 180 and by π.

Figure 4

$$\begin{array}{ccc} \text{Divide both} & \overline{\quad 180° = \pi \text{ radians} \quad} & \text{Divide both} \\ \text{sides by 180} & \downarrow \qquad\qquad\qquad\qquad \downarrow & \text{sides by } \pi \end{array}$$

$$1° = \frac{\pi}{180} \text{ radians} \qquad \left(\frac{180}{\pi}\right)° = 1 \text{ radian}$$

To gain some insight into the relationship between degrees and radians, we can approximate π with 3.14 to obtain the approximate number of degrees in 1 radian.

$$1 \text{ radian} = 1\left(\frac{180}{\pi}\right)^\circ$$

$$\approx 1\left(\frac{180}{3.14}\right)^\circ$$

$$= 57.3^\circ \text{ to the nearest tenth}$$

We see that 1 radian is approximately 57°. A radian is much larger than a degree. Figure 5 illustrates the relationship between 20° and 20 radians.

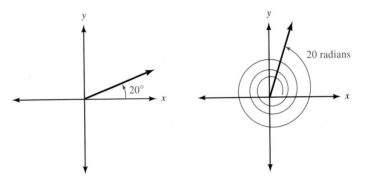

Figure 5

Here are some further conversions between degrees and radians.

Converting from Degrees to Radians

▶ **EXAMPLE 2**   Convert 45° to radians.

Solution   Since $1^\circ = \dfrac{\pi}{180}$ radians, and 45° is the same as $45(1^\circ)$, we have

$$45^\circ = 45\left(\frac{\pi}{180}\right) \text{ radians} = \frac{\pi}{4} \text{ radians}$$

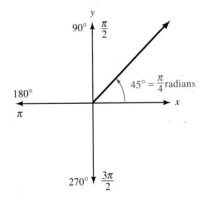

Figure 6

When we leave our answer in terms of π, as in $\pi/4$, we are writing an exact value. If we wanted a decimal approximation, we would substitute 3.14 for π.

Exact value $\dfrac{\pi}{4} \approx \dfrac{3.14}{4} = 0.785$ Approximate value

Note also that if we wanted the radian equivalent of 90°, we could simply multiply $\pi/4$ by 2, since $90° = 2 \times 45°$.

$$90° = 2 \times 45° = 2 \times \dfrac{\pi}{4} = \dfrac{\pi}{2}$$ ◀

▶ **EXAMPLE 3**   Convert 450° to radians.

Solution   Multiplying by $\pi/180$ we have

$$450° = 450\left(\dfrac{\pi}{180}\right) = \dfrac{5\pi}{2} \text{ radians}$$

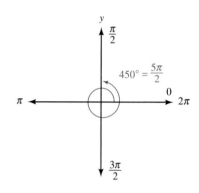

Figure 7

Again, $5\pi/2$ is the exact value. If we wanted a decimal approximation, we would substitute 3.14 for π.

Exact value $\dfrac{5\pi}{2} \approx \dfrac{5(3.14)}{2} = 7.85$ Approximate value   ◀

Converting from Radians to Degrees

▶ **EXAMPLE 4**   Convert $\pi/6$ to degrees.

Solution   To convert from radians to degrees, we multiply by $180/\pi$.

$$\dfrac{\pi}{6} \text{ (radians)} = \dfrac{\pi}{6}\left(\dfrac{180}{\pi}\right)°$$
$$= 30°$$

Note that 60° is twice 30°, so $2(\pi/6) = \pi/3$ must be the radian equivalent of 60°. Figure 8 illustrates.

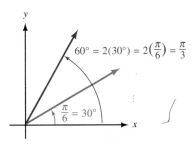

Figure 8   ◄

► **EXAMPLE 5**   Convert $4\pi/3$ to degrees.

Solution   Multiplying by $180/\pi$ we have

$$\frac{4\pi}{3} \text{ (radians)} = \frac{4\pi}{3}\left(\frac{180}{\pi}\right)^\circ$$

$$= 240°$$

Note that the reference angle for the angle shown in Figure 9 can be given in either degrees or radians.

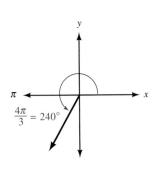

Figure 9

in degrees: $\hat\theta = 240° - 180° = 60°$

in radians: $\hat\theta = \frac{4\pi}{3} - \pi = \frac{4\pi}{3} - \frac{3\pi}{3} = \frac{\pi}{3}$ ◄

As is apparent from the preceding examples, changing from degrees to radians and radians to degrees is simply a matter of multiplying by the appropriate conversion factors.

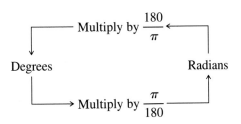

Table 1 shows the most common angles written in both degrees and radians. In each case, the radian measure is given in exact values and approximations accurate to the nearest hundredth of a radian.

Table 1

Degrees	Radians	
	Exact values	*Approximations*
0°	0	0
30°	$\dfrac{\pi}{6}$	0.52
45°	$\dfrac{\pi}{4}$	0.79
60°	$\dfrac{\pi}{3}$	1.05
90°	$\dfrac{\pi}{2}$	1.57
180°	π	3.14
270°	$\dfrac{3\pi}{2}$	4.71
360°	2π	6.28

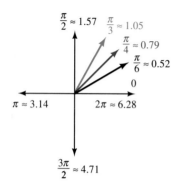

Figure 10

▶ **EXAMPLE 6**   Find $\sin \dfrac{\pi}{6}$.

Solution   Since $\pi/6$ and 30° are equivalent, so are their sines.

$$\sin \frac{\pi}{6} = \frac{1}{2}$$

◀

Calculator Note   To work this problem on a calculator, we must first put the calculator in radian mode. (Consult the manual that came with your calculator to see how to do this.) If your calculator does not have a key labeled π, use 3.1416. Here is the sequence to key in your calculator to work the problem given in Example 6.

$$\boxed{\text{Rad}}\ \ 3.1416\ \ \boxed{\div}\ \ 6\ \ \boxed{=}\ \ \boxed{\sin}$$

▶ **EXAMPLE 7**   Find $4 \sin \dfrac{7\pi}{6}$.

Solution   Since $7\pi/6$ terminates in QIII, its sine will be negative. The reference angle is $7\pi/6 - \pi = \pi/6$.

$$4 \sin \frac{7\pi}{6} = -4 \sin \frac{\pi}{6}$$
$$= -4\left(\frac{1}{2}\right)$$
$$= -2$$

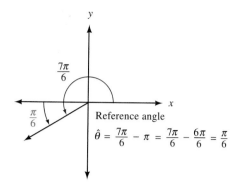

Figure 11

◀

▶ **EXAMPLE 8**   Evaluate $4 \sin (2x + \pi)$ when $x = \pi/6$.

Solution   Substituting $\pi/6$ for x and simplifying, we have

$$4 \sin \left(2 \cdot \frac{\pi}{6} + \pi \right) = 4 \sin \left(\frac{\pi}{3} + \pi \right)$$

$$= 4 \sin \frac{4\pi}{3}$$

$$= 4 \left(-\frac{\sqrt{3}}{2} \right)$$

$$= -2\sqrt{3}$$

◀

PROBLEM SET 3.2

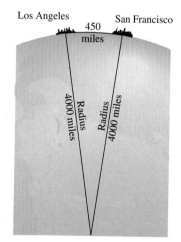

Figure 12

Find the radian measure of angle θ, if θ is a central angle in a circle of radius r, and θ cuts off an arc of length s.

1. $r = 3$ cm, $s = 9$ cm   **2.** $r = 6$ cm, $s = 3$ cm
3. $r = 10$ inches, $s = 5$ inches   **4.** $r = 5$ inches, $s = 10$ inches
5. $r = 4$ inches, $s = 12\pi$ inches **6.** $r = 3$ inches, $s = 12$ inches

7. $r = \dfrac{1}{4}$ cm, $s = \dfrac{1}{2}$ cm   **8.** $r = \dfrac{1}{4}$ cm, $s = \dfrac{1}{8}$ cm

9. Los Angeles and San Francisco are approximately 450 mi apart on the surface of the earth. Assuming that the radius of the earth is 4,000 mi, find the radian measure of the central angle with vertex at the center of the earth that has Los Angeles on one side and San Francisco on the other side. (See Figure 12.)

10. Los Angeles and New York City are approximately 2,500 mi apart on the surface of the earth. Assuming that the radius of the earth is 4,000 mi, find the radian measure of the central angle with vertex at the center of the earth that has Los Angeles on one side and New York City on the other side.

For each angle below:
a. Draw the angle in standard position.
b. Convert to radian measure using exact values.
c. Name the reference angle in both degrees and radians.

11. 30°	**12.** 60°	**13.** 90°	**14.** 270°
15. 260°	**16.** 340°	**17.** −150°	**18.** −210°
19. 420°	**20.** 390°	**21.** −135°	**22.** −120°

For Problems 23–26, use 3.1416 for π unless your calculator has a key marked π.

23. Use a calculator to convert 120° 40′ to radians. Round your answer to the nearest hundredth. (First convert to decimal degrees, then multiply by the appropriate conversion factor to convert to radians.)

24. Use a calculator to convert 256° 20′ to radians to the nearest hundredth of a radian.

25. Use a calculator to convert 1′ (1 minute) to radians to three significant digits.

26. Use a calculator to convert 1° to radians to three significant digits.

27. If a central angle with its vertex at the center of the earth has a measure of 1′, then the arc on the surface of the earth that is cut off by this angle has a measure of 1 nautical mile. Find the number of regular (statute) miles in 1 nautical mile to the nearest hundredth of a mile. (Use 4,000 mi for the radius of the earth.) (See Figure 13.)

28. If two ships are 20 nautical miles apart on the ocean, how many statute miles apart are they? (Use the results of Problem 27 to do the calculations.)

1 nautical mile

1′

Radius of earth: 4000 miles

Figure 13

For each angle below:
a. Convert to degree measure.
b. Draw the angle in standard position.
c. Label the reference angle in both degrees and radians.

29. $\dfrac{\pi}{3}$	**30.** $\dfrac{\pi}{4}$	**31.** $\dfrac{2\pi}{3}$	**32.** $\dfrac{3\pi}{4}$
33. $\dfrac{-7\pi}{6}$	**34.** $\dfrac{-5\pi}{6}$	**35.** $\dfrac{10\pi}{6}$	**36.** $\dfrac{7\pi}{3}$
37. 4π	**38.** 3π	**39.** $\dfrac{\pi}{12}$	**40.** $\dfrac{5\pi}{12}$

Use a calculator to convert each of the following to degree measure to the nearest tenth of a degree.

41. 1	**42.** 2	**43.** 1.3	**44.** 2.4
45. 0.75	**46.** 0.25	**47.** 5	**48.** 6

Give the exact value of each of the following:

49. $\sin\dfrac{4\pi}{3}$	**50.** $\cos\dfrac{4\pi}{3}$	**51.** $\tan\dfrac{\pi}{6}$	**52.** $\cot\dfrac{\pi}{3}$

53. $\sec \dfrac{2\pi}{3}$ **54.** $\csc \dfrac{3\pi}{2}$ **55.** $\csc \dfrac{5\pi}{6}$ **56.** $\sec \dfrac{5\pi}{6}$

57. $4 \sin\left(-\dfrac{\pi}{4}\right)$ **58.** $4 \cos\left(-\dfrac{\pi}{4}\right)$

59. $-\sin \dfrac{\pi}{4}$ **60.** $-\cos \dfrac{\pi}{4}$ **61.** $2 \cos \dfrac{\pi}{6}$ **62.** $2 \sin \dfrac{\pi}{6}$

Evaluate each of the following expressions when x is $\pi/6$. In each case, use exact values.

63. $\sin 2x$ **64.** $\sin 3x$ **65.** $6 \cos 3x$ **66.** $6 \cos 2x$

67. $\sin\left(x + \dfrac{\pi}{2}\right)$ **68.** $\sin\left(x - \dfrac{\pi}{2}\right)$

69. $4 \cos\left(2x + \dfrac{\pi}{3}\right)$ **70.** $4 \cos\left(3x + \dfrac{\pi}{6}\right)$

For the following expressions, find the value of y that corresponds to each value of x, then write your results as ordered pairs (x, y).

71. $y = \sin x$ for $x = 0, \dfrac{\pi}{4}, \dfrac{\pi}{2}, \dfrac{3\pi}{4}, \pi$

72. $y = \cos x$ for $x = 0, \dfrac{\pi}{4}, \dfrac{\pi}{2}, \dfrac{3\pi}{4}, \pi$

73. $y = 2 \sin x$ for $x = 0, \dfrac{\pi}{2}, \pi, \dfrac{3\pi}{2}, 2\pi$

74. $y = \dfrac{1}{2} \cos x$ for $x = 0, \dfrac{\pi}{2}, \pi, \dfrac{3\pi}{2}, 2\pi$

75. $y = \sin 2x$ for $x = 0, \dfrac{\pi}{4}, \dfrac{\pi}{2}, \dfrac{3\pi}{4}, \pi$

76. $y = \cos 3x$ for $x = 0, \dfrac{\pi}{6}, \dfrac{\pi}{3}, \dfrac{\pi}{2}, \dfrac{2\pi}{3}$

77. $y = \sin\left(x - \dfrac{\pi}{2}\right)$ for $x = \dfrac{\pi}{2}, \pi, \dfrac{3\pi}{2}, 2\pi, \dfrac{5\pi}{2}$

78. $y = \cos\left(x - \dfrac{\pi}{6}\right)$ for $x = \dfrac{\pi}{6}, \dfrac{\pi}{3}, \dfrac{2\pi}{3}, \pi, \dfrac{7\pi}{6}$

79. $y = 3 \sin\left(2x + \dfrac{\pi}{2}\right)$ for $x = -\dfrac{\pi}{4}, 0, \dfrac{\pi}{4}, \dfrac{\pi}{2}, \dfrac{3\pi}{4}$

80. $y = 5 \cos\left(2x - \dfrac{\pi}{3}\right)$ for $x = \dfrac{\pi}{6}, \dfrac{\pi}{3}, \dfrac{2\pi}{3}, \pi, \dfrac{7\pi}{6}$

Review Problems

The problems that follow review material we covered in Section 1.3.

Find all six trigonometric functions of θ, if the given point is on the terminal side of θ.

81. $(1, -3)$ **82.** $(-1, 3)$ **83.** (m, n) **84.** (a, b)

85. Find the remaining trigonometric functions of θ, if $\sin \theta = \frac{1}{2}$ and θ terminates in QII.

86. Find the remaining trigonometric functions of θ, if $\cos \theta = -1/\sqrt{2}$ and θ terminates in QII.

87. Find all six trigonometric functions of θ, if the terminal side of θ lies along the line $y = 2x$ in QI.

88. Find the six trigonometric functions of θ, if the terminal side of θ lies along the line $y = 2x$ in QIII.

SECTION 3.3 **Definition III: Circular Functions**

In this section, we will give a third and final definition for the trigonometric functions. Rather than give the new definition first and then show that it does not conflict with our previous definitions, we will do the reverse and show that either of our first two definitions leads to conclusions that imply a third definition.

To begin, recall that the unit circle (Figure 1) is the circle with center at the origin and radius 1. The equation of the unit circle is $x^2 + y^2 = 1$.

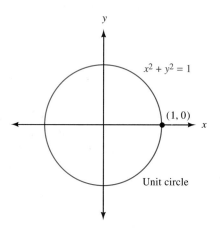

Figure 1

Suppose the terminal side of angle θ, in standard position, intersects the unit circle at point (x, y) as shown in Figure 2.

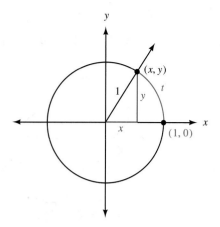

Figure 2

Since the radius of the unit circle is 1, the distance from the origin to the point (x, y) is 1. We can use our first definition for the trigonometric functions from Section 1.3 to write

$$\cos \theta = \frac{x}{r} = \frac{x}{1} = x \quad \text{and} \quad \sin \theta = \frac{y}{r} = \frac{y}{1} = y$$

In other words, the ordered pair (x, y) shown in Figure 2 can be written $(\cos \theta, \sin \theta)$. We can arrive at this same result by applying our second definition for the trigonometric functions (from Section 2.1). Since θ is an acute angle in a right triangle, by Definition II we have

$$\cos \theta = \frac{\text{side adjacent } \theta}{\text{hypotenuse}} = \frac{x}{1} = x$$

$$\sin \theta = \frac{\text{side opposite } \theta}{\text{hypotenuse}} = \frac{y}{1} = y$$

As you can see, both of our first two definitions lead to the conclusion that the point (x, y) shown in Figure 2 can be written as $(\cos \theta, \sin \theta)$.

Next, consider the length of the arc from $(1, 0)$ to (x, y), which we have labeled t in Figure 2. If θ is measured in radians, then by definition

$$\theta = \frac{t}{r} = \frac{t}{1} = t$$

The length of the arc from $(1, 0)$ to (x, y) is exactly the same as the radian measure of angle θ. Therefore, we can write

$$\cos \theta = \cos t = x \quad \text{and} \quad \sin \theta = \sin t = y$$

These results give rise to a third definition for the trigonometric functions.

DEFINITION III   Circular Functions

If (x, y) is any point on the unit circle, and t is the distance from $(1, 0)$ to (x, y) along the circumference of the unit circle, then

$$\cos t = x$$

$$\sin t = y$$

$$\tan t = \frac{y}{x}$$

$$\cot t = \frac{x}{y}$$

$$\csc t = \frac{1}{y}$$

$$\sec t = \frac{1}{x}$$

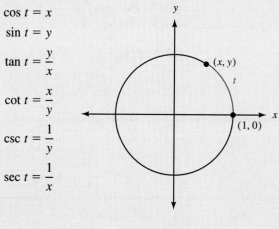

Figure 3

As we travel around the unit circle, starting at the point $(1, 0)$, the points we come across all have coordinates $(\cos t, \sin t)$, where t is the distance we have traveled. When we define the trigonometric functions this way, we call them circular functions because of their relationship to the unit circle.

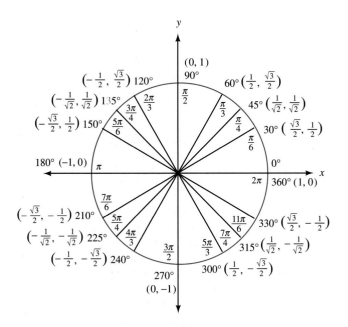

Figure 4

Figure 4 shows an enlarged version of the unit circle with multiples of $\pi/6$ and $\pi/4$ marked off. Each angle is given in both degrees and radians. The radian measure of each angle is the same as the distance from $(1, 0)$ to the point on the terminal side of the angle, as measured along the circumference of the circle. The x- and y-coordinate of each point shown are the cosine and sine, respectively, of the associated angle or distance.

▶ **EXAMPLE 1**   Use Figure 4 to find the six trigonometric functions of $5\pi/6$.

Solution   We obtain cosine and sine directly from Figure 4. The other trigonometric functions of $5\pi/6$ are found by using the ratio and reciprocal identities, rather than the new definition.

$$\sin \frac{5\pi}{6} = \frac{1}{2}$$

$$\cos \frac{5\pi}{6} = -\frac{\sqrt{3}}{2}$$

$$\tan \frac{5\pi}{6} = \frac{\sin 5\pi/6}{\cos 5\pi/6} = \frac{1/2}{-\sqrt{3}/2} = -\frac{1}{\sqrt{3}}$$

$$\cot \frac{5\pi}{6} = \frac{1}{\tan 5\pi/6} = \frac{1}{-1/\sqrt{3}} = -\sqrt{3}$$

$$\sec \frac{5\pi}{6} = \frac{1}{\cos 5\pi/6} = \frac{1}{-\sqrt{3}/2} = -\frac{2}{\sqrt{3}}$$

$$\csc \frac{5\pi}{6} = \frac{1}{\sin 5\pi/6} = \frac{1}{1/2} = 2$$

◀

Figure 4 is helpful in visualizing the relationships among the angles shown and the trigonometric functions of those angles. You may want to make a larger copy of this diagram yourself. In the process of doing so you will become more familiar with the relationship between degrees and radians and the exact values of the angles in the diagram.

▶ **EXAMPLE 2**   Use the unit circle to find all values of t between 0 and 2π for which $\cos t = \frac{1}{2}$.

Solution   We look for all ordered pairs on the unit circle with an x-coordinate of $\frac{1}{2}$. The angles, or distances, associated with these points are the angles for which $\cos t = \frac{1}{2}$. They are $t = \pi/3$ or $60°$ and $t = 5\pi/3$ or $300°$.   ◀

Even and Odd Functions

Recall from algebra the definitions of even and odd functions.

DEFINITION   An *even function* is a function for which

$$f(-x) = f(x) \text{ for all } x \text{ in the domain of } f.$$

An even function is a function for which replacing x with $-x$ leaves the equation that defines the function unchanged. If a function is even, then every time the point (x, y) is on the graph, so is the point $(-x, y)$. The function $f(x) = x^2 + 3$ is an even function since

$$f(-x) = (-x)^2 + 3 = x^2 + 3 = f(x)$$

DEFINITION   An *odd function* is a function for which

$$f(-x) = -f(x) \text{ for all } x \text{ in the domain of } f.$$

An odd function is a function for which replacing x with $-x$ changes the sign of the equation that defines the function. If a function is odd, then every time the point (x, y) is on the graph, so is the point $(-x, -y)$. The function $f(x) = x^3 - x$ is an odd function since

$$f(-x) = (-x)^3 - (-x) = -x^3 + x = -(x^3 - x) = -f(x)$$

From the unit circle it is apparent that sine is an odd function and cosine is an even function. To begin to see that this is true, we locate $\pi/6$ and $-\pi/6$ ($-\pi/6$ is coterminal with $11\pi/6$) on the unit circle and notice that

$$\cos\left(-\frac{\pi}{6}\right) = \frac{\sqrt{3}}{2} = \cos\frac{\pi}{6}$$

and

$$\sin\left(-\frac{\pi}{6}\right) = -\frac{1}{2} = -\sin\frac{\pi}{6}$$

We can generalize this result by drawing an angle θ and its opposite $-\theta$ in standard position and then labeling the points where their terminal sides intersect the unit circle with (x, y) and $(x, -y)$, respectively. (Can you see from Figure 5 why we label these two points in this way? That is, does it make sense that if (x, y) is on the terminal side of θ, then $(x, -y)$ must be on the terminal side of $-\theta$?)

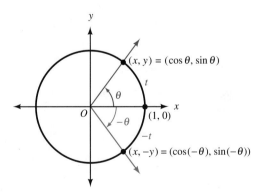

Figure 5

Since, on the unit circle, $\cos \theta = x$ and $\sin \theta = y$, we have

$$\cos(-\theta) = x = \cos \theta$$

indicating that cosine is an even function and

$$\sin(-\theta) = -y = -\sin \theta$$

indicating that sine is an odd function.

Now that we have established that sine is an odd function and cosine is an even function, we can use our ratio and reciprocal identities to find which of the other trigonometric functions are even and which are odd. Example 3 shows how this is done for the cosecant function.

▶ **EXAMPLE 3** Show that cosecant is an odd function.

Solution We must prove that $\csc(-\theta) = -\csc \theta$. That is, we must turn $\csc(-\theta)$ into $-\csc \theta$. Here is how it goes:

$$\csc(-\theta) = \frac{1}{\sin(-\theta)} \qquad \text{Reciprocal identity}$$

$$= \frac{1}{-\sin \theta} \qquad \text{Sine is an odd function}$$

$$= -\frac{1}{\sin \theta} \qquad \text{Algebra}$$

$$= -\csc \theta \qquad \text{Reciprocal identity} \qquad ◀$$

▶ **EXAMPLE 4** Use the even and odd function relationships to find exact values for each of the following.

$$\textbf{a. } \sin(-60°) \qquad \textbf{b. } \cos\left(-\frac{2\pi}{3}\right) \qquad \textbf{c. } \csc(-225°)$$

Solution

$$\textbf{a. } \sin(-60°) = -\sin 60° \qquad \text{Sine is an odd function}$$

$$= -\frac{\sqrt{3}}{2} \qquad \text{Unit circle}$$

$$\textbf{b. } \cos\left(-\frac{2\pi}{3}\right) = \cos\left(\frac{2\pi}{3}\right) \qquad \text{Cosine is an even function}$$

$$= -\frac{1}{2} \qquad \text{Unit circle}$$

c. $\csc(-225°) = \dfrac{1}{\sin(-225°)}$ Reciprocal identity

$= \dfrac{1}{-\sin 225°}$ Sine is an odd function

$= \dfrac{1}{-(-1/\sqrt{2})}$ Unit circle

$= \sqrt{2}$ ◀

There is an interesting diagram we can draw using the ideas given in this section. Figure 6 shows a point $P(x, y)$ that is t units from the point $(1, 0)$ on the circumference of the unit circle. Therefore, $\cos t = x$ and $\sin t = y$. Can you see why QB is labeled tan t? It has to do with the fact that tan t is the ratio of sin t to cos t along with the fact that triangles *POA* and *QOB* are similar.

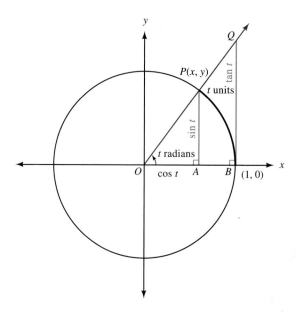

Figure 6

There are many concepts that can be visualized from Figure 6. One of the more important is the variations that occur in sin t, cos t, and tan t as P travels around the unit circle. To illustrate, imagine P traveling once around the unit circle starting at $(1, 0)$ and ending 2π units later at the same point. As P moves from $(1, 0)$ to $(0, 1)$, t increases from 0 to $\pi/2$. At the same time, sin t increases from 0 to 1, and cos t decreases from 1 down to 0. As we continue around the unit circle, sin t and cos t simply oscillate between -1 and 1; furthermore, everywhere sin t is 1 or -1, cos t is 0. We will look at these oscillations in more detail in Chapter 4.

PROBLEM SET 3.3

Use the unit circle to find the six trigonometric functions of each angle.

1. 150°   **2.** 135°   **3.** $11\pi/6$ **4.** $5\pi/3$
5. 180°   **6.** 270°   **7.** $3\pi/4$ **8.** $5\pi/4$

Use the unit circle and the fact that cosine is an even function to find each of the following:

9. $\cos(-60°)$ **10.** $\cos(-120°)$

11. $\cos\left(-\dfrac{5\pi}{6}\right)$ **12.** $\cos\left(-\dfrac{4\pi}{3}\right)$

Use the unit circle and the fact that sine is an odd function to find each of the following:

13. $\sin(-30°)$ **14.** $\sin(-90°)$

15. $\sin\left(-\dfrac{3\pi}{4}\right)$ **16.** $\sin\left(-\dfrac{7\pi}{4}\right)$

Use the unit circle to find all values of θ between 0 and 2π for which

17. $\sin\theta = 1/2$ **18.** $\sin\theta = -1/2$
19. $\cos\theta = -\sqrt{3}/2$ **20.** $\cos\theta = 0$
21. $\tan\theta = -\sqrt{3}$ **22.** $\cot\theta = \sqrt{3}$

23. If angle θ is in standard position and intersects the unit circle at $(1/\sqrt{5}, -2/\sqrt{5})$, find $\sin\theta$, $\cos\theta$, and $\tan\theta$.
24. If angle θ is in standard position and intersects the unit circle at the point $(-1/\sqrt{10}, -3/\sqrt{10})$, find $\sin\theta$, $\cos\theta$, and $\tan\theta$.
25. If $\sin\theta = -1/3$, find $\sin(-\theta)$.   **26.** If $\cos\theta = -1/3$, find $\cos(-\theta)$.

If we start at the point $(1, 0)$ and travel once around the unit circle, we travel a distance of 2π units and arrive back where we started at the point $(1, 0)$. If we continue around the unit circle a second time, we will repeat all the values of x and y that occurred during our first trip around. Use this discussion to evaluate the following expressions:

27. $\sin\left(2\pi + \dfrac{\pi}{2}\right)$ **28.** $\cos\left(2\pi + \dfrac{\pi}{2}\right)$

29. $\sin\left(2\pi + \dfrac{\pi}{6}\right)$ **30.** $\cos\left(2\pi + \dfrac{\pi}{3}\right)$

31. $\sin\dfrac{5\pi}{2}$ **32.** $\cos\dfrac{5\pi}{2}$

33. $\sin\dfrac{13\pi}{6}$ **34.** $\cos\dfrac{7\pi}{3}$

Make a diagram of the unit circle with an angle θ in quadrant I and its supplement $180° - \theta$ in quadrant II. Label the point on the terminal side of θ and the unit circle with (x, y) and the point on the terminal side of $180° - \theta$ and the unit circle with $(-x, y)$. Use the diagram to show that

35. $\sin (180° - \theta) = \sin \theta$ **36.** $\cos (180° - \theta) = -\cos \theta$

37. Show that tangent is an odd function.

38. Show that cotangent is an odd function.

Prove each identity.

39. $\sin (-\theta) \cot (-\theta) = \cos \theta$ **40.** $\cos (-\theta) \tan \theta = \sin \theta$

41. $\sin (-\theta) \sec (-\theta) \cot (-\theta) = 1$ **42.** $\cos (-\theta) \csc (-\theta) \tan (-\theta) = 1$

43. $\csc \theta + \sin (-\theta) = \dfrac{\cos^2 \theta}{\sin \theta}$ **44.** $\sec \theta - \cos (-\theta) = \dfrac{\sin^2 \theta}{\cos \theta}$

45. Redraw the diagram in Figure 6 from this section and label the line segment that corresponds to sec t.

46. Make a diagram similar to the diagram in Figure 6 from this section, but instead of labeling the point $(1, 0)$ with B, label the point $(0, 1)$ with B. Then place Q on the line OP and connect Q to B so that QB is perpendicular to the y-axis. Now, if $P(x, y)$ is t units from $(1, 0)$, label the line segments that correspond to sin t, cos t, cot t, and csc t.

Review Problems

The problems that follow review material we covered in Section 2.3.

Problems 47–54 refer to right triangle ABC in which $C = 90°$. Solve each triangle.

47. $A = 42°$, $c = 36$ **48.** $A = 58°$, $c = 17$

49. $B = 22°$, $b = 320$ **50.** $B = 48°$, $b = 270$

51. $a = 20.5$, $b = 31.4$ **52.** $a = 16.3$, $b = 20.8$

53. $a = 4.37$, $c = 6.21$ **54.** $a = 7.12$, $c = 8.44$

SECTION 3.4 **Arc Length and Area of a Sector**

In Chapter 2, we discussed some of the aspects of the first Ferris wheel, which was built by George Ferris in 1893. In particular, we used right triangle trigonometry to find the height above the ground of a rider on the wheel at any point on the wheel. The first topic we will cover in this section is *arc length*. Our study of arc length will allow us to find the distance traveled by a rider on a Ferris wheel at any point during the ride.

In Section 3.2, we found that if a central angle θ, measured in radians, in a circle of radius r, cuts off an arc of length s, then the relationship between s, r, and θ can be

written as $\theta = \dfrac{s}{r}$

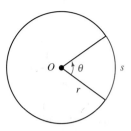

Figure 1

If we multiply both sides of this equation by r, we will obtain the equation that gives arc length s, in terms of r and θ.

$$\theta = \frac{s}{r} \qquad \text{Definition of radian measure}$$

$$r \cdot \theta = r \cdot \frac{s}{r} \qquad \text{Multiply both sides by } r$$

$$r\theta = s$$

Arc Length

If θ (in radians) is a central angle in a circle with radius r, then the length of the arc cut off by θ is given by

$$s = r\theta \qquad (\theta \text{ in radians})$$

▶ **EXAMPLE 1**   Give the length of the arc cut off by a central angle of 2 radians in a circle of radius 4.3 inches.

Solution   We have $\theta = 2$ and $r = 4.3$ inches. Applying the formula $s = r\theta$ gives us

$$s = r\theta$$
$$= 4.3(2)$$
$$= 8.6 \text{ inches}$$

8.6 in

2

4.3 in

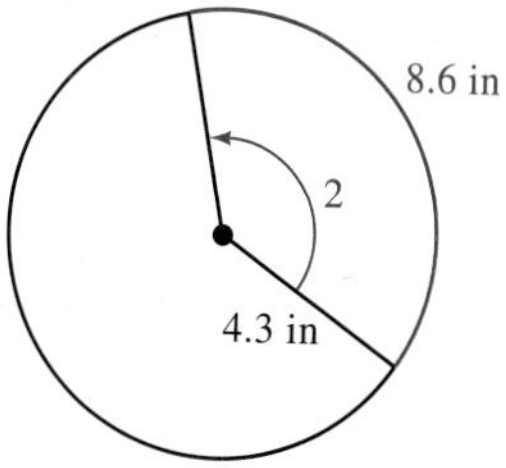

Figure 2

◀

▶ **EXAMPLE 2**   Figure 3 is a model of George Ferris's Ferris wheel. Recall that the diameter of the wheel is 250 feet, and θ is the central angle formed as a rider travels from his or her initial position P_0 to position P_1. Find the distance traveled by the rider if $\theta = 45°$ and if $\theta = 105°$.

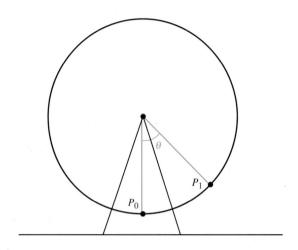

Figure 3

Solution   The formula for arc length, $s = r\theta$, requires θ to be given in radians. Since θ is given in degrees, we must multiply it by $\pi/180$ to convert to radians. Also, since the diameter of the wheel is 250 feet, the radius is 125 feet.

For $\theta = 45°$: $s = r\theta$

$$= 125(45)\left(\frac{\pi}{180}\right)$$

$$= \frac{125\pi}{4} \text{ feet}$$

$$= 98.2 \text{ feet} \qquad \text{To the nearest tenth}$$

For $\theta = 105°$: $s = r\theta$

$$= 125(105)\left(\frac{\pi}{180}\right)$$

$$= \frac{875\pi}{12} \text{ feet}$$

$$= 229.1 \text{ feet} \qquad \text{To the nearest tenth} \qquad ◀$$

▶ **EXAMPLE 3**   The minute hand of a clock is 1.2 centimeters long. To two significant digits, how far does the tip of the minute hand move in 20 minutes?

Solution   We have $r = 1.2$ centimeters. Since we are looking for s, we need to find θ. We can use a proportion to find θ. Since one complete rotation is 60 minutes and 2π radians, we say θ is to 2π as 20 minutes is to 60 minutes, or

$$\text{If} \quad \frac{\theta}{2\pi} = \frac{20}{60} \quad \text{then} \quad \theta = \frac{2\pi}{3}$$

Now we can find s.

$$s = r\theta$$
$$= 1.2\left(\frac{2\pi}{3}\right)$$
$$= \frac{2.4\pi}{3}$$
$$\approx 2.5 \text{ centimeters}$$

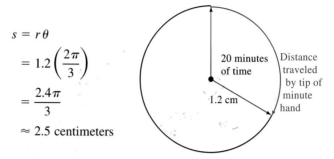

Figure 4

The tip of the minute hand will travel approximately 2.5 centimeters every 20 minutes. ◀

If we are working with relatively small central angles in circles with large radii, we can use the length of the intercepted arc to approximate the length of the associated chord. For example, Figure 5 shows a central angle of $1°$ in a circle of radius 1,800 feet, along with the arc and chord cut off by $1°$. (Figure 5 is not drawn to scale.)

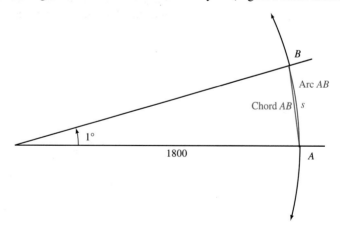

Figure 5

To find the length of arc AB, we convert θ to radians by multiplying by $\pi/180°$. Then we apply the formula $s = r\theta$.

$$s = r\theta = 1,800(1°)\left(\frac{\pi}{180°}\right) = 10\pi \approx 31.4 \text{ feet}$$

If we had carried out the calculation of arc AB to six significant digits, we would have obtained $s = 31.4159$. The length of the chord AB is 31.4155 to six significant digits (found by using the law of sines, which we will cover in Chapter 7). As you can see, the first five digits in each number are the same. It seems reasonable then to approximate the length of chord AB with the length of arc AB.

As our next example illustrates, we can also use the procedure just outlined in the reverse order to find the radius of a circle by approximating arc length with the length of the associated chord.

▶ **EXAMPLE 4**   A person standing on the earth notices that a 747 Jumbo Jet flying overhead subtends an angle of 0.45°. If the length of the jet is 230 feet, find its altitude to the nearest thousand feet.

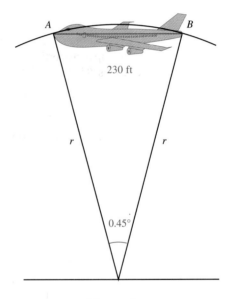

Figure 6

Solution   Figure 6 is a diagram of the situation. Since we are working with a relatively small angle in a circle with a large radius, we use the length of the airplane (chord AB in Figure 6) as an approximation of the length of the arc AB, and r as an approximation for the altitude of the plane.

Since $s = r\theta$, $r = \dfrac{s}{\theta}$

so $r = \dfrac{230}{(0.45°)(\pi/180°)}$ We multiply 0.45° by $\pi/180$ to change to radian measure

$= \dfrac{230(180)}{(0.45)(\pi)}$

$= 29{,}000$ feet to the nearest thousand feet   ◀

Area of a Sector

Next we want to derive the formula for the area of the sector formed by a central angle θ (Figure 7).

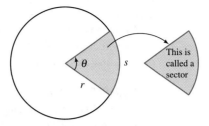

Figure 7

If we let A represent the area of the sector formed by central angle θ, we can find A by setting up a proportion as follows: We say the area A of the sector is to the area of the circle as θ is to one full rotation. That is,

$$\text{Area of sector} \longrightarrow \frac{A}{\pi r^2} = \frac{\theta}{2\pi} \longleftarrow \text{Central angle } \theta$$
$$\text{Area of circle} \longrightarrow \phantom{\frac{A}{\pi r^2} = \frac{\theta}{2\pi}} \longleftarrow \text{One full rotation}$$

We solve for A by multiplying both sides of the proportion by πr^2.

$$\pi r^2 \cdot \frac{A}{\pi r^2} = \frac{\theta}{2\pi} \cdot \pi r^2$$

$$A = \frac{1}{2} r^2 \theta$$

Area of a Sector

If θ (in radians) is a central angle in a circle with radius r, then the area of the sector formed by angle θ is given by

$$A = \frac{1}{2} r^2 \theta \qquad (\theta \text{ in radians})$$

▶ **EXAMPLE 5**   Find the area of the sector formed by a central angle of 1.4 radians in a circle of radius 2.1 meters.

Solution   We have $r = 2.1$ meters and $\theta = 1.4$. Applying the formula for A gives us

$$A = \frac{1}{2} r^2 \theta$$

$$= \frac{1}{2} (2.1)^2 (1.4)$$

$$= 3.1 \text{ meters}^2 \qquad \text{To the nearest tenth}$$

Remember Area is measured in square units. When $r = 2.1$ meters, $r^2 = (2.1 \text{ meters})^2 = 4.41 \text{ meters}^2$. ◀

▶ **EXAMPLE 6**   If the sector formed by a central angle of $15°$ has an area of $\pi/3$ centimeters^2, find the radius of the circle.

Solution   We first convert $15°$ to radians.

$$\theta = 15\left(\frac{\pi}{180}\right) = \frac{\pi}{12}$$

Then we substitute $\theta = \pi/12$ and $A = \pi/3$ into the formula for A and solve for r.

$$A = \frac{1}{2}r^2\theta$$

$$\frac{\pi}{3} = \frac{1}{2}r^2\frac{\pi}{12}$$

$$\frac{\pi}{3} = \frac{\pi}{24}r^2$$

$$r^2 = \frac{\pi}{3} \cdot \frac{24}{\pi}$$

$$r^2 = 8$$
$$r = 2\sqrt{2} \text{ centimeters}$$

Note that we need only use the positive square root of 8, since we know our radius must be measured with positive units. ◀

▶ **EXAMPLE 7**   A lawn sprinkler located at the corner of a yard is set to rotate through $90°$ and project water out 30 feet. To three significant digits, what area of lawn is watered by the sprinkler?

Solution   We have $\theta = 90° = \dfrac{\pi}{2} \approx 1.57$ radians and $r = 30$ feet.

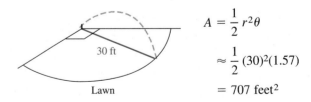

30 ft

Lawn

$$A = \frac{1}{2}r^2\theta$$

$$\approx \frac{1}{2}(30)^2(1.57)$$

$$= 707 \text{ feet}^2$$

◀

Figure 8

PROBLEM SET 3.4

Unless otherwise stated, all answers in this Problem Set that need to be rounded should be rounded to three significant digits.

For each problem below, θ is a central angle in a circle of radius r. In each case, find the length of arc s cut off by θ.

1. $\theta = 2$, $r = 3$ inches
2. $\theta = 3$, $r = 2$ inches
3. $\theta = 1.5$, $r = 1.5$ ft
4. $\theta = 2.4$, $r = 1.8$ ft
5. $\theta = \pi/6$, $r = 12$ cm
6. $\theta = \pi/3$, $r = 12$ cm
7. $\theta = 60°$, $r = 4$ mm
8. $\theta = 30°$, $r = 4$ mm
9. $\theta = 240°$, $r = 10$ inches
10. $\theta = 315°$, $r = 5$ inches

11. The minute hand of a clock is 2.4 cm long. How far does the tip of the minute hand travel in 20 minutes?
12. The minute hand of a clock is 1.2 cm long. How far does the tip of the minute hand travel in 40 minutes?
13. A space shuttle 200 mi above the earth is orbiting the earth once every 6 hours. How far does the shuttle travel in 1 hour? (Assume the radius of the earth is 4,000 mi.) Give your answer as both an exact value and an approximation to three significant digits. (See Figure 9.)
14. How long, in hours, does it take the space shuttle in Problem 13 to travel 8,400 mi? Give both the exact value and an approximate value for your answer.
15. The pendulum on a grandfather clock swings from side to side once every second. If the length of the pendulum is 4 ft and the angle through which it swings is 20°, how far does the tip of the pendulum travel in 1 second?
16. Find the total distance traveled in 1 minute by the tip of the pendulum on the grandfather clock in Problem 15.
17. From the earth, the moon subtends an angle of approximately 0.5°. If the distance to the moon is approximately 240,000 mi, find an approximation for the diameter of the moon accurate to the nearest hundred miles. (See Example 4 and the discussion that precedes it.)
18. If the distance to the sun is approximately 93 million mi, and, from the earth, the sun subtends an angle of approximately 0.5°, estimate the diameter of the sun to the nearest 10,000 miles.

Repeat Example 2 from this section for the following values of θ.

19. $\theta = 30°$
20. $\theta = 60°$
21. $\theta = 220°$
22. $\theta = 315°$

23. In Problem Set 2.3, we mentioned a Ferris wheel built in Vienna in 1897, known as the Great Wheel. The diameter of this wheel is 197 ft. Use Figure 3 from this section as a model of the Great Wheel, and find the distance traveled by a rider in going from initial position P_0 to position P_1 if
   a. θ is 60°
   b. θ is 210°
   c. θ is 285°

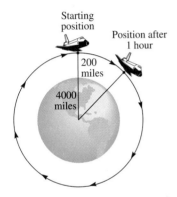

Starting
position
Position after
1 hour
200 miles
4000 miles

Figure 9

24. A Ferris wheel called Colossus that we mentioned in Problem Set 2.3 has a diameter of 165 feet. Using Figure 3 from this section as a model, find the distance traveled by someone starting at initial position P_0 and moving to position P_1 if

    a. θ is $150°$

    b. θ is $240°$

    c. θ is $345°$

In each problem below, θ is a central angle that cuts off an arc of length s. In each case, find the radius of the circle.

25. $\theta = 6, s = 3$ ft   **26.** $\theta = 1, s = 2$ ft

27. $\theta = 1.4, s = 4.2$ inches   **28.** $\theta = 5.1, s = 10.2$ inches

29. $\theta = \pi/4, s = \pi$ cm   **30.** $\theta = 3\pi/4, s = \pi$ cm

31. $\theta = 90°, s = \pi/2$ m   **32.** $\theta = 180°, s = \pi/2$ m

33. $\theta = 225°, s = 4$ km   **34.** $\theta = 150°, s = 5$ km

Find the area of the sector formed by central angle θ in a circle of radius r if

35. $\theta = 2, r = 3$ cm   **36.** $\theta = 3, r = 2$ cm

37. $\theta = 2.4, r = 4$ inches   **38.** $\theta = 1.8, r = 2$ inches

39. $\theta = \pi/5, r = 3$ m   **40.** $\theta = 2\pi/5, r = 3$ m

41. $\theta = 15°, r = 5$ m   **42.** $\theta = 15°, r = 10$ m

43. A central angle of 2 radians cuts off an arc of length 4 inches. Find the area of the sector formed.

44. An arc of length 3 ft is cut off by a central angle of $\pi/4$ radians. Find the area of the sector formed.

45. If the sector formed by a central angle of $30°$ has an area of $\pi/3$ cm², find the radius of the circle.

46. What is the length of the arc cut off by angle θ in Problem 45?

47. A sector of area $2\pi/3$ inches² is formed by a central angle of $45°$. What is the radius of the circle?

48. A sector of area 25 inches² is formed by a central angle of 4 radians. Find the radius of the circle.

49. A lawn sprinkler is located at the corner of a yard. The sprinkler is set to rotate through $90°$ and project water out 60 ft. What is the area of the yard watered by the sprinkler?

50. An automobile windshield wiper 10 inches long rotates through an angle of $60°$. If the rubber part of the blade covers only the last 9 inches of the wiper, find the area of the windshield cleaned by the windshield wiper.

Review Problems

The problems that follow review material we covered in Section 2.4.

51. If a 75-ft flagpole casts a shadow 43 ft long, what is the angle of elevation of the sun from the tip of the shadow?

52. A road up a hill makes an angle of 5° with the horizontal. If the road from the bottom of the hill to the top of the hill is 2.5 mi long, how high is the hill?

53. A person standing 5.2 ft from a mirror notices that the angle of depression from his eyes to the bottom of the mirror is 13°, while the angle of elevation to the top of the mirror is 12°. Find the vertical dimension of the mirror.

54. A boat travels on a course of bearing S 63° 50′ E for 114 mi. How many miles south and how many miles east has the boat traveled?

55. The height of a right circular cone is 35.8 cm. If the diameter of the base is 20.5 cm, what angle does the side of the cone make with the base?

56. A ship leaves the harbor entrance and travels 35 mi in the direction N 42° E. The captain then turns the ship 90° and travels another 24 mi in the direction S 48° E. At that time, how far is the ship from the harbor entrance, and what is the bearing of the ship from the harbor entrance?

57. A man standing on the roof of a building 86.0 ft above the ground looks down to the building next door. He finds the angle of depression to the roof of that building from the roof of his building to be 14.5°, while the angle of depression from the roof of his building to the bottom of the building next door is 43.2°. How tall is the building next door?

58. Two people decide to find the height of a tree. They position themselves 35 ft apart in line with, and on the same side of, the tree. If they find the angles of elevation from the ground where they are standing to the top of the tree are 65° and 44°, how tall is the tree?

SECTION 3.5 ## Velocities

The specifications for the first Ferris wheel indicate that one trip around the wheel took 20 minutes. How fast was a rider traveling around the wheel? There are a number of ways to answer this question. The most intuitive measure of the rate at which the rider is traveling around the wheel is what we call *linear velocity*. The units of linear velocity are miles per hour, feet per second, and so forth. Another way to specify how fast the rider is traveling around the wheel is with what we call *angular velocity*. Angular velocity is given as the amount of central angle through which the rider travels over a given amount of time. The central angle swept out by a rider traveling once around the wheel is 360°, or 2π radians. If one trip around the wheel takes 20 minutes, then the angular velocity of a rider is

$$\frac{2\pi \text{ radians}}{20 \text{ minutes}} = \frac{\pi}{10} \text{ radians per minute}$$

In this section, we will learn more about angular velocity and linear velocity and the relationship between them. Let's start with the formal definition for the linear velocity of a point moving on the circumference of a circle.

DEFINITION   If P is a point on a circle of radius r, and P moves a distance s on the circumference of the circle in an amount of time t, then the *linear velocity, v,* of P is given by the formula

$$v = \frac{s}{t}$$

▶ **EXAMPLE 1**   A point on a circle travels 5 centimeters in 2 seconds. Find the linear velocity of the point.

Solution   Substituting $s = 5$ and $t = 2$ into the equation $v = s/t$ gives us

$$v = \frac{5 \text{ centimeters}}{2 \text{ seconds}}$$

$$= 2.5 \text{ centimeters per second}$$ ◀

Note   In all the examples and problems in this section, we are assuming that the point on the circle moves with uniform circular motion. That is, the velocity of the point is constant.

DEFINITION   If P is a point moving with uniform circular motion on a circle of radius r, and the line from the center of the circle through P sweeps out a central angle θ in an amount of time t, then the *angular velocity, ω* (omega), of P is given by the formula

$$\omega = \frac{\theta}{t} \qquad \text{where } \theta \text{ is measured in radians}$$

▶ **EXAMPLE 2**   A point on a circle rotates through $3\pi/4$ radians in 3 seconds. Give the angular velocity of P.

Solution   Substituting $\theta = 3\pi/4$ and $t = 3$ into the equation $\omega = \theta/t$ gives us

$$\omega = \frac{3\pi/4 \text{ radians}}{3 \text{ seconds}}$$

$$= \frac{\pi}{4} \text{ radians per second}$$ ◀

Note   There are a number of equivalent ways to express the units of velocity. For example, the answer to Example 2 can be expressed in each of the following ways; they are all equivalent.

$$\frac{\pi}{4} \text{ radians per second} = \frac{\pi}{4} \text{ rad/sec} = \frac{\pi \text{ radians}}{4 \text{ seconds}} = \frac{\pi \text{ rad}}{4 \text{ sec}}$$

Likewise, you can express the answer to Example 1 in any of the following ways:

$$2.5 \text{ centimeters per second} = 2.5 \text{ cm/sec} = \frac{2.5 \text{ cm}}{1 \text{ sec}}$$

▶ **EXAMPLE 3**   A bicycle wheel with a radius of 13 inches turns with an angular velocity of 3 radians per second. Find the distance traveled by a point on the bicycle tire in 1 minute.

Solution   We have $\omega = 3$ radians per second, $r = 13$ inches, and $t = 60$ seconds. First we find θ using $\omega = \theta/t$.

$$\text{If}\qquad \omega = \frac{\theta}{t}$$

$$\text{then}\qquad \begin{aligned} \theta &= \omega t \\ &= 3(60) \\ &= 180 \text{ radians} \end{aligned}$$

To find the distance traveled by the point in 60 seconds, we use the formula $s = r\theta$ from Section 3.4, with $r = 13$ inches and $\theta = 180$.

$$\begin{aligned} s &= 13(180) \\ &= 2{,}340 \text{ inches} \end{aligned}$$

If we want this result expressed in feet, we divide by 12.

$$\begin{aligned} s &= \frac{2{,}340}{12} \text{ feet} \\ &= 195 \text{ feet} \end{aligned}$$

A point on the tire of the bicycle will travel 195 feet in 1 minute. If the bicycle were being ridden under these conditions, the rider would travel 195 feet in 1 minute.   ◀

▶ **EXAMPLE 4**   Figure 1 shows a fire truck parked on the shoulder of a freeway next to a long block wall. The red light on the top of the truck is 10 feet from the wall and rotates through one complete revolution every 2 seconds. Find the equations that give the lengths d and l in terms of time t.

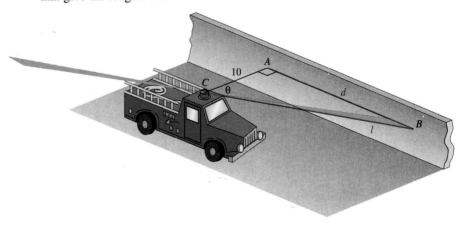

Figure 1

Solution   The angular velocity of the rotating red light is

$$\omega = \frac{\theta}{t} = \frac{2\pi \text{ radians}}{2 \text{ seconds}} = \pi \text{ radians per second}$$

From right triangle *ABC*, we have the following relationships:

$$\tan \theta = \frac{d}{10} \qquad \text{and} \qquad \sec \theta = \frac{l}{10}$$

$$d = 10 \tan \theta \qquad\qquad l = 10 \sec \theta$$

Now, these equations give us *d* and *l* in terms of θ. To write *d* and *l* in terms of *t*, we solve $\omega = \theta/t$ for θ to obtain $\theta = \omega t = \pi t$. Substituting this for θ in each equation, we have *d* and *l* expressed in terms of *t*.

$$d = 10 \tan \pi t \qquad l = 10 \sec \pi t \qquad\qquad \blacktriangleleft$$

The Relationship Between the Two Velocities

To find the relationship between the two kinds of velocities we have developed so far, we can take the equation that relates arc length and central angle measure, $s = r\theta$, and divide both sides by time, *t*.

$$\text{If} \qquad s = r\theta$$

$$\text{then} \qquad \frac{s}{t} = \frac{r\theta}{t}$$

$$\frac{s}{t} = r\frac{\theta}{t}$$

$$v = r\omega$$

Linear velocity is the product of the radius and the angular velocity.

▶ **EXAMPLE 5**   A phonograph record is turning at 45 rpm (revolutions per minute). If the distance from the center of the record to a point on the edge of the record is 3 inches, find the angular velocity and the linear velocity of the point in feet per minute.

Solution   The quantity 45 rpm is another way of expressing the rate at which the point on the record is moving. We can obtain the angular velocity from it by remembering that one complete revolution is equivalent to 2π radians. Therefore,

$$\omega = 45(2\pi) \text{ radians per minute}$$
$$= 90\pi \text{ radians per minute}$$

To find the linear velocity, we multiply ω by the radius.

$$
\begin{aligned}
v &= r\omega \\
&= 3(90\pi) \\
&= 270\pi \text{ inches per minute} &&\text{Exact value} \\
&\approx 270(3.14) \text{ inches per minute} &&\text{Approximate value} \\
&= 848 \text{ inches per minute} &&\text{To 3 significant digits}
\end{aligned}
$$

To convert 848 inches per minute to feet per minute, we write 848 inches/min as

$$\frac{848 \text{ inches}}{1 \text{ min}}$$

then we multiply by the number 1 in the form

$$\frac{1 \text{ ft}}{12 \text{ inches}}$$

By doing so, inches will divide out in much the same way that common factors divide out when we reduce fractions to lowest terms. Here is our work:

$$848 \text{ inches/min} = \frac{848 \text{ inches}}{1 \text{ min}} \cdot \frac{1 \text{ ft}}{12 \text{ inches}} = \frac{848}{12} \text{ ft/min} \approx 70.7 \text{ ft/min}$$

The fraction $\dfrac{1 \text{ ft}}{12 \text{ inches}}$ is called a *conversion factor*. It is simply the number 1 (because 12 inches and 1 foot are equivalent) in a form that allows us to convert from inches to feet by dividing out inches.   ◀

▶ **EXAMPLE 6**   The Ferris wheel shown in Figure 2 is a model of the one we mentioned in the introduction to this section. If the diameter of the wheel is 250 feet and one complete revolution takes 20 minutes, find the linear velocity of a person riding on the wheel. Give the answer in miles per hour.

Solution   We found the angular velocity in the introduction to this section. It is

$$\omega = \frac{\pi}{10} \text{ rad/min}$$

Next, we use the formula $v = r\omega$ to find the linear velocity. That is, we multiply the angular velocity by the radius to find the linear velocity.

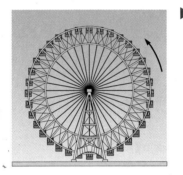

Figure 2

$$v = r\omega$$

$$= (125 \text{ ft}) \left(\frac{\pi}{10} \text{ rad/min} \right)$$

$$= \frac{25\pi}{2} \text{ ft/min}$$

$$\approx 39.27 \text{ ft/min (intermediate answer)}$$

To convert to miles per hour, we use the facts that there are 60 minutes in 1 hour and 5,280 feet in 1 mile.

$$39.27 \text{ ft/min} = \frac{39.27 \text{ ft}}{1 \text{ min}} \cdot \frac{60 \text{ min}}{1 \text{ hr}} \cdot \frac{1 \text{ mi}}{5,280 \text{ ft}}$$

$$= \frac{(39.27)(60) \text{ mi}}{5,280 \text{ hr}}$$

$$\approx 0.45 \text{ mi/hr}$$ ◀

To gain a more intuitive understanding of the relationship between the radius of the circle and the linear velocity of a point on the circle, imagine that a bird is sitting on one of the wires that connects the center of the Ferris wheel in Example 6 to the wheel itself. Imagine further that the bird is sitting exactly halfway between the center of the wheel and a rider on the wheel. How does the linear velocity of the bird compare with the linear velocity of the rider? Since the angular velocity of the bird and the rider are equal (both sweep out the same amount of central angle in a given amount of time), and linear velocity is the product of the radius and the angular velocity, we can simply multiply the linear velocity of the rider by $\frac{1}{2}$ to obtain the linear velocity of the bird. Therefore, the bird is traveling at

$$\frac{1}{2} \cdot 0.45 \text{ mi/hr} = 0.225 \text{ mi/hr}$$

In one revolution around the wheel, the rider will travel a greater distance than the bird in the same amount of time, so the rider's velocity is greater than that of the bird. Twice as great, to be exact.

PROBLEM SET 3.5

In this Problem Set, round any answers that need rounding to three significant digits.

Find the linear velocity of a point moving with uniform circular motion, if the point covers a distance s in an amount of time t, where

1. $s = 3$ ft and $t = 2$ min
2. $s = 10$ ft and $t = 2$ min
3. $s = 12$ cm and $t = 4$ sec
4. $s = 12$ cm and $t = 2$ sec
5. $s = 30$ mi and $t = 2$ hr
6. $s = 100$ mi and $t = 4$ hr

Find the distance s covered by a point moving with linear velocity v for a time t if

7. $v = 20$ ft/sec and $t = 4$ sec
8. $v = 10$ ft/sec and $t = 4$ sec
9. $v = 45$ mph and $t = \frac{1}{2}$ hr
10. $v = 55$ mph and $t = \frac{1}{2}$ hr
11. $v = 21$ mph and $t = 20$ min
12. $v = 63$ mph and $t = 10$ sec

Point P sweeps out central angle θ as it rotates on a circle of radius r as given below. In each case, find the angular velocity of point P.

13. $\theta = 2\pi/3$, $t = 5$ sec
14. $\theta = 3\pi/4$, $t = 5$ sec
15. $\theta = 12$, $t = 3$ min
16. $\theta = 24$, $t = 6$ min
17. $\theta = 8\pi$, $t = 3\pi$ sec
18. $\theta = 12\pi$, $t = 5\pi$ sec
19. $\theta = 45\pi$, $t = 1.2$ hr
20. $\theta = 24\pi$, $t = 1.8$ hr

21. Figure 3 shows a lighthouse that is 100 ft from a long straight wall on the beach. The light in the lighthouse rotates through one complete rotation once every 4 seconds. Find an equation that gives the distance d in terms of time t, then find d when t is $\frac{1}{2}$ second and $\frac{3}{2}$ seconds. What happens when you try $t = 1$ second in the equation? How do you interpret this?

22. Using the diagram in Figure 3, find an equation that expresses l in terms of time t. Find l when t is 0.5 second, 1.0 second, and 1.5 seconds. (Assume the light goes through one rotation every 4 seconds.)

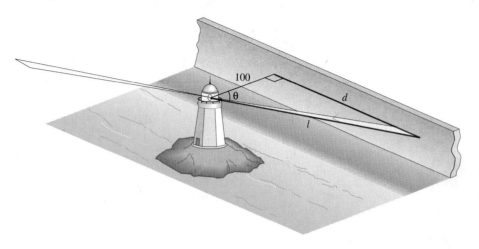

Figure 3

In the problems that follow, point P moves with angular velocity ω on a circle of radius r. In each case, find the distance s traveled by the point in time t.

23. $\omega = 4$ rad/sec, $r = 2$ inches, $t = 5$ sec
24. $\omega = 2$ rad/sec, $r = 4$ inches, $t = 5$ sec
25. $\omega = 3\pi/2$ rad/sec, $r = 4$ m, $t = 30$ sec
26. $\omega = 4\pi/3$ rad/sec, $r = 8$ m, $t = 20$ sec
27. $\omega = 15$ rad/sec, $r = 5$ ft, $t = 1$ min
28. $\omega = 10$ rad/sec, $r = 6$ ft, $t = 2$ min

For each of the following problems, find the angular velocity associated with the given rpm.

29. 10 rpm   30. 20 rpm   31. $33\frac{1}{3}$ rpm   32. $16\frac{2}{3}$ rpm
33. 5.8 rpm   34. 7.2 rpm

For each problem below, a point is rotating with uniform circular motion on a circle of radius r.

35. Find v if $r = 2$ inches and $\omega = 5$ rad/sec.
36. Find v if $r = 8$ inches and $\omega = 4$ rad/sec.
37. Find ω if $r = 6$ cm and $v = 3$ cm/sec.
38. Find ω if $r = 3$ cm and $v = 8$ cm/sec.
39. Find v if $r = 4$ ft and the point rotates at 10 rpm.
40. Find v if $r = 1$ ft and the point rotates at 20 rpm.

41. The earth rotates through one complete revolution every 24 hours. Since the axis of rotation is perpendicular to the equator, you can think of a person standing on

the equator as standing on the edge of a disc that is rotating through one complete revolution every 24 hours. Find the angular velocity of a person standing on the equator.

42. Assuming the radius of the earth is 4,000 mi, use the information from Problem 41 to find the linear velocity of a person standing on the equator.

43. A boy is twirling a model airplane on a string 5 ft long. If he twirls the plane at 0.5 rpm, how far does the plane travel in 2 minutes?

44. A mixing blade on a food processor extends out 3 inches from its center. If the blade is turning at 600 rpm, what is the linear velocity of the tip of the blade in feet per minute?

45. A gasoline-driven lawnmower has a blade that extends out 1 ft from its center. The tip of the blade is traveling at the speed of sound, which is 1,100 ft/sec. Through how many rpm is the blade turning?

46. A $3\frac{1}{2}$-inch diskette, when placed in the disk drive of a computer, rotates at 300 rpm. Find the linear velocity of a point 1.5 inches from the center of the diskette.

47. A $5\frac{1}{4}$-inch floppy diskette, when placed in the disk drive of a computer, rotates at 360 rpm. Find the linear velocity of a point 2.5 inches from the center of the diskette. Then find the linear velocity of a point 1.25 inches from the center of the diskette.

48. A 5-inch fixed disk in a computer rotates at 3,600 rpm. Find the linear velocity of a point 2 inches from the center of the disk. Then find the linear velocity of a point 1 inch from the center.

49. Figure 4 is a model of the Ferris wheel known as the Reisenrad, or Great Wheel, that was built in Vienna in 1897. The diameter of the wheel is 197 ft, and one complete revolution takes 15 minutes. Find the linear velocity of a person riding on the wheel. Give your answer in miles per hour.

50. Use Figure 4 as a model of the Ferris wheel called Colossus that was built in St. Louis in 1986. The diameter of the wheel is 165 ft. A brochure that gives some statistics associated with Colossus indicates that it rotates at 1.5 rpm. The same brochure also indicates that a rider on the wheel is traveling at 10 mph. Explain why these two numbers, 1.5 rpm and 10 mph, cannot both be correct.

51. How far does the tip of a 12-cm minute hand on a clock travel in 1 day?

52. How far does the tip of a 10-cm hour hand on a clock travel in 1 day?

53. A woman rides a bicycle for 1 hour and travels 16 km (about 10 mi). Find the angular velocity of the wheel if the radius is 30 cm.

54. Find the number of rpm for the wheel in Problem 53.

Figure 4

Review Problems

The problems that follow review material we covered in Section 2.5.

55. A boat is crossing a river that runs due west. The heading of the boat is due north; it is moving through the water at 9.50 mph. If the current is running at a constant 4.25 mph, find the true course of the boat.

56. A ship is 4° off course. How far off course is it after traveling for 75 mi?

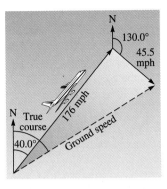

Figure 5

57. A plane has an airspeed of 176 mph and a heading of 40.0°. The air currents are moving at a constant 45.5 mph at 130.0°. Find the ground speed and true course of the plane. (See Figure 5.)

58. A plane has an airspeed of 255 mph and a heading of 130°. The air currents are moving at a constant 25.5 mph at 40.0°. Find the ground speed and true course of the plane.

59. Find the magnitude of the horizontal and vertical components of a velocity vector of 68 ft/sec with angle of elevation 37°.

60. The magnitude of the horizontal component of a vector is 75, while the magnitude of its vertical component is 45. What is the magnitude of the vector?

61. A ship sails for 85.5 mi on a 237.3° course. How far west and how far south has the boat traveled?

62. A plane flying with a constant ground speed of 285.5 mph flies for 2 hours on a course with bearing N 48.7° W. How far north and how far west does the plane fly?

Research Projects

63. We mentioned in Chapter 2 that the Ferris wheel called the Reisenrad that was built in Vienna in 1897 is still in operation today. A brochure that gives some statistics associated with the Reisenrad indicates that passengers riding it travel at 2 ft, 6 inches per second. The Orson Welles movie *The Third Man* contains a scene in which Orson Welles rides the Reisenrad through one complete revolution. Play *The Third Man* on a VCR, so you can view the Reisenrad in operation. Then devise a method of using the movie to estimate the angular velocity of the wheel. Give a detailed account of the procedure you use to arrive at your estimate. Finally, use your results either to prove or to disprove the claim that passengers travel at 2 ft, 6 inches per second on the Reisenrad.

64. Jim Rizzoli owns, maintains, and races an alcohol dragster. On board the dragster is a computer that compiles data into a number of different categories during each of Jim's races. The table below shows some of the data from a race Jim was in during the 1993 Winternationals.

Time in Seconds	Speed in Miles/Hour	Front Axle RPM
0	0	0
1	72.7	1,107
2	129.9	1,978
3	162.8	2,486
4	192.2	2,919
5	212.4	3,233
6	228.1	3,473

(problem continues on p. 154)

The front wheels, at each end of the front axle, are 22.07 inches in diameter. Derive a formula that will convert rpm's from a 22.07-inch wheel into miles per hour. Test your formula on all six rows of the table. Explain any discrepancies between the table values and the values obtained from your formula.

CHAPTER 3 SUMMARY

Reference Angle [3.1]

The *reference angle* $\hat{\theta}$ for any angle θ in standard position is the positive acute angle between the terminal side of θ and the x-axis.

A trigonometric function of an angle and its reference angle differ at most in sign. We find trigonometric functions for angles between $0°$ and $360°$ by first finding the reference angle. We then find the value of the trigonometric function of the reference angle and use the quadrant in which the angle terminates to assign the correct sign.

Examples

1. $30°$ is the reference angle for $30°$, $150°$, $210°$, and $330°$.

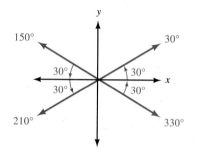

$$\sin 150° = \sin 30° = \frac{1}{2}$$

$$\sin 210° = -\sin 30° = -\frac{1}{2}$$

$$\sin 330° = -\sin 30° = -\frac{1}{2}$$

2.

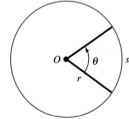

Radian Measure [3.2]

In a circle with radius r, if central angle θ cuts off an arc of length s, then the radian measure of θ is given by

$$\theta = \frac{s}{r}$$

3. Radians to degrees

$$\frac{4\pi}{3} \text{ radians} = \frac{4\pi}{3}\left(\frac{180}{\pi}\right)°$$
$$= 240°$$

Radians and Degrees [3.2]

Changing from degrees to radians and radians to degrees is simply a matter of multiplying by the appropriate conversion factor.

Degrees to radians

$$450° = 450\left(\frac{\pi}{180}\right)$$

$$= \frac{5\pi}{2} \text{ radians}$$

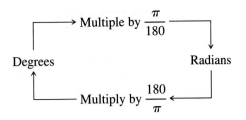

The Unit Circle [3.3]

The unit circle is the circle with its center at the origin and a radius of 1. The equation of the unit circle is $x^2 + y^2 = 1$. Because the radius of the unit circle is 1, any point (x, y) on the circle is such that

$$x = \cos \theta \text{ and } y = \sin \theta$$

5.   Tan θ is an odd function.

$$\tan(-\theta) = \frac{\sin(-\theta)}{\cos(-\theta)}$$

$$= \frac{-\sin \theta}{\cos \theta}$$

$$= -\frac{\sin \theta}{\cos \theta}$$

$$= -\tan \theta$$

Even and Odd Functions [3.3]

An *even function* is a function for which

$$f(-x) = f(x) \text{ for all } x \text{ in the domain of } f$$

and an *odd function* is a function for which

$$f(-x) = -f(x) \text{ for all } x \text{ in the domain of } f$$

Cosine is an even function, and sine is an odd function. That is,

$$\cos(-\theta) = \cos \theta \qquad \text{Cosine is an even function}$$

and

$$\sin(-\theta) = -\sin \theta \qquad \text{Sine is an odd function}$$

6.   The arc cut off by 2.5 radians in a circle of radius 4 inches is

$$s = 4(2.5) = 10 \text{ inches}$$

Arc Length [3.4]

If s is an arc cut off by a central angle θ, measured in radians, in a circle of radius r, then

$$s = r\theta$$

7.   The area of the sector formed by a central angle of 2.5 radians in a circle of radius 4 inches is

$$A = \frac{1}{2}(4)^2(2.5) = 20 \text{ inches}^2$$

Area of a Sector [3.4]

The area of the sector formed by a central angle θ in a circle of radius r is

$$A = \frac{1}{2}r^2\theta$$

where θ is measured in radians.

8. If a point moving at a uniform speed on a circle travels 12 cm every 3 seconds, then the linear velocity of the point is

$$v = \frac{12 \text{ cm}}{3 \text{ sec}} = 4 \text{ cm/sec}$$

9. If a point moving at uniform speed on a circle of radius 4 inches rotates through $3\pi/4$ radians every 3 seconds, then the angular velocity of the point is

$$\omega = \frac{3\pi/4 \text{ rad}}{3 \text{ sec}} = \frac{\pi}{4} \text{ rad/sec}$$

The linear velocity of the same point is given by

$$v = 4\left(\frac{\pi}{4}\right) = \pi \text{ inches/sec}$$

Linear Velocity [3.5]

If P is a point on a circle of radius r, and P moves a distance s on the circumference of the circle, in an amount of time t, then the *linear velocity, v,* of P is given by the formula

$$v = \frac{s}{t}$$

Angular Velocity [3.5]

If P is a point moving with uniform circular motion on a circle of radius r, and the line from the center of the circle through P sweeps out a central angle θ, in an amount of time t, then the *angular velocity, ω,* of P is given by the formula

$$\omega = \frac{\theta}{t} \qquad \text{where } \theta \text{ is measured in radians}$$

The relationship between linear velocity and angular velocity is given by the formula

$$v = r\omega$$

CHAPTER 3 TEST

Draw each of the following angles in standard position and then name the reference angle:

1. 235°

2. 117.8°

3. 410° 20′

4. −225°

Use a calculator to find each of the following:

5. cot 320°

6. cot (−25°)

7. csc (−236.7°)

8. sec 322.3°

9. sec 140° 20′

10. csc 188° 50′

Use a calculator to find θ, if θ is between 0° and 360° and

11. $\sin \theta = 0.1045$ with θ in QII

12. $\cos \theta = -0.4772$ with θ in QIII

13. $\cot \theta = 0.9659$ with θ in QIII

14. $\sec \theta = 1.545$ with θ in QIV

Give the exact value of each of the following:

15. sin 225°

16. cos 135°

17. tan 330°

18. sec 390°

Convert each of the following to radian measure. Write each answer as an exact value.

19. 250°

20. −390°

Convert each of the following to degree measure:

21. $4\pi/3$

22. $7\pi/12$

Give the exact value of each of the following:

23. $\sin \dfrac{2\pi}{3}$

24. $\cos \dfrac{2\pi}{3}$

25. $4 \cos \left(-\dfrac{3\pi}{4} \right)$

26. $2 \cos \left(-\dfrac{5\pi}{3} \right)$

27. $\sec \dfrac{5\pi}{6}$

28. $\csc \dfrac{5\pi}{6}$

29. Evaluate $2 \cos \left(3x - \dfrac{\pi}{2} \right)$ when x is $\dfrac{\pi}{3}$.

30. Evaluate $4 \sin \left(2x + \dfrac{\pi}{4} \right)$ when x is $\dfrac{\pi}{4}$.

31. Show that cotangent is an odd function.
32. Prove the identity $\sin(-\theta) \sec(-\theta) \cot(-\theta) = 1$.

For Problems 33 and 34, θ is a central angle in a circle of radius r. In each case, find the length of arc s cut off by θ.

33. $\theta = \pi/6$, $r = 12$ m

34. $\theta = 60°$, $r = 6$ ft

In Problems 35 and 36, θ is a central angle that cuts off an arc of length s. In each case, find the radius of the circle.

35. $\theta = \pi/4$, $s = \pi$ cm

36. $\theta = 2\pi/3$, $s = \pi/4$ cm

Find the area of the sector formed by central angle θ in a circle of radius r if

37. $\theta = 90°$, $r = 4$ inches

38. $\theta = 2.4$, $r = 3$ cm

39. The minute hand of a clock is 2 cm long. How far does the tip of the minute hand travel in 30 minutes?
40. A central angle of 4 radians cuts off an arc of length 8 inches. Find the area of the sector formed.

Find the distance s covered by a point moving with linear velocity v for a time t if

41. $v = 30$ ft/sec and $t = 3$ sec

42. $v = 66$ ft/sec and $t = 1$ min

In the problems that follow, point P moves with angular velocity ω on a circle of radius r. In each case, find the distance s traveled by the point in time t.

43. $\omega = 4$ rad/sec, $r = 3$ inches, $t = 6$ sec
44. $\omega = 3\pi/4$ rad/sec, $r = 8$ ft, $t = 20$ sec

For Problems 45 and 46, find the angular velocity associated with the given rpm.

45. 6 rpm

46. 2 rpm

For each problem below, a point is rotating with uniform circular motion on a circle of radius r.

47. Find ω if $r = 10$ cm and $v = 5$ cm/sec.
48. Find ω if $r = 3$ cm and $v = 5$ cm/sec.
49. Find v if $r = 2$ ft and the point rotates at 20 rpm.
50. Find v if $r = 1$ ft and the point rotates at 10 rpm.
51. A belt connects a pulley of radius 8 cm to a pulley of radius 6 cm. Each point on the belt is traveling at 24 cm/sec. Find the angular velocity of each pulley (see Figure 1).

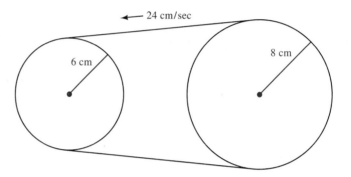

Figure 1

52. A propeller with radius 1.50 ft is rotating at 900 rpm. Find the linear velocity of the tip of the propeller. Give the exact value and an approximation to three significant digits.

4

Graphing and Inverse Functions

ı ı ı ı ı ı ı ı ı ı

▶ **To the Student**

In this chapter, we will consider the graphs of the trigonometric functions. We will begin by graphing $y = \sin x$ and $y = \cos x$ and then proceed to more complicated graphs. To begin with, we will make tables of values of x and y that satisfy the equations and then use the information in these tables to sketch the associated graphs. Our ultimate goal, however, is to graph trigonometric functions of the form $y = A \sin (Bx + C)$ and $y = A \cos (Bx + C)$ without the use of tables. Equations of this form have many applications. For instance, they can be used as mathematical models for sound waves and to describe such things as the variations in electrical current in an alternating circuit.

Basic Graphs

The Sine Graph

To graph the equation $y = \sin x$, we begin by making a table of values of x and y that satisfy the equation and then use the information in the table to sketch the graph. To make it easy on ourselves, we will let x take on values that are multiples of $\pi/4$. As an aid in sketching the graphs, we will approximate $1/\sqrt{2}$ with 0.7.

Table 1

x	$y = \sin x$	(x, y)
0	$y = \sin 0 = 0$	$(0, 0)$
$\dfrac{\pi}{4}$	$y = \sin \dfrac{\pi}{4} = \dfrac{1}{\sqrt{2}} \approx 0.7$	$\left(\dfrac{\pi}{4}, 0.7\right)$
$\dfrac{\pi}{2}$	$y = \sin \dfrac{\pi}{2} = 1$	$\left(\dfrac{\pi}{2}, 1\right)$
$\dfrac{3\pi}{4}$	$y = \sin \dfrac{3\pi}{4} = \dfrac{1}{\sqrt{2}} \approx 0.7$	$\left(\dfrac{3\pi}{4}, 0.7\right)$
π	$y = \sin \pi = 0$	$(\pi, 0)$
$\dfrac{5\pi}{4}$	$y = \sin \dfrac{5\pi}{4} = -\dfrac{1}{\sqrt{2}} \approx -0.7$	$\left(\dfrac{5\pi}{4}, -0.7\right)$
$\dfrac{3\pi}{2}$	$y = \sin \dfrac{3\pi}{2} = -1$	$\left(\dfrac{3\pi}{2}, -1\right)$
$\dfrac{7\pi}{4}$	$y = \sin \dfrac{7\pi}{4} = -\dfrac{1}{\sqrt{2}} \approx -0.7$	$\left(\dfrac{7\pi}{4}, -0.7\right)$
2π	$y = \sin 2\pi = 0$	$(2\pi, 0)$

Graphing each ordered pair and then connecting them with a smooth curve, we obtain the following graph:

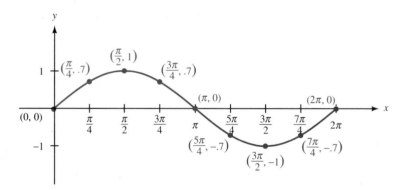

Figure 1

To further justify the graph in Figure 1, we could find additional ordered pairs that satisfy the equation. For example, we could continue our table by letting x take on multiples of $\pi/6$ and $\pi/3$. If we were to do so, we would find that any new ordered pair that satisfied the equation $y = \sin x$ would be such that its graph would lie on the curve in Figure 1. Figure 2 shows the curve in Figure 1 again, but this time with all the ordered pairs with x-coordinates that are multiples of $\pi/6$ or $\pi/3$ and the corresponding y-coordinates that satisfy the equation $y = \sin x$.

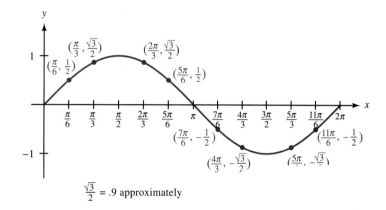

$$\frac{\sqrt{3}}{2} = .9 \text{ approximately}$$

Figure 2

▶ **EXAMPLE 1** Use the graph of $y = \sin x$ in Figure 2 to find all values of x between 0 and 2π for which $\sin x = \frac{1}{2}$.

Solution We locate $\frac{1}{2}$ on the y-axis and draw a horizontal line through it. We follow this line to the points where it intersects the graph of $y = \sin x$. The values of x just below these points of intersection are the values of x for which $\sin x = \frac{1}{2}$. As Figure 3 indicates, they are $\pi/6$ and $5\pi/6$.

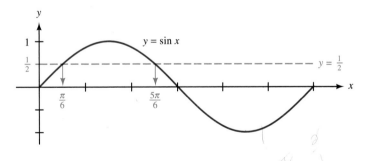

Figure 3

Graphing $y = \sin x$ Using the Unit Circle

We can also obtain the graph of the sine function by using the unit circle definition we introduced in Section 3.3 (Definition III): If the point (x, y) is t units from $(1, 0)$ along the circumference of the unit circle, then $\sin t = y$. Therefore, if we start at the point (1, 0) and travel once around the unit circle (a distance of 2π units), we can find the value of y in the equation $y = \sin t$ by simply keeping track of the y-coordinates of the points that are t units from (1, 0).

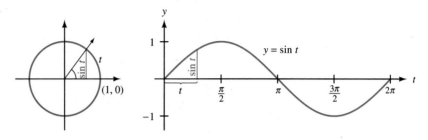

unit circle

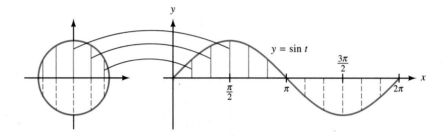

Figure 4

Extending the Sine Graph

Figures 1 through 4 each show one complete cycle of $y = \sin x$. We can extend the graph of $y = \sin x$ to the right of $x = 2\pi$ by realizing that, once we go past $x = 2\pi$, we will begin to name angles that are coterminal with the angles between 0 and 2π. Because of this, we will start to repeat the values of $\sin x$. Likewise, if we let x take on values to the left of $x = 0$, we will simply get the values of $\sin x$ between 0 and 2π in the reverse order. Figure 5 shows the graph of $y = \sin x$ extended beyond the interval from $x = 0$ to $x = 2\pi$.

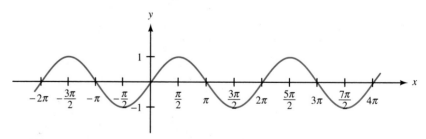

Figure 5

Our graph of $y = \sin x$ never goes above 1 or below -1, and it repeats itself every 2π units on the x-axis. This gives rise to the following two definitions.

DEFINITION   Period

For any function $y = f(x)$, the smallest positive number p for which

$$f(x + p) = f(x) \qquad \text{for all } x$$

is called the *period* of $f(x)$. In the case of $y = \sin x$, the period is $p = 2\pi$ since 2π is the smallest positive number for which

$$\sin (x + p) = \sin x \text{ for all } x$$

DEFINITION   Amplitude

If the greatest value of y is M and the least value of y is m, then the *amplitude* of the graph of y is defined to be

$$A = \frac{1}{2}|M - m|$$

In the case of $y = \sin x$, the amplitude is 1 because $\frac{1}{2}|1 - (-1)| = \frac{1}{2}(2) = 1$.

Along with the definitions for period and amplitude, we need to review the definitions for domain and range from algebra.

DEFINITION   Domain

The *domain* for the function $y = f(x)$ is the set of values that x can assume. For the function $y = \sin x$, the domain is all real numbers, since there is no value of x for which $\sin x$ is undefined.

DEFINITION   Range

The *range* for the function $y = f(x)$ is the set of values that y assumes. From the graph of $y = \sin x$ shown in Figure 5, we find that the range is the set

$$\{y \mid -1 \le y \le 1\}$$

The Cosine Graph

The graph of $y = \cos x$ has the same general shape as the graph of $y = \sin x$.

▶ **EXAMPLE 2**   Sketch the graph of $y = \cos x$.

Solution   We can arrive at the graph by making a table of convenient values of x and y.

Table 2

x	$y = \cos x$	(x, y)
0	$y = \cos 0 = 1$	$(0, 1)$
$\dfrac{\pi}{4}$	$y = \cos \dfrac{\pi}{4} = \dfrac{1}{\sqrt{2}}$	$\left(\dfrac{\pi}{4}, \dfrac{1}{\sqrt{2}}\right)$
$\dfrac{\pi}{2}$	$y = \cos \dfrac{\pi}{2} = 0$	$\left(\dfrac{\pi}{2}, 0\right)$
$\dfrac{3\pi}{4}$	$y = \cos \dfrac{3\pi}{4} = -\dfrac{1}{\sqrt{2}}$	$\left(\dfrac{3\pi}{4}, -\dfrac{1}{\sqrt{2}}\right)$
π	$y = \cos \pi = -1$	$(\pi, -1)$
$\dfrac{5\pi}{4}$	$y = \cos \dfrac{5\pi}{4} = -\dfrac{1}{\sqrt{2}}$	$\left(\dfrac{5\pi}{4}, -\dfrac{1}{\sqrt{2}}\right)$
$\dfrac{3\pi}{2}$	$y = \cos \dfrac{3\pi}{2} = 0$	$\left(\dfrac{3\pi}{2}, 0\right)$
$\dfrac{7\pi}{4}$	$y = \cos \dfrac{7\pi}{4} = \dfrac{1}{\sqrt{2}}$	$\left(\dfrac{7\pi}{4}, \dfrac{1}{\sqrt{2}}\right)$
2π	$y = \cos 2\pi = 1$	$(2\pi, 1)$

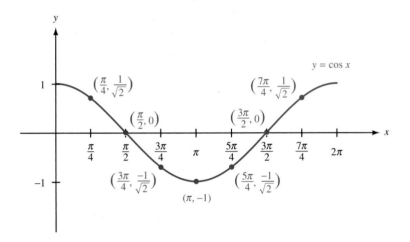

Figure 6 ◀

▶ **EXAMPLE 3** Find all values of x for which $\cos x = -1$.

Solution We draw a horizontal line through $y = -1$ and notice where it intersects the graph of $y = \cos x$. The x-coordinates of those points are solutions to $\cos x = -1$. (Figure 7 is the graph of $y = \cos x$ extended beyond the interval from $x = 0$ to $x = 2\pi$.)

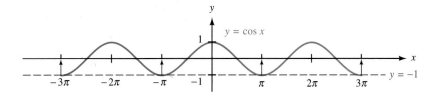

Figure 7

Figure 7 indicates that all solutions to $\cos x = -1$ are

$$x = \ldots, -3\pi, -\pi, \pi, 3\pi, \ldots$$

Since each pair of consecutive solutions differs by 2π, we can write the solutions in a more compact form as

$$x = \pi + 2k\pi \qquad \text{where } k \text{ is an integer} \qquad \blacktriangleleft$$

The Tangent Graph

Table 3 lists some solutions to $y = \tan x$ between $x = 0$ and $x = 2\pi$. Note that, at $\pi/2$ and $3\pi/2$, tangent is undefined. If we think of $\tan x$ as $(\sin x)/(\cos x)$, it must be undefined at these values of x since $\cos x$ is 0 at multiples of $\pi/2$ and division by 0 is undefined.

Table 3

x	$\tan x$	x	$\tan x$
0	0	$\dfrac{5\pi}{4}$	1
$\dfrac{\pi}{4}$	1	$\dfrac{4\pi}{3}$	$\sqrt{3} \approx 1.7$
$\dfrac{\pi}{3}$	$\sqrt{3} \approx 1.7$	$\dfrac{3\pi}{2}$	undefined
$\dfrac{\pi}{2}$	undefined	$\dfrac{5\pi}{3}$	$-\sqrt{3} \approx -1.7$
$\dfrac{2\pi}{3}$	$-\sqrt{3} \approx -1.7$	$\dfrac{7\pi}{4}$	-1
$\dfrac{3\pi}{4}$	-1	2π	0
π	0		

The entries in Table 3 for which $y = \tan x$ is undefined correspond to values of x for which there is no corresponding value of y. That is, there will be no point on the

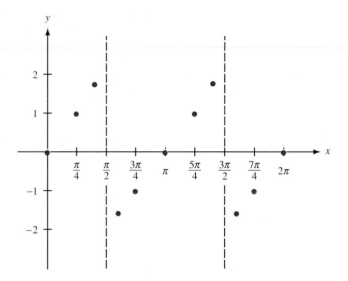

Figure 8

graph with an x-coordinate of $\pi/2$ or $3\pi/2$. To help us remember this, we have drawn dotted vertical lines through $x = \pi/2$ and $x = 3\pi/2$. These vertical lines are called *asymptotes*. Our graph will never cross or touch these lines. Figure 8 shows the information we have so far. If we were to use a calculator or table to find other values of tan x close to the asymptotes in Figure 8, we would find that tan x would become very large as we got close to the left side of an asymptote and very small as we got close to the right side of an asymptote. For example, if we were to think of the numbers on the x-axis as degrees rather than radians, $x = 85°$ would be found just to the left of the asymptote at $\pi/2$ and tan 85° would be approximately 11. Likewise tan 89° would be approximately 57, and tan 89.9° would be about 573. As you can see,

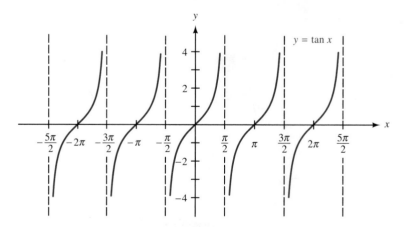

Figure 9

as x moves closer to $\pi/2$ from the left side of $\pi/2$, tan x gets larger and larger. In Figure 9, we connect the points from Table 3 in the manner appropriate to this discussion and then extend this graph to the right of 2π and to the left of 0 as we did with our sine and cosine curves.

As Figure 9 indicates, the period of $y = \tan x$ is π. The tangent function has no amplitude since there is no largest or smallest value of y on the graph of $y = \tan x$.

▶ **EXAMPLE 4** Find x, if tan $x = -1$ and $-\pi/2 \leq x \leq 3\pi/2$.

Solution Graphing $y = \tan x$ between $x = -\pi/2$ and $x = 3\pi/2$ and then drawing a horizontal line through $y = -1$, we have the graph shown in Figure 10. The values of x between $x = -\pi/2$ and $x = 3\pi/2$ that satisfy the equation tan $x = -1$ are $x = -\pi/4$ and $3\pi/4$.   ◀

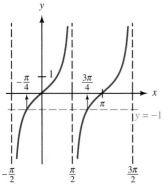

Figure 10

The Cosecant Graph

▶ **EXAMPLE 5** Sketch the graph of $y = \csc x$.

Solution Instead of making a table of values to help us graph $y = \csc x$, we can use the fact that csc x and sin x are reciprocal. Figure 11 shows how this looks graphically.

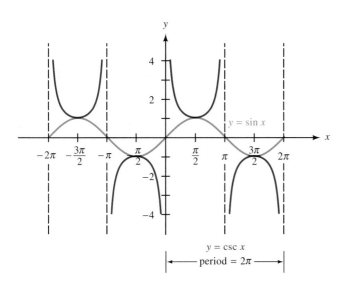

When sin x is	csc x will be
1	1
$\frac{1}{2}$	2
$\frac{1}{3}$	3
$\frac{1}{4}$	4
0	undefined
$-\frac{1}{4}$	-4
$-\frac{1}{3}$	-3
$-\frac{1}{2}$	-2
-1	-1

Figure 11

From the graph, we see that the period of $y = \csc x$ is 2π and, as was the case with $y = \tan x$, there is no amplitude.   ◀

The Cotangent and Secant Graphs

In Problem Set 4.1, you will be asked to graph $y = \cot x$ and $y = \sec x$. These graphs are shown in Figures 12 and 13 for reference.

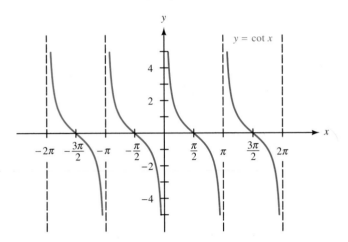

Figure 12

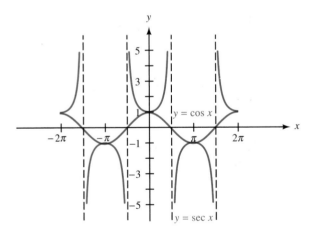

Figure 13

We end this section with a summary of the important facts associated with the graphs of our trigonometric functions.

Functions	One Cycle of the Graph	Domain	Range	Amplitude	Period
$y = \sin x$		all real numbers	$-1 \leq y \leq 1$	1	2π
$y = \cos x$		all real numbers	$-1 \leq y \leq 1$	1	2π
$y = \tan x$		all real numbers except $x = \dfrac{\pi}{2} + k\pi$ where k is an integer	all real numbers	none	π
$y = \cot x$		all real numbers except $x = k\pi$ where k is an integer	all real numbers	none	π
$y = \sec x$		all real numbers except $x = \dfrac{\pi}{2} + k\pi$ where k is an integer	$y \leq -1 \quad$ or $\quad y \geq 1$	none	2π
$y = \csc x$		all real numbers except $x = k\pi$ where k is an integer	$y \leq -1 \quad$ or $\quad y \geq 1$	none	2π

PROBLEM SET 4.1

Make a table of values for Problems 1 through 6 using multiples of $\pi/4$ for x. Then use the entries in the table to sketch the graph of each function for x between 0 and 2π.

1. $y = \cos x$ **2.** $y = \cot x$ **3.** $y = \csc x$ **4.** $y = \sin x$

5. $y = \tan x$ **6.** $y = \sec x$

Use the graphs you found in Problems 1 through 6 to find all values of x between 0 and 2π for which each of the following is true:

7. $\cos x = \frac{1}{2}$ **8.** $\sin x = \frac{1}{2}$

9. $\csc x = -1$ **10.** $\csc x = 1$

11. $\cos x = -1/\sqrt{2}$ **12.** $\cot x = -1$

13. $\tan x = 1$ **14.** $\sin x = -\sqrt{3}/2$

15. $\cos x = \sqrt{3}/2$ **16.** $\sin x = 1/\sqrt{2}$

Sketch the graphs of each of the following between $x = -4\pi$ and $x = 4\pi$ by extending the graphs you made in Problems 1 through 6:

17. $y = \sin x$ **18.** $y = \cos x$ **19.** $y = \sec x$ **20.** $y = \csc x$

21. $y = \cot x$ **22.** $y = \tan x$

Use the graphs you found in Problems 17 through 22 to find all values of x between $x = 0$ and $x = 4\pi$ for which each of the following is true:

23. $\sin x = \sqrt{3}/2$ **24.** $\cos x = 1/\sqrt{2}$

25. $\sec x = -1$ **26.** $\csc x = 1$

27. $\sin x = -\frac{1}{2}$ **28.** $\cos x = -\frac{1}{2}$

29. $\cot x = 1$ **30.** $\tan x = -1$

31. $\sin x = 1/\sqrt{2}$ **32.** $\cos x = -\sqrt{3}/2$

Find all values of x for which the following are true:

33. $\cos x = 0$ **34.** $\sin x = 0$ **35.** $\sin x = 1$ **36.** $\cos x = 1$

37. $\tan x = 0$ **38.** $\cot x = 0$

Give the amplitude and period of each of the following graphs:

39.

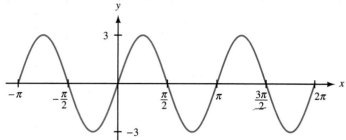

40.

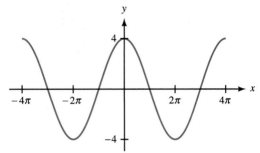

41.

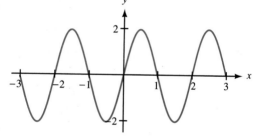

42.

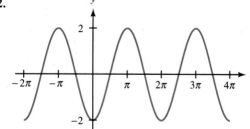

43.

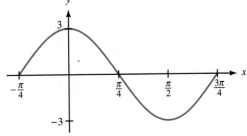

44.

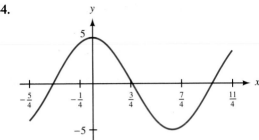

45. Sketch the graph of $y = 2 \sin x$ from $x = 0$ to $x = 2\pi$ by making a table using multiples of $\pi/2$ for x. What is the amplitude of the graph you obtain?

46. Sketch the graph of $y = \frac{1}{2} \cos x$ from $x = 0$ to $x = 2\pi$ by making a table using multiples of $\pi/2$ for x. What is the amplitude of the graph you obtain?

47. Make a table using multiples of $\pi/4$ for x to sketch the graph of $y = \sin 2x$ from $x = 0$ to $x = 2\pi$. After you have obtained the graph, state the number of complete cycles your graph goes through between 0 and 2π.

48. Make a table using multiples of $\pi/6$ and $\pi/3$ to sketch the graph of $y = \sin 3x$ from $x = 0$ to $x = 2\pi$. After you have obtained the graph, state the number of complete cycles your graph goes through between 0 and 2π.

Review Problems

The problems that follow review material we covered in Sections 1.5, 3.2, and 3.3.

Prove the following identities.

49. $\cos \theta \tan \theta = \sin \theta$

50. $\sin \theta \tan \theta + \cos \theta = \sec \theta$

51. $(1 + \sin \theta)(1 - \sin \theta) = \cos^2 \theta$

52. $(\sin \theta + \cos \theta)^2 = 1 + 2 \sin \theta \cos \theta$

53. $\csc \theta + \sin (-\theta) = \dfrac{\cos^2 \theta}{\sin \theta}$

54. $\sec \theta - \cos (-\theta) = \dfrac{\sin^2 \theta}{\cos \theta}$

Write each of the following in degrees.

55. $\dfrac{\pi}{3}$

56. $\dfrac{\pi}{6}$

57. $\dfrac{\pi}{4}$

58. $\dfrac{\pi}{2}$

59. $\dfrac{2\pi}{3}$

60. $\dfrac{5\pi}{3}$

61. $\dfrac{11\pi}{6}$

62. $\dfrac{7\pi}{6}$

| SECTION 4.2 | **Amplitude and Period** |

In Section 4.1, the graphs of $y = \sin x$ and $y = \cos x$ were shown to have a period of 2π and an amplitude of 1. In this section, we will extend our work with graphing to include a more detailed look at amplitude and period.

▶ **EXAMPLE 1**   Sketch the graph of $y = 2 \sin x$, if $0 \le x \le 2\pi$.

Solution   The coefficient 2 on the right side of the equation will simply multiply each value of $\sin x$ by a factor of 2. Therefore, the values of y in $y = 2 \sin x$ should all be twice the corresponding values of y in $y = \sin x$. Table 1 contains some values for $y = 2 \sin x$. Figure 1 shows the graphs of $y = \sin x$ and $y = 2 \sin x$. (We are including the graph of $y = \sin x$ simply for reference and comparison. With both graphs to look at, it is easier to see what change is brought about by the coefficient 2.)

Table 1

x	$y = 2 \sin x$	(x, y)
0	$y = 2 \sin 0 = 2(0) = 0$	$(0, 0)$
$\dfrac{\pi}{2}$	$y = 2 \sin \dfrac{\pi}{2} = 2(1) = 2$	$\left(\dfrac{\pi}{2}, 2\right)$
π	$y = 2 \sin \pi = 2(0) = 0$	$(\pi, 0)$
$\dfrac{3\pi}{2}$	$y = 2 \sin \dfrac{3\pi}{2} = 2(-1) = -2$	$\left(\dfrac{3\pi}{2}, -2\right)$
2π	$y = 2 \sin 2\pi = 2(0) = 0$	$(2\pi, 0)$

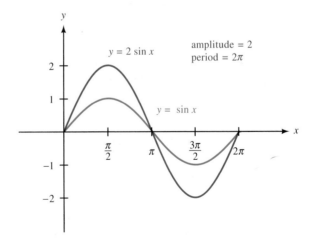

Figure 1

The coefficient 2 in $y = 2 \sin x$ changes the amplitude from 1 to 2 but does not affect the period. That is, we can think of the graph of $y = 2 \sin x$ as if it were the graph of $y = \sin x$ with the amplitude extended to 2 instead of 1.   ◀

▶ **EXAMPLE 2**   Sketch one complete cycle of the graph of $y = \frac{1}{2}\cos x$.

Solution   Table 2 gives us some points on the curve $y = \frac{1}{2}\cos x$. Figure 2 shows the graphs of both $y = \frac{1}{2}\cos x$ and $y = \cos x$ on the same set of axes, from $x = 0$ to $x = 2\pi$.

Table 2

	$y = \dfrac{1}{2}\cos x$	(x, y)
0	$y = \dfrac{1}{2}\cos 0 = \dfrac{1}{2}(1) = \dfrac{1}{2}$	$\left(0, \dfrac{1}{2}\right)$
$\dfrac{\pi}{2}$	$y = \dfrac{1}{2}\cos\dfrac{\pi}{2} = \dfrac{1}{2}(0) = 0$	$\left(\dfrac{\pi}{2}, 0\right)$
π	$y = \dfrac{1}{2}\cos\pi = \dfrac{1}{2}(-1) = -\dfrac{1}{2}$	$\left(\pi, -\dfrac{1}{2}\right)$
$\dfrac{3\pi}{2}$	$y = \dfrac{1}{2}\cos\dfrac{3\pi}{2} = \dfrac{1}{2}(0) = 0$	$\left(\dfrac{3\pi}{2}, 0\right)$
2π	$y = \dfrac{1}{2}\cos 2\pi = \dfrac{1}{2}(1) = \dfrac{1}{2}$	$\left(2\pi, \dfrac{1}{2}\right)$

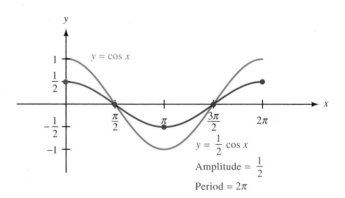

Figure 2

The coefficient $\frac{1}{2}$ in $y = \frac{1}{2}\cos x$ determines the amplitude of the graph.   ◀

Generalizing the results of these first two examples, we can say that if A is a positive number, then the graphs of $y = A\sin x$ and $y = A\cos x$ will have amplitude A.

▶ **EXAMPLE 3**   Graph $y = 3 \cos x$ and $y = \frac{1}{4} \sin x$, if $0 \leq x \leq 2\pi$.

Solution   The amplitude for $y = 3 \cos x$ is 3, while the amplitude for $y = \frac{1}{4} \sin x$ is $\frac{1}{4}$. The graphs are shown in Figure 3.

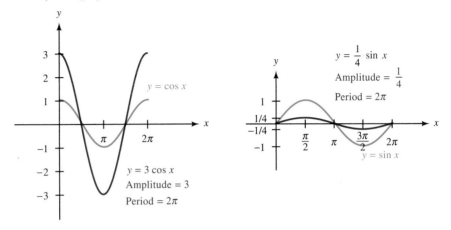

Figure 3   ◀

▶ **EXAMPLE 4**   Graph $y = \sin 2x$, if $0 \leq x \leq 2\pi$.

Solution   To see how the coefficient 2 in $y = \sin 2x$ affects the graph, we can make a table in which the values of x are multiples of $\pi/4$. (Multiples of $\pi/4$ are convenient because the coefficient 2 divides the 4 in $\pi/4$ exactly.) Table 3 shows the values of x and y, while Figure 4 contains the graphs of $y = \sin x$ and $y = \sin 2x$.

Table 3

x	$y = \sin 2x$	(x, y)
0	$y = \sin 2 \cdot 0 = 0$	$(0, 0)$
$\dfrac{\pi}{4}$	$y = \sin 2 \cdot \dfrac{\pi}{4} = \sin \dfrac{\pi}{2} = 1$	$\left(\dfrac{\pi}{4}, 1\right)$
$\dfrac{\pi}{2}$	$y = \sin 2 \cdot \dfrac{\pi}{2} = \sin \pi = 0$	$\left(\dfrac{\pi}{2}, 0\right)$
$\dfrac{3\pi}{4}$	$y = \sin 2 \cdot \dfrac{3\pi}{4} = \sin \dfrac{3\pi}{2} = -1$	$\left(\dfrac{3\pi}{4}, -1\right)$
π	$y = \sin 2 \cdot \pi = 0$	$(\pi, 0)$
$\dfrac{5\pi}{4}$	$y = \sin 2 \cdot \dfrac{5\pi}{4} = \sin \dfrac{5\pi}{2} = 1$	$\left(\dfrac{5\pi}{4}, 1\right)$
$\dfrac{3\pi}{2}$	$y = \sin 2 \cdot \dfrac{3\pi}{2} = \sin 3\pi = 0$	$\left(\dfrac{3\pi}{2}, 0\right)$
$\dfrac{7\pi}{4}$	$y = \sin 2 \cdot \dfrac{7\pi}{4} = \sin \dfrac{7\pi}{2} = -1$	$\left(\dfrac{7\pi}{4}, -1\right)$
2π	$y = \sin 2 \cdot 2\pi = \sin 4\pi = 0$	$(2\pi, 0)$

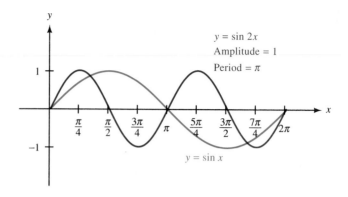

$y = \sin 2x$
Amplitude $= 1$
Period $= \pi$

$y = \sin x$

Figure 4

The graph of $y = \sin 2x$ has a period of π. It goes through two complete cyles in 2π units on the x-axis.   ◄

▶ **EXAMPLE 5**   Graph $y = \sin 3x$ from $x = 0$ to $x = 2\pi$.

Solution   To see the effect of the coefficient 3 on the graph, it is convenient to use a table in which the values of x are multiples of $\pi/6$, because 3 divides 6 exactly.

Table 4

x	$y = \sin 3x$	(x, y)
0	$y = \sin 3 \cdot 0 = \sin 0 = 0$	$(0, 0)$
$\dfrac{\pi}{6}$	$y = \sin 3 \cdot \dfrac{\pi}{6} = \sin \dfrac{\pi}{2} = 1$	$\left(\dfrac{\pi}{6}, 1\right)$
$\dfrac{\pi}{3}$	$y = \sin 3 \cdot \dfrac{\pi}{3} = \sin \pi = 0$	$\left(\dfrac{\pi}{3}, 0\right)$
$\dfrac{\pi}{2}$	$y = \sin 3 \cdot \dfrac{\pi}{2} = \sin \dfrac{3\pi}{2} = -1$	$\left(\dfrac{\pi}{2}, -1\right)$
$\dfrac{2\pi}{3}$	$y = \sin 3 \cdot \dfrac{2\pi}{3} = \sin 2\pi = 0$	$\left(\dfrac{2\pi}{3}, 0\right)$

The information in Table 4 indicates the period of $y = \sin 3x$ is $2\pi/3$. The graph will go through three complete cycles in 2π units on the x-axis. Figure 5 shows the graph of $y = \sin 3x$ and the graph of $y = \sin x$, on the interval $0 \le x \le 2\pi$.

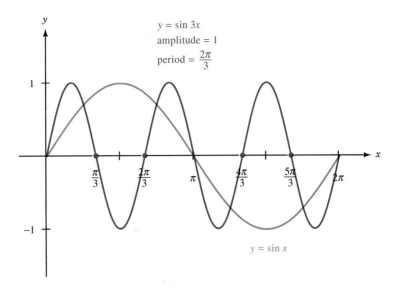

Figure 5

Table 5 summarizes the information obtained from Examples 4 and 5.

Table 5

Equation	Number of Cycles Every 2π Units	Period	
$y = \sin x$	1	2π	
$y = \sin 2x$	2	π	
$y = \sin 3x$	3	$\dfrac{2\pi}{3}$	
$y = \sin Bx$	B	$\dfrac{2\pi}{B}$	B is positive

▶ **EXAMPLE 6**   Graph one complete cycle of $y = \cos \frac{1}{2} x$.

Solution   The coefficient of x is $\frac{1}{2}$. The graph will go through $\frac{1}{2}$ of a complete cycle every 2π units. The period will be

$$\text{Period} = \frac{2\pi}{1/2} = 4\pi$$

Figure 6 shows the graph.

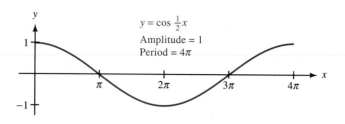

Figure 6

Continuing to generalize from the examples, we have the following summary.

Amplitude and Period for Sine and Cosine Curves

If B is a positive number, the graphs of $y = A \sin Bx$ and $y = A \cos Bx$ will have

$$\text{Amplitude} = |A|$$

$$\text{Period} = \frac{2\pi}{B}$$

In the next three examples, we use this information about amplitude and period to graph one complete cycle of some sine and cosine curves, then we extend these graphs to cover more than one complete cycle.

▶ **EXAMPLES**   Graph one complete cycle of each of the following equations and then extend the graph to cover the given interval.

7. $y = 3 \sin 2x, \ -\pi \le x \le 2\pi$

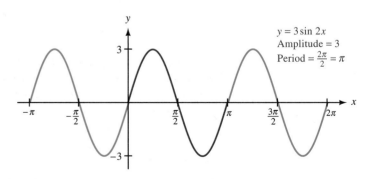

Figure 7

8. $y = 4 \cos \frac{1}{2} x,\ -4\pi \le x \le 4\pi$

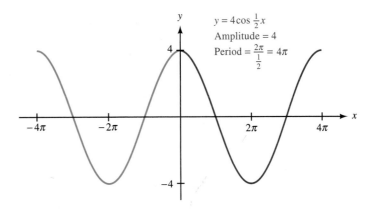

Figure 8

9. $y = 2 \sin \pi x,\ -3 \le x \le 3$

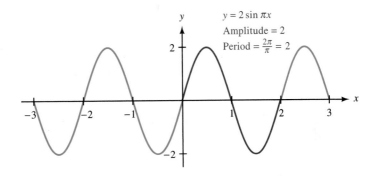

Figure 9   ◀

Note that, on the graphs in Figures 7, 8, and 9, the axes have not been labeled proportionally. Instead, they are labeled so that the amplitude and period are easy to read. As you can see, once we have the graph of one complete cycle of a curve, it is easy to extend the curve to any interval of interest.

Reflecting About the *x*-Axis

So far in this section, all of the coefficients *A* and *B* we have encountered have been positive. If we are given an equation to graph in which *B* is negative, we can use the properties of even and odd functions to rewrite the equation with B positive. For example,

$$y = 3 \sin (-2x) \text{ is equivalent to } y = -3 \sin 2x$$
because sine is an odd function.

$$y = 3 \cos (-2x) \text{ is equivalent to } y = 3 \cos 2x$$
because cosine is an even function.

So we do not need to worry about negative values of B; we simply make them positive by using the properties of even and odd functions and then graph as usual. To see how a negative value of A affects graphing, we will graph $y = -2 \cos x$.

▶ **EXAMPLE 10**   Graph $y = -2 \cos x$, from $x = -2\pi$ to $x = 4\pi$.

Solution   Each value of y on the graph of $y = -2 \cos x$ will be the opposite of the corresponding value of y on the graph of $y = 2 \cos x$. The result is that the graph of $y = -2 \cos x$ is the reflection of the graph of $y = 2 \cos x$ about the x-axis. Figure 10 shows the extension of one complete cycle of $y = -2 \cos x$ to the interval $-2\pi \leq x \leq 4\pi$.

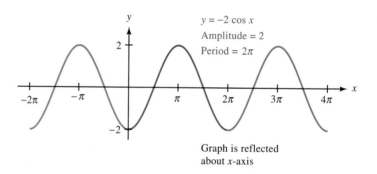

$y = -2 \cos x$
Amplitude = 2
Period = 2π

Graph is reflected
about x-axis

Figure 10   ◀

Summary

The graphs of $y = A \sin Bx$ and $y = A \cos Bx$, where B is a positive number, will be reflected about the x-axis if A is negative.

Vertical Translations

Recall from algebra the relationship between the graphs of $y = x^2$ and $y = x^2 - 3$. Figures 11 and 12 show the graphs. The graph of $y = x^2 - 3$ has the same shape as the graph of $y = x^2$ but with its vertex (and all other points) moved down three units. If we were to graph $y = x^2 + 2$, the graph would have the same shape as the graph of $y = x^2$, but it would be moved up two units. In general, the graph of $y = f(x) + k$ is

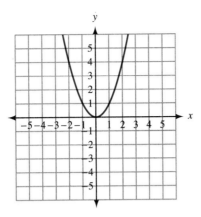

Figure 11

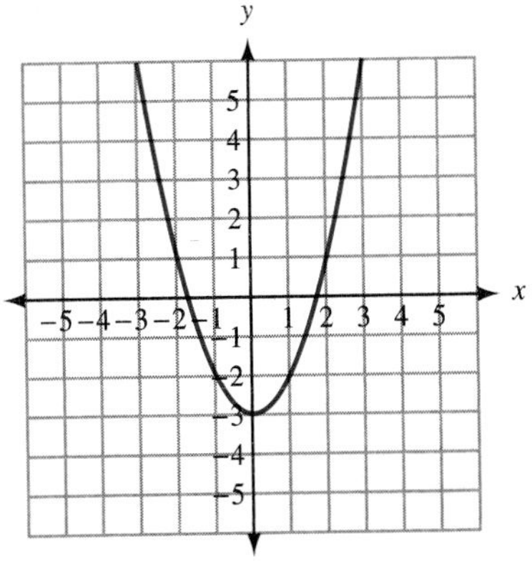

Figure 12

the graph of $y = f(x)$ translated k units vertically. If k is a positive number, the translation is up. If k is a negative number, the translation is down.

▶ **EXAMPLE 11**   Use the results of Examples 7 and 9 to graph each of the following equations.

a. $y = -4 + 3 \sin 2x$ **b.** $y = 3 + 2 \sin \pi x$

Solution   Each equation has the form $y = k + f(x)$ indicating that the vertical translation is k. (Since addition is a commutative operation, it makes no difference whether our equations have the form $y = k + f(x)$ or $y = f(x) + k$; both have the same meaning.) The graph of $y = -4 + 3 \sin 2x$ is the graph of $y = 3 \sin 2x$ with all points moved down four units. Likewise, the graph of $y = 3 + 2 \sin \pi x$ is the graph of $y = 2 \sin \pi x$ with all points moved up three units. The graphs are shown in Figures 13 and 14.

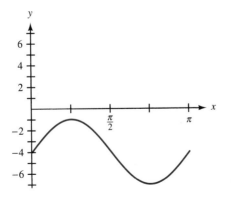

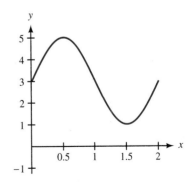

Figure 13

Figure 14

▶ **EXAMPLE 12**   Graph $y = \csc 2x$ from $x = 0$ to $x = 2\pi$.

Solution   Since $\csc 2x$ is the reciprocal of $\sin 2x$, the graph of $y = \csc 2x$ will have the same period as $y = \sin 2x$. That period is $2\pi/2 = \pi$. Also, because of this reciprocal relationship, the graph of $y = \csc 2x$ will be undefined, and a vertical asymptote will occur, whenever the graph of $y = \sin 2x$ is 0. We graphed $y = \sin 2x$ in Example 4. Figure 15 shows that graph again, along with the graph of $y = \csc 2x$, from $x = 0$ to $x = 2\pi$.

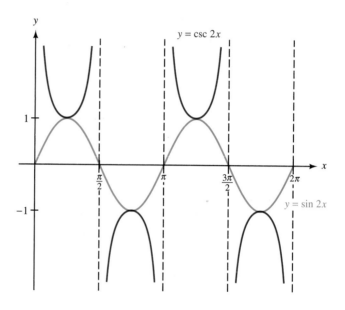

Figure 15   ◀

▶ **EXAMPLE 13**   Graph $y = 3 \csc 2x$ from $x = 0$ to $x = 2\pi$.

Solution   If we look back to Figure 15, we can see that the range of y values on the graph of $y = \csc 2x$ is $y \geq 1$ and $y \leq -1$. Since all the y values on the graph of $y = 3 \csc 2x$ will be three times as large or small as the y values on $y = \csc 2x$, it is reasonable to assume the range of y values on the graph of $y = 3 \csc 2x$ will be $y \geq 3$ and $y \leq -3$. Following this kind of reasoning, we have the graph shown in Figure 16.

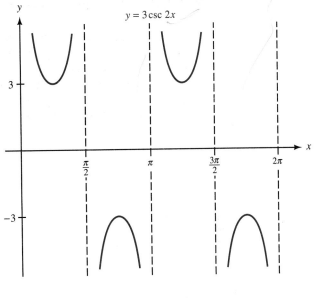

$y = 3 \csc 2x$

Figure 16

Since the graphs of $y = A \csc Bx$ and $y = A \sec Bx$ are similar, we use the information from Examples 12 and 13 to generalize as follows:

Range and Period for Secant and Cosecant Graphs

If B is a positive number, the graphs of $y = A \sec Bx$ and $y = A \csc Bx$ will have:

$$\text{Period} = \frac{2\pi}{B}$$

$$\text{Range: } y \geq |A| \text{ and } y \leq -|A|$$

If A is negative, the graphs will be reflected about the x-axis.

For our next example, we look at the graph of one of the equations we found in Example 4 of Section 3.5. Example 14 gives the main facts from that example.

▶ **EXAMPLE 14**   Figure 17 shows a fire truck parked on the shoulder of a freeway next to a long block wall. The red light on the top of the truck is 10 feet from the wall and rotates through one complete revolution every 2 seconds. Graph the equation that gives the length d in terms of time t from $t = 0$ to $t = 2$.

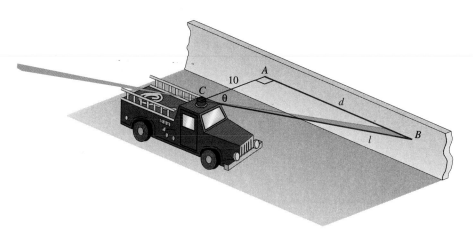

Figure 17

Solution   From Example 4 in Section 3.5 we know that

$$d = 10 \tan \pi t$$

To graph this equation between $t = 0$ and $t = 2$, we construct a table of values in which t assumes all multiples of $\frac{1}{4}$ from $t = 0$ to $t = 2$. Plotting the points given in Table 6 and then connecting them with a smooth tangent curve, we have the graph in Figure 18.

Table 6

t	$d = 10 \tan \pi t$	d
0	$d = 10 \tan \pi \cdot 0 = 10 \tan 0 = 0$	0
$\frac{1}{4}$	$d = 10 \tan \pi \cdot \frac{1}{4} = 10 \tan \frac{\pi}{4} = 10$	10
$\frac{1}{2}$	$d = 10 \tan \pi \cdot \frac{1}{2} = 10 \tan \frac{\pi}{2}$	undefined
$\frac{3}{4}$	$d = 10 \tan \pi \cdot \frac{3}{4} = 10 \tan \frac{3\pi}{4} = -10$	-10
1	$d = 10 \tan \pi \cdot 1 = 10 \tan \pi = 0$	0
$\frac{5}{4}$	$d = 10 \tan \pi \cdot \frac{5}{4} = 10 \tan \frac{5\pi}{4} = 10$	10
$\frac{3}{2}$	$d = 10 \tan \pi \cdot \frac{3}{2} = 10 \tan \frac{3\pi}{2}$	undefined
$\frac{7}{4}$	$d = 10 \tan \pi \cdot \frac{7}{4} = 10 \tan \frac{7\pi}{4} = -10$	-10
2	$d = 10 \tan \pi \cdot 2 = 10 \tan 2\pi = 0$	0

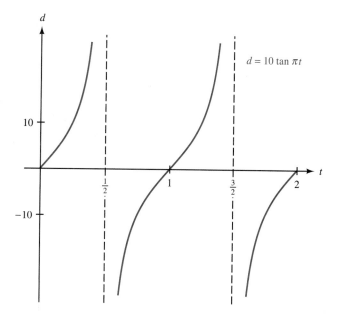

$d = 10 \tan \pi t$

Figure 18

Note that the period of the function $d = 10 \tan \pi t$ is 1. Since the period of $y = \tan x$ is π, we can conclude that the new period 1 must come from dividing the normal period π by π to get 1. Also, the number 10 in the equation $d = 10 \tan \pi t$ causes the tangent graph to rise faster above the horizontal axis and fall faster below it.

◀

We generalize the results of Example 14 as follows.

Period and Rate of Increase for Tangent and Cotangent Graphs

If B is a positive number, the graphs of $y = A \tan Bx$ and $y = A \cot Bx$ will have:

$$\text{Period} = \frac{\pi}{B}$$

Each graph will rise and fall at a faster rate than the corresponding graphs of $y = \tan x$ and $y = \cot x$ if $|A|$ is greater than 1 and at a slower rate if $|A|$ is between 0 and 1.

If A is negative, the graphs will be reflected about the x-axis.

In Section 1.5, we found that the expression $\sqrt{x^2 + 9}$ could be rewritten without a square root by making the substitution $x = 3 \tan \theta$. Then we noted that the substitution itself was questionable because we did not know at that time if every real number x could be written as $3 \tan \theta$, for some value of θ. We can clear up this point by looking at the graph of $y = 3 \tan x$.

▶ **EXAMPLE 15** Graph $y = 3 \tan x$ for $-\dfrac{\pi}{2} < x < \dfrac{\pi}{2}$.

Solution The graph is shown in Figure 19, along with the graph of $y = \tan x$ over the same interval. Note that the coefficient 3 affects the graph by making it rise and fall faster than the graph of $y = \tan x$.

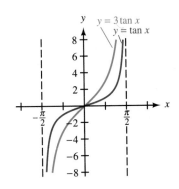

Figure 19

As you can see from the graph, the range for $y = 3 \tan x$ is all real numbers. This means that for any real number y there is a value of x for which $y = 3 \tan x$. In other words, any real number can be written as $3 \tan \theta$, which takes care of the question we had about the substitution we made in Section 1.5. ◀

PROBLEM SET 4.2

Graph one complete cycle of each of the following. In each case, label the axes accurately and identify the amplitude and period for each graph.

1. $y = 6 \sin x$

2. $y = 6 \cos x$

3. $y = \sin 2x$

4. $y = \sin \dfrac{1}{2} x$

5. $y = \cos \dfrac{1}{3} x$

6. $y = \cos 3x$

7. $y = \dfrac{1}{3} \sin x$

8. $y = \dfrac{1}{2} \cos x$

9. $y = \sin \pi x$

10. $y = \cos \pi x$

11. $y = \sin \dfrac{\pi}{2} x$

12. $y = \cos \dfrac{\pi}{2} x$

Graph one complete cycle for each of the following. In each case, label the axes so that the amplitude and period are easy to read.

13. $y = 4 \sin 2x$

15. $y = 2 \cos 4x$

17. $y = 3 \sin \dfrac{1}{2} x$

19. $y = \dfrac{1}{2} \cos 3x$

21. $y = \dfrac{1}{2} \sin \dfrac{\pi}{2} x$

14. $y = 2 \sin 4x$

16. $y = 3 \cos 2x$

18. $y = 2 \sin \dfrac{1}{3} x$

20. $y = \dfrac{1}{2} \sin 3x$

22. $y = 2 \sin \dfrac{\pi}{2} x$

Use your answers for Problems 13 through 22 for reference, and graph one complete cycle of each of the following equations.

23. $y = 4 + 4 \sin 2x$

25. $y = -3 + 2 \cos 4x$

27. $y = 1 + 3 \sin \dfrac{1}{2} x$

29. $y = -1 + \dfrac{1}{2} \cos 3x$

31. $y = \dfrac{1}{2} + \dfrac{1}{2} \sin \dfrac{\pi}{2} x$

24. $y = -2 + 2 \sin 4x$

26. $y = 4 + 3 \cos 2x$

28. $y = -1 + 2 \sin \dfrac{1}{3} x$

30. $y = 1 + \dfrac{1}{2} \sin 3x$

32. $y = -2 + 2 \sin \dfrac{\pi}{2} x$

Graph each of the following over the given interval. Label the axes so that the amplitude and period are easy to read.

33. $y = 2 \sin \pi x, \ -4 \le x \le 4$

35. $y = 3 \sin 2x, \ -\pi \le x \le 2\pi$

37. $y = -3 \cos \dfrac{1}{2} x, \ -2\pi \le x \le 6\pi$

39. $y = -2 \sin (-3x), \ 0 \le x \le 2\pi$

34. $y = 3 \cos \pi x, \ -2 \le x \le 4$

36. $y = -3 \sin 2x, \ -2\pi \le x \le 2\pi$

38. $y = 3 \cos \dfrac{1}{2} x, \ -4\pi \le x \le 4\pi$

40. $y = -2 \cos (-3x), \ 0 \le x \le 2\pi$

41. The current in an alternating circuit varies in intensity with time. If I represents the intensity of the current and t represents time, then the relationship between I and t is given by

$$I = 20 \sin 120\pi t$$

where I is measured in amperes and t is measured in seconds. Find the maximum value of I and the time it takes for I to go through one complete cycle.

42. A weight is hung from a spring and set in motion so that it moves up and down continuously. The velocity v of the weight at any time t is given by the equation

$$v = 3.5 \cos 2\pi t$$

where v is measured in meters per second and t is measured in seconds. Find the maximum velocity of the weight and the amount of time it takes for the weight to move from its lowest position to its highest position.

Graph each of the following from $x = 0$ to $x = 2\pi$.

43. $y = \csc 3x$ **44.** $y = \sec 3x$

45. $y = 2 \csc 3x$ **46.** $y = 2 \sec 3x$

47. $y = -2 \csc 3x$ **48.** $y = -2 \sec 3x$

49. $y = 3 \sec \dfrac{1}{2} x$ **50.** $y = 3 \csc \dfrac{1}{2} x$

51. Since the period for $y = \tan x$ is π, the graph of $y = \tan x$ will go through one complete cycle every π units. Through how many cycles will the graph of $y = \tan 2x$ go every π units? What is the period of $y = \tan 2x$? Sketch the graph of $y = \tan 2x$, from $x = -\pi/4$ to $x = 3\pi/4$.

52. Through how many complete cycles will the graph of $y = \tan \frac{1}{2} x$ go every π units? What is the period of this graph? Sketch the graph from $x = -\pi$ to $x = 3\pi$.

53. Figure 20 shows a lighthouse that is 100 ft from a long straight wall on the beach. The light in the lighthouse rotates through one complete rotation once every 4 seconds. In Problem 21 of Problem Set 3.5, you found the equation that gives d in terms of t to be $d = 100 \tan \dfrac{\pi}{2} t$. Graph this equation by making a table in which t assumes all multiples of $\frac{1}{2}$ from $t = 0$ to $t = 4$.

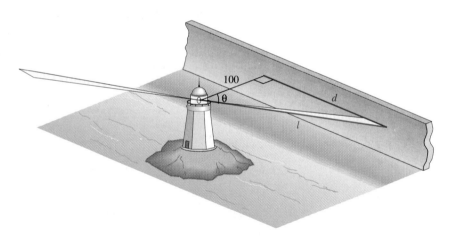

Figure 20

54. Sketch the graph of $y = -\tan x$, for $-\pi/2 \le x \le 3\pi/2$.

Graph each of the following from $x = 0$ to $x = \pi$.

55. $y = \tan 3x$ **56.** $y = \cot 3x$

57. $y = \cot 2x$ **58.** $y = \tan 4x$

59. $y = -\cot 2x$ **60.** $y = -\tan 4x$

61. Referring to Example 14 from this section, the equation that gives l in terms of time t is $l = 10 \sec \pi t$. Graph this equation by making a table of values in which t takes on multiples of $\frac{1}{4}$ starting at $t = 0$ and ending at $t = 2$.

62. In Figure 20, the equation that gives l in terms of t is $l = 100 \sec \dfrac{\pi}{2} t$. Graph this equation from $t = 0$ to $t = 4$.

Review Problems

The problems that follow review material we covered in Section 3.2. Reviewing these problems will help you with the next section.

Evaluate each of the following if x is $\pi/2$ and y is $\pi/6$.

63. $\sin\left(x + \dfrac{\pi}{2}\right)$

64. $\sin\left(x - \dfrac{\pi}{2}\right)$

65. $\cos\left(y - \dfrac{\pi}{6}\right)$

66. $\cos\left(y + \dfrac{\pi}{6}\right)$

67. $\sin(x + y)$

68. $\cos(x + y)$

69. $\sin x + \sin y$

70. $\cos x + \cos y$

Convert each of the following to radians without using a calculator.

71. $45°$ **72.** $30°$ **73.** $60°$ **74.** $90°$

75. $150°$ **76.** $300°$ **77.** $225°$ **78.** $120°$

SECTION 4.3   ## Phase Shift

In this section, we will consider equations of the form

$$y = A \sin (Bx + C) \qquad \text{and} \qquad y = A \cos (Bx + C) \qquad \text{where } B > 0$$

We already know how the coefficients A and B affect the graphs of these equations. The only thing we have left to do is discover what effect C has on the graphs. We will start our investigation with a couple of equations in which A and B are equal to 1.

▶ **EXAMPLE 1**   Graph $y = \sin\left(x + \dfrac{\pi}{2}\right)$, if $-\dfrac{\pi}{2} \le x \le \dfrac{3\pi}{2}$.

Solution   Since we have not graphed an equation of this form before, it is a good idea to begin by making a table (Table 1). In this case, multiples of $\pi/2$ will be the most convenient replacements for x in the table. Also, if we start with $x = -\pi/2$, our first value of y will be 0.

Table 1

x	$y = \sin\left(x + \dfrac{\pi}{2}\right)$	(x, y)
$-\dfrac{\pi}{2}$	$y = \sin\left(-\dfrac{\pi}{2} + \dfrac{\pi}{2}\right) = \sin 0 = 0$	$\left(-\dfrac{\pi}{2}, 0\right)$
0	$y = \sin\left(0 + \dfrac{\pi}{2}\right) = \sin\dfrac{\pi}{2} = 1$	$(0, 1)$
$\dfrac{\pi}{2}$	$y = \sin\left(\dfrac{\pi}{2} + \dfrac{\pi}{2}\right) = \sin \pi = 0$	$\left(\dfrac{\pi}{2}, 0\right)$
π	$y = \sin\left(\pi + \dfrac{\pi}{2}\right) = \sin\dfrac{3\pi}{2} = -1$	$(\pi, -1)$
$\dfrac{3\pi}{2}$	$y = \sin\left(\dfrac{3\pi}{2} + \dfrac{\pi}{2}\right) = \sin 2\pi = 0$	$\left(\dfrac{3\pi}{2}, 0\right)$

Graphing these points and then drawing the sine curve that connects them gives us the graph of $y = \sin(x + \pi/2)$, as shown in Figure 1. Figure 1 also includes the graph of $y = \sin x$ for reference; we are trying to discover how the graphs of $y = \sin(x + \pi/2)$ and $y = \sin x$ differ.

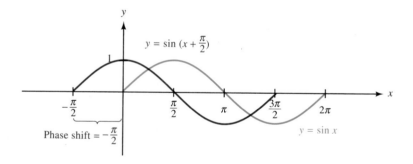

Figure 1

It seems that the graph of $y = \sin(x + \pi/2)$ is shifted $\pi/2$ units to the left of the graph of $y = \sin x$. We say the graph of $y = \sin(x + \pi/2)$ has a *phase shift* of $-\pi/2$, where the negative sign indicates the shift is to the left (in the negative direction). ◀

From the results in Example 1, we would expect the graph of $y = \sin(x - \pi/2)$ to have a phase shift of $+\pi/2$. That is, we expect the graph of $y = \sin(x - \pi/2)$ to be shifted $\pi/2$ units to the *right* of the graph of $y = \sin x$.

▶ **EXAMPLE 2**   Graph one complete cycle of $y = \sin\left(x - \dfrac{\pi}{2}\right)$.

Solution   Proceeding as we did in Example 1, we make a table (Table 2) using multiples of $\pi/2$ for x, and then use the information in the table to sketch the graph. In this example, we start with $x = \pi/2$, since this value of x will give us $y = 0$.

Table 2

x	$y = \sin\left(x - \dfrac{\pi}{2}\right)$	(x, y)
$\dfrac{\pi}{2}$	$y = \sin\left(\dfrac{\pi}{2} - \dfrac{\pi}{2}\right) = \sin 0 = 0$	$\left(\dfrac{\pi}{2}, 0\right)$
π	$y = \sin\left(\pi - \dfrac{\pi}{2}\right) = \sin\dfrac{\pi}{2} = 1$	$(\pi, 1)$
$\dfrac{3\pi}{2}$	$y = \sin\left(\dfrac{3\pi}{2} - \dfrac{\pi}{2}\right) = \sin \pi = 0$	$\left(\dfrac{3\pi}{2}, 0\right)$
2π	$y = \sin\left(2\pi - \dfrac{\pi}{2}\right) = \sin\dfrac{3\pi}{2} = -1$	$(2\pi, -1)$
$\dfrac{5\pi}{2}$	$y = \sin\left(\dfrac{5\pi}{2} - \dfrac{\pi}{2}\right) = \sin 2\pi = 0$	$\left(\dfrac{5\pi}{2}, 0\right)$

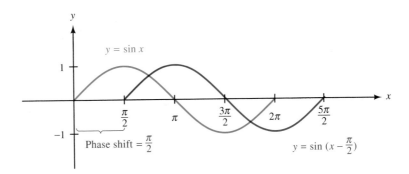

Figure 2

The graph of $y = \sin(x - \pi/2)$, as we expected, is a sine curve shifted $\pi/2$ units to the right of the graph of $y = \sin x$ (Figure 2). The phase shift, in this case, is $+\pi/2$.   ◀

Before we write any conclusions about phase shift, we should look at another example in which A and B are not 1.

▶ **EXAMPLE 3**   Graph $y = 3 \sin\left(2x + \dfrac{\pi}{2}\right)$, if $-\dfrac{\pi}{4} \le x \le \dfrac{3\pi}{4}$.

Solution   We know the coefficient $B = 2$ will change the period from 2π to π. Because the period is smaller (π instead of 2π), we should use values of x in our table (Table 3) that are closer together, like multiples of $\pi/4$ instead of $\pi/2$.

Table 3

x	$y = 3 \sin\left(2x + \dfrac{\pi}{2}\right)$	(x, y)
$-\dfrac{\pi}{4}$	$y = 3 \sin\left[2\left(-\dfrac{\pi}{4}\right) + \dfrac{\pi}{2}\right] = 3 \sin 0 = 0$	$\left(-\dfrac{\pi}{4}, 0\right)$
0	$y = 3 \sin\left(2 \cdot 0 + \dfrac{\pi}{2}\right) = 3 \sin \dfrac{\pi}{2} = 3$	$(0, 3)$
$\dfrac{\pi}{4}$	$y = 3 \sin\left(2 \cdot \dfrac{\pi}{4} + \dfrac{\pi}{2}\right) = 3 \sin \pi = 0$	$\left(\dfrac{\pi}{4}, 0\right)$
$\dfrac{\pi}{2}$	$y = 3 \sin\left(2 \cdot \dfrac{\pi}{2} + \dfrac{\pi}{2}\right) = 3 \sin \dfrac{3\pi}{2} = -3$	$\left(\dfrac{\pi}{2}, -3\right)$
$\dfrac{3\pi}{4}$	$y = 3 \sin\left(2 \cdot \dfrac{3\pi}{4} + \dfrac{\pi}{2}\right) = 3 \sin 2\pi = 0$	$\left(\dfrac{3\pi}{4}, 0\right)$

The amplitude and period are as we would expect. The phase shift, however, is half of $-\pi/2$. The phase shift, $-\pi/4$, comes from the ratio $-C/B$, or in this case

$$\frac{-\pi/2}{2} = -\frac{\pi}{4}$$

The phase shift in the equation $y = A \sin(Bx + C)$ depends on both B and C. In this example, the period is half of the period of $y = \sin x$, and the phase shift is half of the phase shift of $y = \sin(x + \pi/2)$, which was found in Example 1 (see Figure 3).

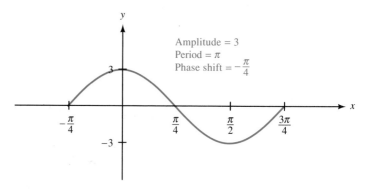

Figure 3

Although all of the examples we have completed so far in this section have been sine curves, the results also apply to cosine curves. Here is a summary.

Amplitude, Period, and Phase Shift

The graphs of $y = A \sin (Bx + C)$ and $y = A \cos (Bx + C)$, where $B > 0$, will have the following characteristics:

$$1. \ \text{Amplitude} = |A|$$

$$2. \ \text{Period} = \frac{2\pi}{B}$$

$$3. \ \text{Phase shift} = -\frac{C}{B}$$

If $A < 0$, the graphs will be reflected about the x-axis.

The information on amplitude, period, and phase shift allows us to sketch sine and cosine curves without having to make tables.

▶ **EXAMPLE 4**   Graph one complete cycle of $y = 2 \sin (3x + \pi)$.

Solution   Here is a detailed list of steps to use in graphing sine and cosine curves for which B is positive.

Step 1: Use A, B, and C to find the amplitude, period, and phase shift.

$$\text{Amplitude} = |A| = 2$$

$$\text{Period} = \frac{2\pi}{B} = \frac{2\pi}{3}$$

$$\text{Phase shift} = -\frac{C}{B} = -\frac{\pi}{3}$$

Step 2: On the x-axis, label the starting point, ending point, and the point halfway between them for each cycle of the curve in question. The starting point is the phase shift. The ending point is the phase shift plus the period. It is also a good idea to label the points one-fourth and three-fourths of the way between the starting point and ending point.

Step 3: Label the y-axis with the amplitude and the opposite of the amplitude. It is okay if the units on the x-axis and the y-axis are not proportional. That is, one unit on the y-axis can be a different length from one unit on the x-axis. The idea is to make the graph easy to read.

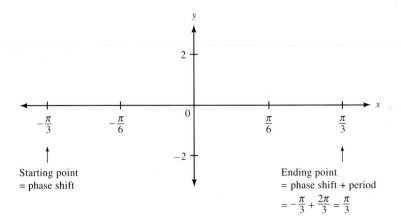

Figure 4

Step 4: Sketch in the curve in question, keeping in mind that the graph will be reflected about the *x*-axis if *A* is negative. In this case, we want a sine curve that will be 0 at the starting point and 0 at the ending point.

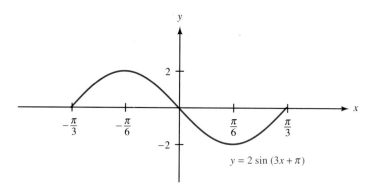

$y = 2 \sin (3x + \pi)$

Figure 5 ◀

The steps listed in Example 4 may seem complicated at first. With a little practice they do not take much time at all, especially when compared with the time it would take to make a table. Also, once we have graphed one complete cycle of the curve, it would be fairly easy to extend the graph in either direction.

▶ **EXAMPLE 5** Graph $y = 2 \cos (3x + \pi)$ from $x = -2\pi/3$ to $x = 2\pi/3$.

Solution *A, B,* and *C* are the same here as they were in Example 4. We use the same labeling on the axes as we used in Example 4, but we draw in a cosine curve instead of a sine curve and then extend it to cover the interval $-2\pi/3 \le x \le 2\pi/3$.

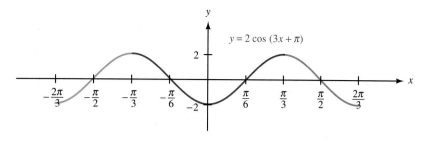

Figure 6

▶ **EXAMPLE 6**   Graph $y = -2 \cos (3x + \pi)$ for $-2\pi/3 \le x \le 2\pi/3$.

Solution   The graph will be the graph found in Example 5 reflected about the x-axis, since the only difference in the two equations is that A is negative here, while it was positive in Example 5.

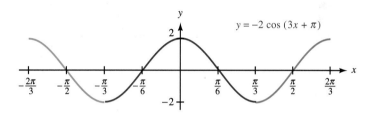

Figure 7

More on Labeling the x-Axis

Now that we have been through a few examples in detail, we can be more specific about how we label the x-axis. Suppose the line in Figure 8 represents the x-axis, and the five points we have labeled with the letters a through e are the five points we will use to graph one complete cycle of a sine or cosine curve. Because these points are equally spaced along the x-axis, we can find their coordinates in the following manner (and the following order).

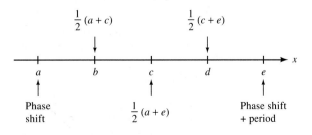

Figure 8

1. The coordinate of the first point, a, is always the phase shift.
2. The coordinate of the last point, e, is the sum of the phase shift and the period.
3. The center point, c, is the average of the first point and the last point. That is,

$$c = \frac{a + e}{2} = \frac{1}{2}(a + e)$$

4. The point b is the average of the first point and the center point. That is,

$$b = \frac{a + c}{2} = \frac{1}{2}(a + c)$$

5. The point d is the average of the center point and the last point. In symbols,

$$d = \frac{c + e}{2} = \frac{1}{2}(c + e)$$

▶ **EXAMPLE 7**   Graph one complete cycle of $y = 4 \sin \left(2x - \dfrac{\pi}{3}\right)$.

Solution   In this case, $A = 4$, $B = 2$, and $C = -\dfrac{\pi}{3}$. These values give us a sine curve with

$$\text{Amplitude} = 4$$

$$\text{Period} = \frac{2\pi}{2} = \pi$$

$$\text{Phase Shift} = \frac{-(-\pi/3)}{2} = \frac{\pi}{6}$$

Because the amplitude is 4, we label the y-axis with 4 and -4. The x-axis is labeled according to the discussion that preceded this example, as shown in Figure 9.

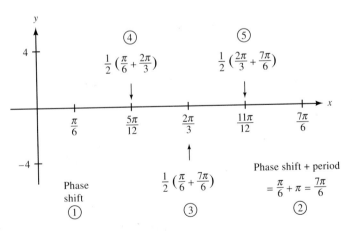

Figure 9

We finish the graph by drawing in a sine curve that starts at $x = \pi/6$ and ends at $x = 7\pi/6$, as shown in Figure 10.

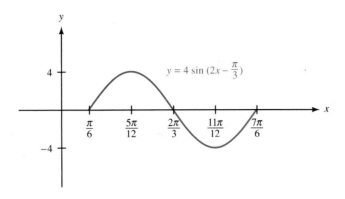

Figure 10 ◀

▶ **EXAMPLE 8** Graph $y = -4 \cos\left(2x - \dfrac{3\pi}{2}\right)$ from $x = 0$ to $x = 2\pi$.

Solution We use A, B, and C to help us sketch one complete cycle of the curve, then we extend the resulting graph to cover the interval $0 \le x \le 2\pi$.

$$\text{Amplitude} = 4 \qquad \text{Period} = \frac{2\pi}{2} = \pi \qquad \text{Phase shift} = -\frac{-3\pi/2}{2} = \frac{3\pi}{4}$$

One complete cycle will start at $3\pi/4$ on the x-axis. It will end π units later at $3\pi/4 + \pi = 7\pi/4$. Since A is negative, the graph is reflected about the x-axis.

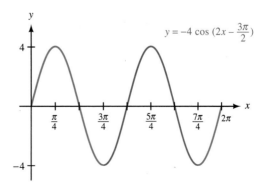

Figure 11 ◀

▶ **EXAMPLE 9**   Graph $y = 5 \sin \left(\pi x + \dfrac{\pi}{4} \right)$, $-\dfrac{5}{4} \le x \le \dfrac{11}{4}$.

Solution   In this example, $A = 5$, $B = \pi$, and $C = \pi/4$. The graph will have the following characteristics:

Amplitude $= 5$,   Period $= \dfrac{2\pi}{\pi} = 2$,   and   Phase shift $= -\dfrac{\pi/4}{\pi} = -\dfrac{1}{4}$

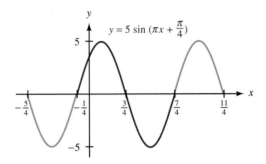

Figure 12   ◀

Vertical Translations

As was the case in Section 4.2, adding a constant k to an equation containing trigonometric functions translates the graph vertically up or down. That is, the graph of $y = k + A \sin (Bx + C)$ will have the same shape (amplitude, period, phase shift, and reflection, if indicated) as $y = A \sin (Bx + C)$ but will be translated k units vertically from the graph of $y = A \sin (Bx + C)$.

▶ **EXAMPLE 10**   Use the graph shown in Example 9 to graph one complete cycle of each of the following:

a. $y = 3 + 5 \sin \left(\pi x + \dfrac{\pi}{4} \right)$ **b.** $y = -5 + 5 \sin \left(\pi x + \dfrac{\pi}{4} \right)$

Solution   The amplitude, period, and phase shift for each graph will be the same as those shown on the graph in Figure 12. The graph of the first equation will be shifted up three units from the graph shown in Figure 12, while the graph of our second equation will be shifted down five units from the graph shown in Figure 12. One complete cycle of each will start at $x = -\frac{1}{4}$ and end two units later at $x = \frac{7}{4}$. The graphs are shown in Figures 13 and 14.

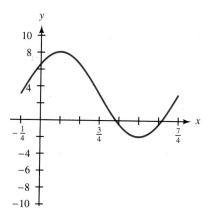

Figure 13

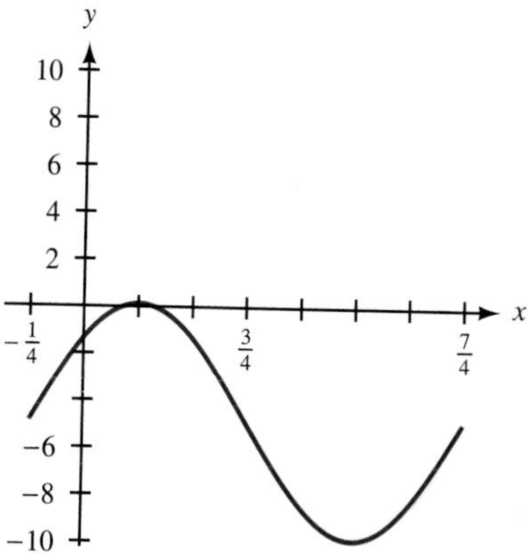

Figure 14   ◀

PROBLEM SET 4.3

For each equation, first identify the phase shift and then sketch one complete cycle of the graph. In each case, graph $y = \sin x$ on the same coordinate system.

1. $y = \sin\left(x + \dfrac{\pi}{4}\right)$

2. $y = \sin\left(x + \dfrac{\pi}{6}\right)$

3. $y = \sin\left(x - \dfrac{\pi}{4}\right)$

4. $y = \sin\left(x - \dfrac{\pi}{6}\right)$

5. $y = \sin\left(x + \dfrac{\pi}{3}\right)$

6. $y = \sin\left(x - \dfrac{\pi}{3}\right)$

For each equation, identify the phase shift and then sketch one complete cycle of the graph. In each case, graph $y = \cos x$ on the same coordinate system.

7. $y = \cos\left(x - \dfrac{\pi}{2}\right)$

8. $y = \cos\left(x + \dfrac{\pi}{2}\right)$

9. $y = \cos\left(x + \dfrac{\pi}{3}\right)$

10. $y = \cos\left(x - \dfrac{\pi}{4}\right)$

For each equation, identify the amplitude, period, and phase shift. Then label the axes accordingly and sketch one complete cycle of the curve.

11. $y = \sin(2x - \pi)$

12. $y = \sin(2x + \pi)$

13. $y = \sin\left(\pi x + \dfrac{\pi}{2}\right)$

14. $y = \sin\left(\pi x - \dfrac{\pi}{2}\right)$

15. $y = -\cos\left(2x + \dfrac{\pi}{2}\right)$

16. $y = -\cos\left(2x - \dfrac{\pi}{2}\right)$

17. $y = 2 \sin \left(\dfrac{1}{2}x + \dfrac{\pi}{2} \right)$

18. $y = 3 \cos \left(\dfrac{1}{2}x + \dfrac{\pi}{3} \right)$

19. $y = \dfrac{1}{2} \cos \left(3x - \dfrac{\pi}{2} \right)$

20. $y = \dfrac{4}{3} \cos \left(3x + \dfrac{\pi}{2} \right)$

21. $y = 3 \sin \left(\dfrac{\pi}{3}x - \dfrac{\pi}{3} \right)$

22. $y = 3 \cos \left(\dfrac{\pi}{3}x - \dfrac{\pi}{3} \right)$

Use your answers for Problems 11 through 20 for reference, and graph one complete cycle of each of the following equations.

23. $y = 1 + \sin (2x - \pi)$

24. $y = -1 + \sin (2x + \pi)$

25. $y = -3 + \sin \left(\pi x + \dfrac{\pi}{2} \right)$

26. $y = 3 + \sin \left(\pi x - \dfrac{\pi}{2} \right)$

27. $y = 2 - \cos \left(2x + \dfrac{\pi}{2} \right)$

28. $y = -2 - \cos \left(2x - \dfrac{\pi}{2} \right)$

29. $y = -2 + 2 \sin \left(\dfrac{1}{2}x + \dfrac{\pi}{2} \right)$

30. $y = 3 + 3 \cos \left(\dfrac{1}{2}x + \dfrac{\pi}{3} \right)$

31. $y = \dfrac{3}{2} + \dfrac{1}{2} \cos \left(3x - \dfrac{\pi}{2} \right)$

32. $y = \dfrac{2}{3} + \dfrac{4}{3} \cos \left(3x + \dfrac{\pi}{2} \right)$

Graph each of the following equations over the given interval. In each case, be sure to label the axes so that the amplitude, period, and phase shift are easy to read.

33. $y = 4 \cos \left(2x - \dfrac{\pi}{2} \right), \ -\dfrac{\pi}{4} \le x \le \dfrac{3\pi}{2}$

34. $y = 3 \sin \left(2x - \dfrac{\pi}{3} \right), \ -\dfrac{5\pi}{6} \le x \le \dfrac{7\pi}{6}$

35. $y = -4 \cos \left(2x - \dfrac{\pi}{2} \right), \ -\dfrac{\pi}{4} \le x \le \dfrac{3\pi}{2}$

36. $y = -3 \sin \left(2x - \dfrac{\pi}{3} \right), \ -\dfrac{5\pi}{6} \le x \le \dfrac{7\pi}{6}$

37. $y = \dfrac{2}{3} \sin \left(3x + \dfrac{\pi}{2} \right), \ -\pi \le x \le \pi$

38. $y = \dfrac{3}{4} \sin \left(3x - \dfrac{\pi}{2} \right), \ -\dfrac{\pi}{2} \le x \le \dfrac{3\pi}{2}$

39. $y = -\dfrac{2}{3} \sin \left(3x + \dfrac{\pi}{2} \right), \ -\pi \le x \le \pi$

40. $y = -\dfrac{3}{4} \sin \left(3x - \dfrac{\pi}{2} \right), \ -\dfrac{\pi}{2} \le x \le \dfrac{3\pi}{2}$

Sketch one complete cycle of each of the following by first graphing the appropriate sine or cosine curve and then using the reciprocal relationships. In each case, be sure to include the asymptotes on your graph.

41. $y = \csc\left(x + \dfrac{\pi}{4}\right)$

42. $y = \sec\left(x + \dfrac{\pi}{4}\right)$

43. $y = 2\sec\left(2x - \dfrac{\pi}{2}\right)$

44. $y = 2\csc\left(2x - \dfrac{\pi}{2}\right)$

45. $y = 3\csc\left(2x + \dfrac{\pi}{3}\right)$

46. $y = 3\sec\left(2x - \dfrac{\pi}{3}\right)$

The periods for $y = \tan x$ and $y = \cot x$ are π. The graphs of $y = \tan(Bx + C)$ and $y = \cot(Bx + C)$ will have periods $= \pi/B$ and phase shifts $= -C/B$, for $B > 0$. Sketch the graph of each equation below. Don't limit your graphs to one complete cycle, and be sure to show the asymptotes with each graph.

47. $y = \tan\left(x + \dfrac{\pi}{4}\right)$

48. $y = \tan\left(x - \dfrac{\pi}{4}\right)$

49. $y = \cot\left(x - \dfrac{\pi}{4}\right)$

50. $y = \cot\left(x + \dfrac{\pi}{4}\right)$

51. $y = \tan\left(2x - \dfrac{\pi}{2}\right)$

52. $y = \tan\left(2x + \dfrac{\pi}{2}\right)$

Review Problems

The following problems review material we covered in Section 3.4.

53. Find the length of arc cut off by a central angle of $\pi/6$ radians in a circle of radius 10 cm.

54. How long is the arc cut off by a central angle of $90°$ in a circle with radius 22 cm?

55. The minute hand of a clock is 2.6 cm long. How far does the tip of the minute hand travel in 30 minutes?

56. The hour hand of a clock is 3 inches long. How far does the tip of the hand travel in 5 hours?

57. Find the radius of a circle if a central angle of 6 radians cuts off an arc of length 4 ft.

58. In a circle, a central angle of $135°$ cuts off an arc of length 75 m. Find the radius of the circle.

59. Find the area of the sector formed by a central angle of $45°$ in a circle of radius 8 inches.

60. An arc of length 4 ft is cut off by a central angle of $\pi/8$. Find the area of the sector formed by the angle.

SECTION 4.4   # Finding an Equation from Its Graph

In this section, we will reverse what we have done in previous sections of this chapter and produce an equation that describes a graph, rather than a graph that describes an equation. Let's start with an example from algebra.

▶ **EXAMPLE 1**   Find the equation of the line shown in Figure 1.

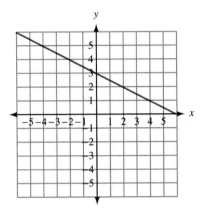

Figure 1

Solution   From algebra we know that the equation of any straight line (except a vertical one) can be written in slope-intercept form as

$$y = mx + b$$

where m is the slope of the line, and b is its y-intercept.

Since the line in Figure 1 crosses the y-axis at 3, we know the y-intercept b is 3. To find the slope of the line, we find the ratio of the vertical change to the horizontal change between any two points on the line (sometimes called rise/run). From Figure 1 we see that this ratio is $-1/2$. Therefore, $m = -1/2$. The equation of our line must be

$$y = -\frac{1}{2}x + 3$$

◀

▶ **EXAMPLE 2**   One cycle of the graph of a trigonometric function is shown in Figure 2. Find an equation to match the graph.

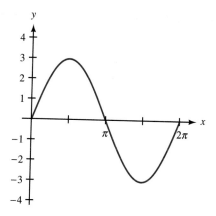

Figure 2

Solution   The graph is a sine curve with an amplitude of 3, period 2π, and no phase shift. The equation is

$$y = 3 \sin x \qquad 0 \le x \le 2\pi$$

◀

▶ **EXAMPLE 3**   Find the equation of the graph shown in Figure 3.

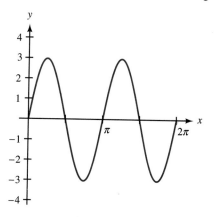

Figure 3

Solution   Again, we have a sine curve, so we know the equation will have the form

$$y = k + A \sin (Bx + C)$$

From Figure 3 we see that the amplitude is 3, which means that $A = 3$. There is no phase shift, nor is there any vertical translation of the graph. Therefore, both C and k are 0.

To find B, we notice that the period is π. Since the formula for the period is $2\pi/B$, we have

$$\pi = \frac{2\pi}{B}$$

which means that B is 2. Our equation must be

$$y = 0 + 3 \sin (2x + 0)$$

which simplifies to

$$y = 3 \sin 2x \qquad \text{for} \qquad 0 \leq x \leq 2\pi$$ ◀

▶ **EXAMPLE 4**   Find the equation of the graph shown in Figure 4.

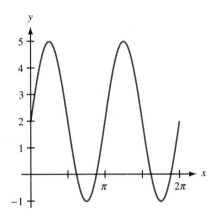

Figure 4

Solution   The graph in Figure 4 has the same shape (amplitude, period, and phase shift) as the graph shown in Figure 3. In addition, it has undergone a vertical shift up of two units; therefore, the equation is

$$y = 2 + 3 \sin 2x \qquad \text{for} \qquad 0 \leq x \leq 2\pi$$ ◀

▶ **EXAMPLE 5**   Find the equation of the graph shown in Figure 5.

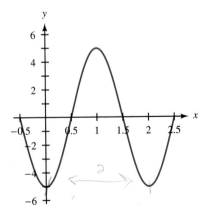

Figure 5

Solution   If we look at the graph from $x = 0$ to $x = 2$, it looks like a cosine curve that has been reflected about the x-axis. The general form for a cosine curve is

$$y = k + A \cos (Bx + C)$$

From Figure 5 we see the amplitude is 5. Since the graph has been reflected about the x-axis, $A = -5$. The period is 2, giving us an equation to solve for B

$$\text{Period} = \frac{2\pi}{B} = 2 \quad \Rightarrow \quad B = \pi$$

There is no horizontal or vertical translation of the curve (if we assume it is a cosine curve), so C and k are both 0. The equation that describes this graph is

$$y = -5 \cos \pi x \qquad -0.5 \le x \le 2.5 \qquad \blacktriangleleft$$

▶ **EXAMPLE 6**   Find the equation of the curve shown in Figure 6.

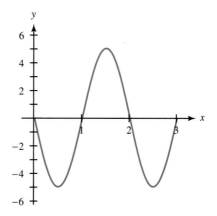

Figure 6

Solution   The graph has the same shape as the graph shown in Figure 5 except it has been moved to the right 0.5 units. Using the results from Example 5, we know that $B = \pi$. To find C, we use the formula for phase shift:

$$\text{Phase shift} = -\frac{C}{B}$$

$$0.5 = -\frac{C}{\pi}$$

Solving for C, we have $C = -\dfrac{\pi}{2}$.

The equation is

$$y = -5 \cos \left(\pi x - \frac{\pi}{2} \right) \qquad \blacktriangleleft$$

In our next example, we use a table of values for two variables to obtain a graph. From the graph, we find the equation.

▶ **EXAMPLE 7** The Ferris wheel built by George Ferris that we encountered in Chapters 2 and 3 is shown in Figure 7. Recall that the diameter is 250 feet and it rotates through one complete revolution every 20 minutes.

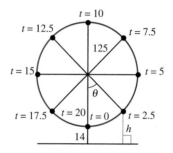

Figure 7

First, make a table that shows the rider's height h above the ground for each of the eight values of t shown in Figure 7. Then graph the ordered pairs indicated by the table. Finally, use the graph to find an equation that gives h as a function of t.

Solution Here is the reasoning we use to find a value of h for each of the given values of t: When $t = 0$, the rider is at the start of the ride, 14 feet above the ground. When $t = 10$, the rider is at the top of the wheel, which is 264 feet above the ground. At times $t = 5$ and $t = 15$, the rider is even with the center of the wheel, which is 139 feet above the ground. The other four values of h are found using right triangle trigonometry, as we did in Chapter 2. Here is our table of corresponding values of t and h:

Table 1

t	h
0 min	14 ft
2.5 min	51 ft
5 min	139 ft
7.5 min	227 ft
10 min	264 ft
12.5 min	227 ft
15 min	139 ft
17.5 min	51 ft
20 min	14 ft

If we graph the points (t, h) on a rectangular coordinate system and then connect them with a smooth curve, we produce the diagram shown in Figure 8. The curve is a cosine curve that has been reflected about the t-axis and then shifted up vertically.

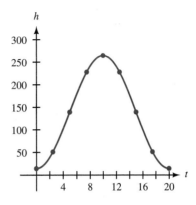

Figure 8

Since the curve is a cosine curve, the equation will have the form

$$h = k + A \cos (Bt + C)$$

To find the equation for the curve in Figure 8, we must find values for k, A, B, and C. We begin by finding C.

Since the curve starts at $t = 0$, the phase shift is 0, which gives us

$$C = 0$$

The amplitude, half the difference between the highest point and the lowest point on the graph, must be 125 (also equal to the radius of the wheel). However, since the graph is a *reflected* cosine curve, we have

$$A = -125$$

The period is 20 minutes, so we find B with the equation

$$20 = \frac{2\pi}{B} \quad \Rightarrow \quad B = \frac{\pi}{10}$$

The amount of the vertical translation, k, is the distance the center of the wheel is off the ground. Therefore,

$$k = 139$$

The equation that gives the height of the rider at any time t during the ride is

$$h = 139 - 125 \cos \frac{\pi}{10} t \qquad \blacktriangleleft$$

PROBLEM SET 4.4

Find the equation of each of the following lines. Write your answers in slope-intercept form, $y = mx + b$.

1.

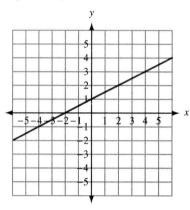

2.

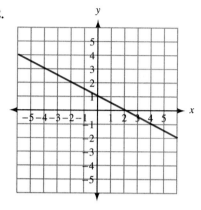

3.

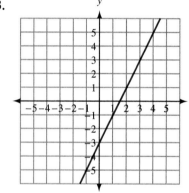

4.

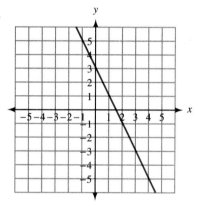

Each graph below is one complete cycle of the graph of an equation containing a trigonometric function. In each case, find an equation to match the graph.

5.

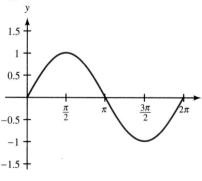

6.

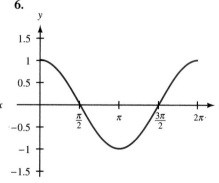

7.

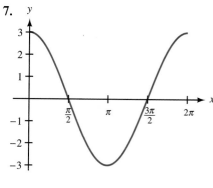

8.

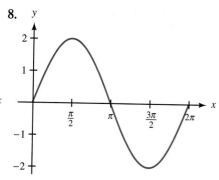

9.

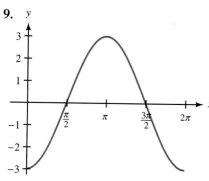

10.

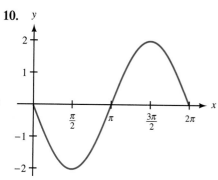

11.

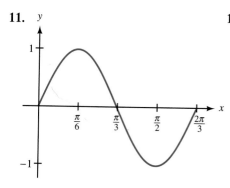

12.

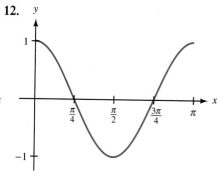

13.

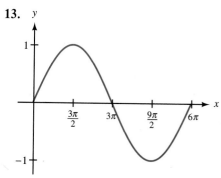

14.

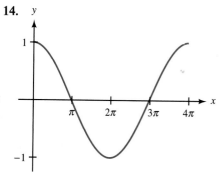

15.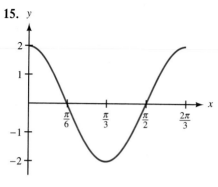

16.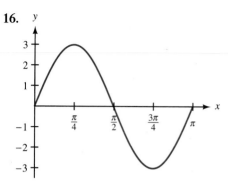

17.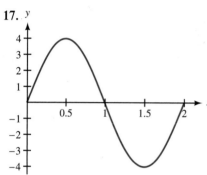

18.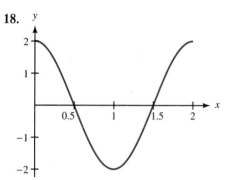

19.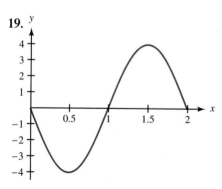

20.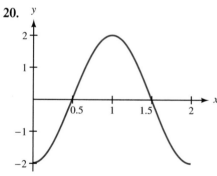

21.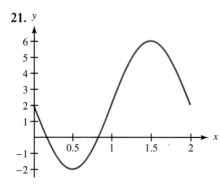

22.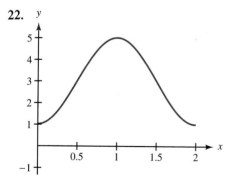

23.

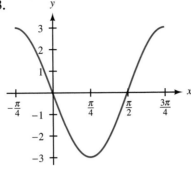

24.

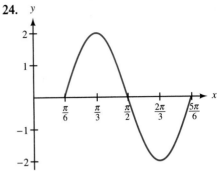

25.

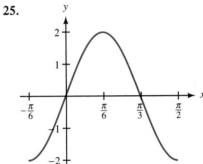

26.

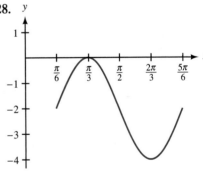

27.

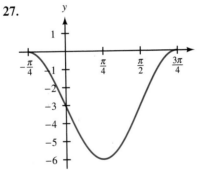

28.

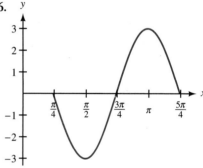

29.

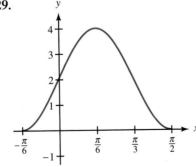

30.

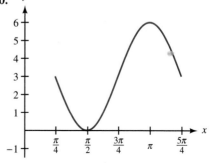

31. Figure 9 is a model of the Ferris wheel known as the Reisenrad that we encountered in Chapters 2 and 3. Recall that the diameter of the wheel is 197 ft, and one complete revolution takes 15 minutes. The bottom of the wheel is 12 ft above the ground. Complete the table next to Figure 9, then plot the points (t, h) from the table. Finally, connect the points with a smooth curve, and use the curve to find an equation that will give a passenger's height above the ground at any time t during the ride.

Table 2

t	h
0 min	
1.875 min	
3.75 min	
5.625 min	
7.5 min	
9.375 min	
11.25 min	
13.125 min	
15 min	

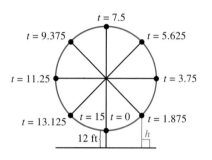

Figure 9

32. In Chapters 2 and 3, we worked some problems involving the Ferris wheel called Colossus that was built in St. Louis in 1986. The diameter of the wheel is 165 ft, it rotates at 1.5 rpm, and the bottom of the wheel is 9 ft above the ground. Find an equation that gives a passenger's height above the ground at any time t during the ride. Assume the passenger starts the ride at the bottom of the wheel.

Review Problems

The following problems review material we covered in Section 3.1.

Name the reference angle for each angle below.

33. $321°$

34. $148°$

35. $236°$

36. $-125°$

37. $-276°$

38. $450°$

Use your calculator to find θ to the nearest tenth of a degree if $0° < \theta < 360°$ and

39. $\sin \theta = 0.7455$ with θ in QII

40. $\cos \theta = 0.7455$ with θ in QIV

41. $\csc \theta = -2.3228$ with θ in QIII

42. $\sec \theta = -3.1416$ with θ in QIII

43. $\cot \theta = -0.2089$ with θ in QIV

44. $\tan \theta = 0.8156$ with θ in QIII

SECTION 4.5

Graphing Combinations of Functions

In this section, we will graph equations of the form $y = y_1 + y_2$, where y_1 and y_2 are algebraic or trigonometric functions of x. For instance, the equation $y = 1 + \sin x$ can be thought of as the sum of the two functions $y_1 = 1$ and $y_2 = \sin x$. That is,

$$\text{if} \quad y_1 = 1 \quad \text{and} \quad y_2 = \sin x,$$
$$\text{then} \quad y = y_1 + y_2$$

Using this kind of reasoning, the graph of $y = 1 + \sin x$ is obtained by adding each value of y_2 in $y_2 = \sin x$ to the corresponding value of y_1 in $y_1 = 1$. Graphically, we can show this by adding the values of y from the graph of y_2 to the corresponding values of y from the graph of y_1 (Figure 1).

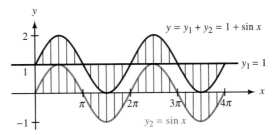

Figure 1

Although in actual practice you may not draw in the little vertical lines we have shown here, they do serve the purpose of allowing us to visualize the idea of adding the y-coordinates on one graph to the corresponding y-coordinates on another graph.

▶ **EXAMPLE 1**   Graph $y = \frac{1}{3}x - \sin x$ between $x = 0$ and $x = 4\pi$.

Solution   We can think of the equation $y = \frac{1}{3}x - \sin x$ as the sum of the equations $y_1 = \frac{1}{3}x$ and $y_2 = -\sin x$. Graphing each of these two equations on the same set of axes and then adding the values of y_2 to the corresponding values of y_1, we have the graph shown in Figure 2.

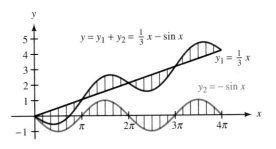

Figure 2   ◀

For the rest of the examples in this section, we will not show the vertical lines used to visualize the process of adding y-coordinates. Sometimes the graphs become too confusing to read when the vertical lines are included. It is the idea behind the vertical lines that is important, not the lines themselves. (And remember, the alternative to graphing these types of equations by adding y-coordinates is to make a table. If you try using a table on some of the examples that follow, you will see that the method being presented here is much faster.)

▶ **EXAMPLE 2** Graph $y = 2 \sin x + \cos 2x$ for x between 0 and 4π.

Solution We can think of y as the sum of y_1 and y_2, where

$$y_1 = 2 \sin x \qquad \text{(amplitude 2, period } 2\pi\text{)}$$

and

$$y_2 = \cos 2x \qquad \text{(amplitude 1, period } \pi\text{)}$$

The graphs of y_1, y_2, and $y = y_1 + y_2$ are shown in Figure 3.

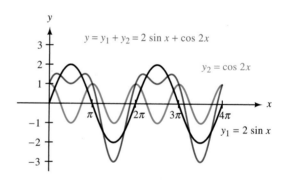

Figure 3

▶ **EXAMPLE 3** Graph $y = \cos x + \cos 2x$ for $0 \le x \le 4\pi$.

Solution We let $y = y_1 + y_2$, where

$$y_1 = \cos x \qquad \text{(amplitude 1, period } 2\pi\text{)}$$

and

$$y_2 = \cos 2x \qquad \text{(amplitude 1, period } \pi\text{)}$$

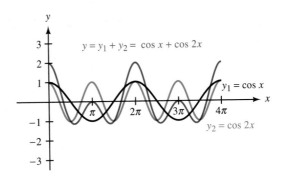

Figure 4   ◀

▶ **EXAMPLE 4**   Graph $y = \sin x + \cos x$ for x between 0 and 4π.

Solution   We let $y_1 = \sin x$ and $y_2 = \cos x$ and graph y_1, y_2, and $y = y_1 + y_2$.

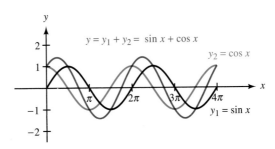

Figure 5

The graph of $y = \sin x + \cos x$ has amplitude $\sqrt{2}$. If we were to extend the graph to the left, we would find it crossed the x-axis at $-\pi/4$. It would then be apparent that the graph of $y = \sin x + \cos x$ is the same as the graph of $y = \sqrt{2} \sin (x + \pi/4)$; both are sine curves with amplitude $\sqrt{2}$ and phase shift $-\pi/4$.   ◀

PROBLEM SET 4.5

Use addition of y-coordinates to sketch the graph of each of the following between $x = 0$ and $x = 4\pi$.

1. $y = 1 + \sin x$
3. $y = 2 - \cos x$
5. $y = 4 + 2 \sin x$

2. $y = 1 + \cos x$
4. $y = 2 - \sin x$
6. $y = 4 + 2 \cos x$

7. $y = \dfrac{1}{3}x - \cos x$ **8.** $y = \dfrac{1}{2}x - \sin x$

9. $y = \dfrac{1}{2}x - \cos x$ **10.** $y = \dfrac{1}{3}x + \cos x$

Sketch the graph of each equation from $x = 0$ to $x = 8$.

11. $y = x + \sin \pi x$ **12.** $y = x + \cos \pi x$

Sketch the graph from $x = 0$ to $x = 4\pi$.

13. $y = 3 \sin x + \cos 2x$ **14.** $y = 3 \cos x + \sin 2x$
15. $y = 2 \sin x - \cos 2x$ **16.** $y = 2 \cos x - \sin 2x$

17. $y = \sin x + \sin \dfrac{x}{2}$ **18.** $y = \cos x + \cos \dfrac{x}{2}$

19. $y = \sin x + \sin 2x$ **20.** $y = \cos x + \cos 2x$

21. $y = \cos x + \dfrac{1}{2}\sin 2x$ **22.** $y = \sin x + \dfrac{1}{2}\cos 2x$

23. $y = \sin x - \cos x$ **24.** $y = \cos x - \sin x$

25. Make a table using multiples of $\pi/2$ for x between 0 and 4π to help sketch the graph of $y = x \sin x$.
26. Sketch the graph of $y = x \cos x$.

Review Problems

The following problems review material we covered in Section 3.5.

27. A point moving on the circumference of a circle covers 5 ft every 20 seconds. Find the linear velocity of the point.
28. A point is moving on the circumference of a circle. Every 30 seconds the point covers 15 cm. Find the linear velocity of the point.
29. A point is moving with a linear velocity of 20 ft/sec on the circumference of a circle. How far does the point move in 1 minute?
30. A point moves at 65 m/sec on the circumference of a circle. How far does the point travel in 1 minute?
31. A point is moving with an angular velocity of 3 rad/sec on a circle of radius 6 m. How far does the point travel in 10 seconds?
32. Convert 30 revolutions per minute (rpm) to angular velocity in radians/second.
33. Convert 120 rpm to angular velocity in radians/second.
34. A point is rotating at 5 rpm on a circle of radius 6 inches. What is the linear velocity of the point?
35. How far does the tip of a 10-cm minute hand on a clock travel in 2 hours?
36. How far does the tip of an 8-cm hour hand on a clock travel in 1 day?

SECTION 4.6 # Inverse Trigonometric Functions

We begin this section with a review of the basic concepts associated with relations and functions. First, let us review the definitions for relations and functions, along with the definitions for domain and range.

> DEFINITION   Relation, Domain, Range, Function
>
> A *relation* is a set of ordered pairs. The set of all first coordinates forms the *domain* of the relation, while the *range* is the set of all second coordinates. A *function* is a relation in which no ordered pairs have the same first coordinates. The formal definition is stated this way: A function is a rule or correspondence that pairs each element in one set, called the domain, with exactly one element from a second set, called the range.

Relations and functions are both sets of ordered pairs. What makes them different from each other is that the ordered pairs of a function will never repeat their first coordinates. Every function is a relation, but not all relations are functions.

Vertical Line Test

If we have the graph of a relation, we can check it visually to see if it is also a function by using the vertical line test. If any vertical line can be found that crosses a graph in more than one place, then that graph cannot be the graph of a function. (If a vertical line crosses a graph in two places, then we have two points with the same first coordinates.)

Next, we review two main topics from inverse functions. The first is the process by which you obtain the equation of the inverse of a function from the function itself, and the second is the relationship between the graph of a function and the graph of its inverse. These two topics are reviewed in Examples 1 and 2.

▶ **EXAMPLE 1**   If the function f is defined by $f(x) = 2x - 3$, find the equation that represents the inverse of f.

Solution   Since the inverse of f is obtained by interchanging the components of all the ordered pairs belonging to f, and each ordered pair in f satisfies the equation $y = 2x - 3$, we simply exchange x and y in the equation $y = 2x - 3$ to get the formula for the inverse of f:

$$x = 2y - 3$$

We now solve this equation for y in terms of x:

$$x + 3 = 2y$$

$$\frac{x + 3}{2} = y$$

$$y = \frac{x + 3}{2}$$

The last line gives the equation that defines the inverse of f. Let's compare the graphs of f and its inverse as given above. (See Figure 1.)

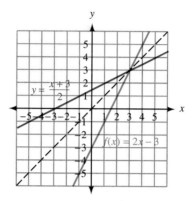

Figure 1

The graphs of f and its inverse have symmetry about the line $y = x$. This is a reasonable result, since the one function was obtained from the other by interchanging x and y in the equation. The ordered pairs (a, b) and (b, a) always have symmetry about the line $y = x$.   ◀

▶ **EXAMPLE 2**   Write an equation for the inverse of the function $y = x^2 - 4$ and then graph both the function and its inverse.

Solution   To find the inverse of $y = x^2 - 4$, we exchange x and y in the equation and then solve for y.

$$\text{The inverse of } y = x^2 - 4$$
$$\text{is } x = y^2 - 4 \qquad \text{Exchange } x \text{ and } y$$
$$\text{or } y^2 - 4 = x$$
$$y^2 = x + 4 \qquad \text{Add 4 to both sides}$$
$$y = \pm\sqrt{x + 4} \qquad \text{Take the square root of both sides}$$

The inverse of the function $y = x^2 - 4$ is given by the equation $y = \pm\sqrt{x + 4}$. The graph of $y = x^2 - 4$ is a parabola that crosses the x-axis at -2 and 2 and has its vertex at $(0, -4)$. To graph its inverse, we reflect the graph of $y = x^2 - 4$ about the line $y = x$. Figure 2 shows both graphs.

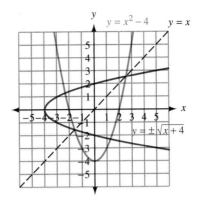

Figure 2

From Figure 2 we see that $y = \pm\sqrt{x + 4}$ is not a function because we can find a vertical line that will cross the graph of $y = \pm\sqrt{x + 4}$ in more than one place, indicating that some values of x correspond to more than one value of y. ◀

The three important points about inverse functions illustrated in Examples 1 and 2 are as follows:

1. The equation of the inverse is found by exchanging x and y in the equation of the function.
2. The graph of the inverse can be found by reflecting the graph of the function about the line $y = x$.
3. A graph is not the graph of a function if a vertical line crosses it in more than one place.

Note   Again, this discussion is meant to be a review. If it is not making any sense, then you should read up on functions and their inverses and then come back and try it again.

Comparing the graphs from Examples 1 and 2, we observe that the inverse of a function is not always a function.

Inverse Function Notation

If $y = f(x)$ is a function, and the inverse of f is also a function, then the inverse of f can be denoted by $y = f^{-1}(x)$. To illustrate, in Example 1 we found the inverse of $f(x) = 2x - 3$ was the function $y = \dfrac{x + 3}{2}$. We can write this inverse function with inverse function notation as

$$f^{-1}(x) = \frac{x + 3}{2}$$

The inverse of the function in Example 2 is not itself a function, so we do not use the notation $f^{-1}(x)$ to represent it.

The Inverse Sine Relation

To find the inverse of $y = \sin x$, we interchange x and y to obtain

$$x = \sin y$$

This is the equation of the inverse sine relation.

To graph $x = \sin y$, we simply reflect the graph of $y = \sin x$ about the line $y = x$, as shown in Figure 3.

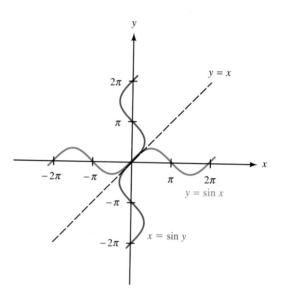

Figure 3

As you can see from the graph, $x = \sin y$ is a relation but not a function. For every value of x in the domain, there are many values of y. The graph of $x = \sin y$ fails the vertical line test.

Inverse Trigonometric Functions

In order that the function $y = \sin x$ have an inverse that is also a function, it is necessary to restrict the values that y can assume. The interval we restrict it to is the interval $-\pi/2 \le y \le \pi/2$. Figure 4 contains the graph of $x = \sin y$ with the restricted interval showing. It is apparent from Figure 4 that if $x = \sin y$ is restricted to the interval $-\pi/2 \le y \le \pi/2$, then each value of x is associated with exactly one

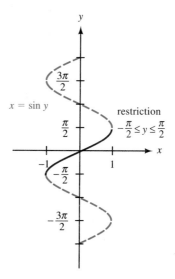

Figure 4

value of y, and we have a function rather than just a relation. (That is, on the interval $-\pi/2 \le y \le \pi/2$, the graph of $x = \sin y$ is such that no vertical line crosses it in more than one place.) The equation $x = \sin y$, and the restriction $-\pi/2 \le y \le \pi/2$, together form the inverse sine function. To designate this function we use the following notation.

Notation

The notation used to indicate the inverse sine function is as follows:

	Notation	**Meaning**
Inverse sine function	$y = \sin^{-1} x$ or $y = \arcsin x$	$x = \sin y$ and $-\dfrac{\pi}{2} \le y \le \dfrac{\pi}{2}$

In this section, we will limit our discussion of inverse trigonometric functions to the inverses of the three major functions: sine, cosine, and tangent. The other three inverse trigonometric functions can be handled with the use of the reciprocal identities.

Figure 5 shows the graphs of $y = \cos^{-1} x$ and $y = \tan^{-1} x$ and the restrictions that will allow them to become functions instead of just relations.

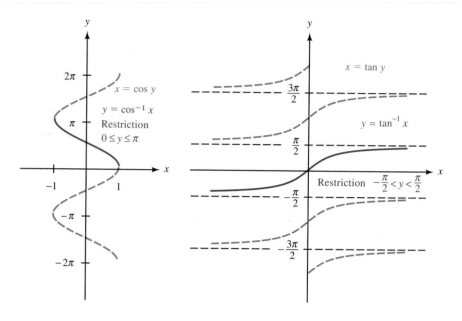

Figure 5

To summarize, here is the definition for the three major inverse trigonometric functions.

DEFINITION   Inverse Trigonometric Functions

The inverse functions for $y = \sin x$, $y = \cos x$, and $y = \tan x$ are as follows:

Inverse Function	**Meaning**
$y = \sin^{-1} x$ or $y = \arcsin x$	$x = \sin y$ and $-\dfrac{\pi}{2} \leq y \leq \dfrac{\pi}{2}$
In words: y is the angle between $-\pi/2$ and $\pi/2$, inclusive, whose sine is x.	
$y = \cos^{-1} x$ or $y = \arccos x$	$x = \cos y$ and $0 \leq y \leq \pi$
In words: y is the angle between 0 and π, inclusive, whose cosine is x.	
$y = \tan^{-1} x$ or $y = \arctan x$	$x = \tan y$ and $-\dfrac{\pi}{2} < y < \dfrac{\pi}{2}$
In words: y is the angle between $-\pi/2$ and $\pi/2$ whose tangent is x.	

Note   The notation $\sin^{-1} x$ is not to be interpreted as meaning the reciprocal of $\sin x$. That is,

$$\sin^{-1} x \neq \frac{1}{\sin x}$$

If we want the reciprocal of $\sin x$, we use $\csc x$ or $(\sin x)^{-1}$, but never $\sin^{-1} x$.

▶ **EXAMPLE 3**   Evaluate in radians without using a calculator or tables.

$$\textbf{a. } \sin^{-1} \frac{1}{2} \qquad\qquad \textbf{b. } \arccos \left(-\frac{\sqrt{3}}{2} \right) \qquad\qquad \textbf{c. } \tan^{-1}(-1)$$

Solution

a. The angle between $-\pi/2$ and $\pi/2$ whose sine is $\frac{1}{2}$ is $\pi/6$.

$$\sin^{-1} \frac{1}{2} = \frac{\pi}{6}$$

b. The angle between 0 and π with a cosine of $-\sqrt{3}/2$ is $5\pi/6$.

$$\arccos \left(-\frac{\sqrt{3}}{2} \right) = \frac{5\pi}{6}$$

c. The angle between $-\pi/2$ and $\pi/2$ the tangent of which is -1 is $-\pi/4$.

$$\tan^{-1}(-1) = -\frac{\pi}{4}$$ ◀

Note   In part c of Example 3, it would be incorrect to give the answer as $7\pi/4$. It is true that $\tan 7\pi/4 = -1$, but $7\pi/4$ is not between $-\pi/2$ and $\pi/2$. There is a difference.

▶ **EXAMPLE 4**   Use a calculator to evaluate each expression to the nearest tenth of a degree.

a. $\arcsin (0.5075)$ **b.** $\arcsin (-0.5075)$
c. $\cos^{-1} (0.6428)$ **d.** $\cos^{-1} (-0.6428)$
e. $\arctan (4.474)$ **f.** $\arctan (-4.474)$

Solution   The easiest method of evaluating these expressions is to use a calculator. Make sure the calculator is set to the degree mode, and then enter the number and push the appropriate button. Scientific calculators are programmed so that the restrictions on the inverse trigonometric functions are automatic. We just push the buttons and the rest is taken care of.

a. $\arcsin (0.5075) = 30.5°$
b. $\arcsin (-0.5075) = -30.5°$ $\Big\}$ Reference angle $30.5°$

c. $\cos^{-1} (0.6428) = 50.0°$
d. $\cos^{-1} (-0.6428) = 130.0°$ $\Big\}$ Reference angle $50°$

$$\left.\begin{array}{l}\textbf{e. } \arctan\,(4.474) = 77.4° \\ \textbf{f. } \arctan\,(-4.474) = -77.4°\end{array}\right\} \quad \text{Reference angle } 77.4°$$ ◄

▶ **EXAMPLE 5** Evaluate $\sin\,(\tan^{-1}\frac{3}{4})$ without using a calculator.

Solution We begin by letting $\theta = \tan^{-1}\,(\frac{3}{4})$. (Remember, $\tan^{-1} x$ is the angle whose tangent is x.) Then we have

$$\text{If } \theta = \tan^{-1}\frac{3}{4}, \text{ then } \tan\theta = \frac{3}{4} \text{ and } 0° < \theta < 90°$$

We can draw a triangle in which one of the acute angles is θ. Since $\tan\theta = \frac{3}{4}$, we label the side opposite θ with 3 and the side adjacent to θ with 4. The hypotenuse is found by applying the Pythagorean Theorem.

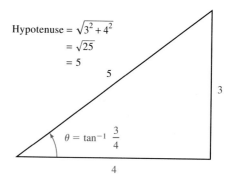

$$\text{Hypotenuse} = \sqrt{3^2 + 4^2}$$
$$= \sqrt{25}$$
$$= 5$$

Figure 6

From Figure 6 we find $\sin\theta$ using the ratio of the side opposite θ to the hypotenuse.

$$\sin\left(\tan^{-1}\frac{3}{4}\right) = \sin\theta = \frac{3}{5}$$ ◄

Calculator Note If we were to do the same problem with the aid of a calculator, the sequence would look like this:

$$3 \;\boxed{\div}\; 4 \;\boxed{=}\; \boxed{\tan^{-1}} \;\boxed{\sin}$$

The display would read 0.6, which is $\frac{3}{5}$.

Although it is a lot easier to use a calculator on problems like the one in Example 5, solving it without a calculator will be of more use to you in the future.

▶ **EXAMPLE 6**   Write the expression $\sin(\cos^{-1} x)$ as an equivalent expression in x only. (Assume x is positive.)

Solution   We let $\theta = \cos^{-1} x$, then $\cos \theta = x$. To help visualize the problem, we draw a right triangle with an acute angle of θ and label it so that $\cos \theta = x$. This is accomplished by labeling the side adjacent to θ with x and the hypotenuse with 1. That way, the ratio of the side adjacent to θ to the hypotenuse is $x/1 = x$. The side opposite θ is found by applying the Pythagorean Theorem.

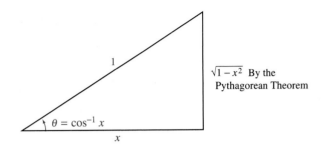

Figure 7

Finding $\sin \theta$ is simply a matter of taking the ratio of the opposite side to the hypotenuse.

$$\sin(\cos^{-1} x) = \sin \theta = \frac{\sqrt{1 - x^2}}{1} = \sqrt{1 - x^2}$$

Note that, since we assumed x was positive, $\cos^{-1} x$ is between $0°$ and $90°$, so that it can be represented as an acute angle in a right triangle.   ◀

PROBLEM SET 4.6

For each equation below, write an equation for its inverse, and then sketch the graph of the equation and its inverse on the same coordinate system.

1. $y = x^2 + 4$
2. $y = x^2 - 3$
3. $y = 2x^2$
4. $y = \dfrac{1}{2}x^2$
5. $y = 3x - 2$
6. $y = 2x + 3$
7. $x^2 + y^2 = 9$
8. $x^2 + y^2 = 4$
9. $y = 3^x$
10. $y = 4^x$

11. Graph $y = \cos x$ between -2π and 2π, and then reflect the graph about the line $y = x$ to obtain the graph of $x = \cos y$.

12. Graph $y = \sin x$ between $-\pi/2$ and $\pi/2$, and then reflect the graph about the line $y = x$ to obtain the graph of $y = \sin^{-1} x$ between $-\pi/2$ and $\pi/2$.
13. Graph $y = \tan x$ for x between $-3\pi/2$ and $3\pi/2$, and then reflect the graph about the line $y = x$ to obtain the graph of $x = \tan y$.
14. Graph $y = \cot x$ for x between 0 and 2π, and then reflect the graph about the line $y = x$ to obtain the graph of $x = \cot y$.

The domain of a function is the set of all first coordinates (the set of all values that x can assume). The range is the set of all second coordinates (all the values that y can assume).

15. What is the domain of the function $y = \sin^{-1} x$?
16. What is the range of the function $y = \sin^{-1} x$?
17. What is the range of $y = \arccos x$?
18. What is the domain of $y = \arccos x$?

Evaluate each expression without using a calculator, and write your answers in radians.

19. $\sin^{-1}\left(\dfrac{\sqrt{3}}{2}\right)$ 20. $\cos^{-1}\left(\dfrac{1}{2}\right)$

21. $\cos^{-1}(-1)$ 22. $\cos^{-1}(0)$
23. $\tan^{-1}(1)$ 24. $\tan^{-1}(0)$

25. $\arccos\left(-\dfrac{1}{\sqrt{2}}\right)$ 26. $\arccos(1)$

27. $\sin^{-1}\left(-\dfrac{1}{2}\right)$ 28. $\sin^{-1}\left(\dfrac{1}{\sqrt{2}}\right)$

29. $\arctan(\sqrt{3})$ 30. $\arctan\left(\dfrac{1}{\sqrt{3}}\right)$

31. $\arcsin(0)$ 32. $\arcsin\left(-\dfrac{\sqrt{3}}{2}\right)$

33. $\tan^{-1}\left(-\dfrac{1}{\sqrt{3}}\right)$ 34. $\tan^{-1}(-\sqrt{3})$

35. $\cos^{-1}\left(-\dfrac{1}{2}\right)$ 36. $\sin^{-1}(1)$

37. $\arccos\left(\dfrac{\sqrt{3}}{2}\right)$ 38. $\arcsin(-1)$

Use a calculator to evaluate each expression to the nearest tenth of a degree.

39. $\sin^{-1}(0.1702)$ 40. $\sin^{-1}(-0.1702)$
41. $\cos^{-1}(-0.8425)$ 42. $\cos^{-1}(0.8425)$
43. $\tan^{-1}(0.3799)$ 44. $\tan^{-1}(-0.3799)$
45. $\arcsin(0.9627)$ 46. $\arccos(0.9627)$
47. $\cos^{-1}(-0.4664)$ 48. $\sin^{-1}(-0.4664)$

49. arctan (-2.748) **50.** arctan (-0.3640)

51. $\sin^{-1}(-0.7660)$ **52.** $\cos^{-1}(-0.7660)$

Evaluate without using a calculator.

53. $\cos\left(\tan^{-1}\dfrac{3}{4}\right)$ **54.** $\csc\left(\tan^{-1}\dfrac{3}{4}\right)$

55. $\tan\left(\sin^{-1}\dfrac{3}{5}\right)$ **56.** $\tan\left(\cos^{-1}\dfrac{3}{5}\right)$

57. $\sec\left(\cos^{-1}\dfrac{1}{\sqrt{5}}\right)$ **58.** $\sin\left(\cos^{-1}\dfrac{1}{\sqrt{5}}\right)$

59. $\sin\left(\cos^{-1}\dfrac{1}{2}\right)$ **60.** $\cos\left(\sin^{-1}\dfrac{1}{2}\right)$

61. $\cot\left(\tan^{-1}\dfrac{1}{2}\right)$ **62.** $\cot\left(\tan^{-1}\dfrac{1}{3}\right)$

Evaluate without using a calculator.

63. $\sin\left(\sin^{-1}\dfrac{3}{5}\right)$ **64.** $\cos\left(\cos^{-1}\dfrac{3}{5}\right)$

65. $\cos\left(\cos^{-1}\dfrac{1}{2}\right)$ **66.** $\sin\left(\sin^{-1}\dfrac{1}{\sqrt{2}}\right)$

67. $\tan\left(\tan^{-1}\dfrac{1}{2}\right)$ **68.** $\tan\left(\tan^{-1}\dfrac{3}{4}\right)$

For each expression below, write an equivalent expression that involves x only. (Assume x is positive.)

69. $\cos(\cos^{-1}x)$ **70.** $\sin(\sin^{-1}x)$
71. $\cos(\sin^{-1}x)$ **72.** $\tan(\cos^{-1}x)$
73. $\sin(\tan^{-1}x)$ **74.** $\cos(\tan^{-1}x)$

75. $\sin\left(\cos^{-1}\dfrac{1}{x}\right)$ **76.** $\cos\left(\sin^{-1}\dfrac{1}{x}\right)$

77. $\sec\left(\cos^{-1}\dfrac{1}{x}\right)$ **78.** $\csc\left(\sin^{-1}\dfrac{1}{x}\right)$

Review Problems

The problems that follow review material we covered in Sections 4.2 and 4.3.

Graph each of the following equations over the indicated interval. Be sure to label the x- and y-axes so that the amplitude and period are easy to see.

79. $y = 4 \sin 2x$, one complete cycle
80. $y = 2 \sin 4x$, one complete cycle
81. $y = 2 \sin \pi x$, $-4 \le x \le 4$
82. $y = 3 \cos \pi x$, $-2 \le x \le 4$

83. $y = -3 \cos \dfrac{1}{2} x$, $\quad -2\pi \le x \le 6\pi$ **84.** $y = -3 \sin 2x$, $\quad -2\pi \le x \le 2\pi$

Graph one complete cycle of each of the following equations. Be sure to label the x- and y-axes so that the amplitude, period, and phase shift for each graph are easy to see.

85. $y = \sin\left(x - \dfrac{\pi}{4}\right)$ **86.** $y = \sin\left(x + \dfrac{\pi}{6}\right)$

87. $y = \cos\left(2x + \dfrac{\pi}{2}\right)$ **88.** $y = \sin(2x + \pi)$

89. $y = 3 \sin\left(2x - \dfrac{\pi}{3}\right)$ **90.** $y = 3 \cos\left(2x - \dfrac{\pi}{3}\right)$

CHAPTER 4 SUMMARY

Examples

1. Since

$$\sin(x + 2\pi) = \sin x,$$

the function $y = \sin x$ is periodic with period 2π. Likewise, since

$$\tan(x + \pi) = \tan x,$$

the function $y = \tan x$ is periodic with period π.

Periodic Functions [4.1]

A function $y = f(x)$ is said to be periodic with period p if p is the smallest positive number such that $f(x + p) = f(x)$ for all x in the domain of f.

Basic Graphs [4.1]

The graphs of $y = \sin x$ and $y = \cos x$ are both periodic with period 2π. The amplitude of each graph is 1. The sine curve passes through 0 on the y-axis, while the cosine curve passes through 1 on the y-axis.

The graphs of $y = \csc x$ and $y = \sec x$ are also periodic with period 2π. We graph them by using the fact that they are reciprocals of sine and cosine. Since there is no largest or smallest value of y, we say the secant and cosecant curves have no amplitude.

The graphs of $y = \tan x$ and $y = \cot x$ are periodic with period π. The tangent curve passes through the origin, while the cotangent is undefined when x is 0. There is no amplitude for either graph.

2.

a.

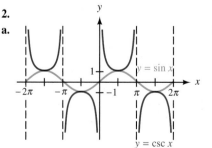

b.

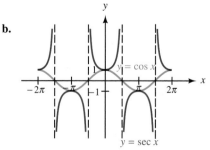

c.

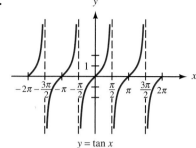

$$y = \tan x$$

d.

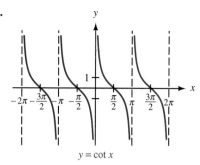

$$y = \cot x$$

3.

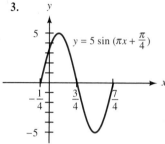

$$A = \tfrac{1}{2}\left|5 - (-5)\right| = \tfrac{1}{2}(10) = 5$$

4. The phase shift for the graph in Example 3 is $-\tfrac{1}{4}$.

5.

a.

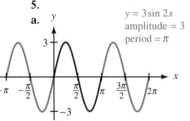

$y = 3 \sin 2x$
amplitude $= 3$
period $= \pi$

b.

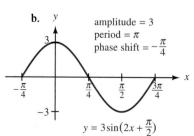

amplitude $= 3$
period $= \pi$
phase shift $= -\tfrac{\pi}{4}$

$$y = 3 \sin\left(2x + \tfrac{\pi}{2}\right)$$

Amplitude [4.1, 4.2]

The *amplitude* A of a curve is half the absolute value of the difference between the largest value of y, denoted by M, and the smallest value of y, denoted by m.

$$A = \frac{1}{2}\left|M - m\right|$$

Phase Shift [4.3]

The *phase shift* for a sine or cosine curve is the distance the curve has moved right or left from the curve $y = \sin x$ or $y = \cos x$. For example, we usually think of the graph of $y = \sin x$ as starting at the origin. If we graph another sine curve that starts at $\pi/4$, then we say this curve has a phase shift of $\pi/4$.

Graphing Sine and Cosine Curves [4.2, 4.3]

The graphs of $y = A \sin (Bx + C)$ and $y = A \cos (Bx + C)$, where $B > 0$, will have the following characteristics:

$$\text{Amplitude} = |A|$$

$$\text{Period} = \frac{2\pi}{B}$$

$$\text{Phase shift} = -\frac{C}{B}$$

To graph one of these curves, we first find the phase shift and label that point on the x-axis (this will be our starting point). We then add the period to the phase shift and mark the result on the x-axis (this is our ending point). We mark the y-axis with the amplitude. Finally, we sketch in one complete cycle of the curve in question, keeping in mind that, if A is negative, the graph must be reflected about the x-axis.

6.

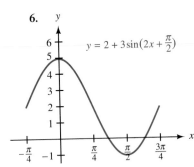

$y = 2 + 3\sin(2x + \frac{\pi}{2})$

Vertical Translations [4.2, 4.3]

Adding a constant k to an equation containing trigonometric functions translates the graph vertically up or down. For example, the graph of $y = k + A \sin(Bx + C)$ will have the same shape (amplitude, period, phase shift, and reflection, if indicated) as $y = A \sin(Bx + C)$ but will be translated k units vertically from the graph of $y = A \sin(Bx + C)$.

7.

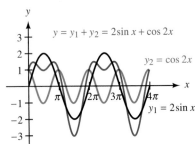

$y = y_1 + y_2 = 2\sin x + \cos 2x$

$y_2 = \cos 2x$

$y_1 = 2\sin x$

Graphing by Addition of y-Coordinates [4.5]

To graph equations of the form $y = y_1 + y_2$, where y_1 and y_2 are algebraic or trigonometric functions of x, we graph y_1 and y_2 separately on the same coordinate system and then add the two graphs to obtain the graph of y.

8. Evaluate in radians without using a calculator.

a. $\sin^{-1}\dfrac{1}{2}$

The angle between $-\pi/2$ and $\pi/2$ whose sine is $\frac{1}{2}$ is $\pi/6$.

$$\sin^{-1}\frac{1}{2} = \frac{\pi}{6}$$

b. $\arccos\left(-\dfrac{\sqrt{3}}{2}\right)$

The angle between 0 and π with a cosine of $-\sqrt{3}/2$ is $5\pi/6$.

$$\arccos\left(-\frac{\sqrt{3}}{2}\right) = \frac{5\pi}{6}$$

Inverse Trigonometric Functions [4.6]

Inverse Function	Meaning
$y = \sin^{-1} x$ or $y = \arcsin x$	$x = \sin y$ and $-\dfrac{\pi}{2} \le y \le \dfrac{\pi}{2}$
In words: y is the angle between $-\pi/2$ and $\pi/2$, inclusive, whose sine is x.	
$y = \cos^{-1} x$ or $y = \arccos x$	$x = \cos y$ and $0 \le y \le \pi$
In words: y is the angle between 0 and π, inclusive, whose cosine is x.	
$y = \tan^{-1} x$ or $y = \arctan x$	$x = \tan y$ and $-\dfrac{\pi}{2} < y < \dfrac{\pi}{2}$
In words: y is the angle between $-\pi/2$ and $\pi/2$ whose tangent is x.	

CHAPTER 4 TEST

Graph each of the following between $x = -4\pi$ and $x = 4\pi$.

1. $y = \sin x$
2. $y = \cos x$
3. $y = \tan x$
4. $y = \sec x$

5. How many complete cycles of the graph of the equation $y = \sin x$ are shown in your answer to Problem 1?
6. How many complete cycles of the graph of the equation $y = \tan x$ are shown in your answer to Problem 3?
7. Use your answer to Problem 4 to find all values of x between -4π and 4π for which $\sec x = -1$.
8. Use your answer to Problem 2 to find all values of x between -4π and 4π for which $\cos x = \frac{1}{2}$.

For each equation below, first identify the amplitude and period and then use this information to sketch one complete cycle of the graph.

9. $y = \cos \pi x$ 　　　　　　　　　　 10. $y = -3 \cos x$

Graph each of the following on the given interval.

11. $y = 2 + 3 \sin 2x, \ -\pi \le x \le 2\pi$ 　　 12. $y = 2 \sin \pi x, \ -4 \le x \le 4$

For each equation below, identify the amplitude, period, and phase shift and then use this information to sketch one complete cycle of the graph.

13. $y = \sin \left(x + \dfrac{\pi}{4} \right)$ 　　　　　 14. $y = \cos \left(x - \dfrac{\pi}{2} \right)$

15. $y = 3 \sin \left(2x - \dfrac{\pi}{3} \right)$ 　　　　 16. $y = -3 + 3 \sin \left(\dfrac{\pi}{3} x - \dfrac{\pi}{3} \right)$

17. $y = \csc \left(x + \dfrac{\pi}{4} \right)$ 　　　　　 18. $y = \tan \left(2x - \dfrac{\pi}{2} \right)$

Graph each of the following on the given interval.

19. $y = 2 \sin (3x - \pi), \ -\dfrac{\pi}{3} \le x \le \dfrac{5\pi}{3}$

20. $y = 2 \sin \left(\dfrac{\pi}{2} x - \dfrac{\pi}{4} \right), \ -\dfrac{1}{2} \le x \le \dfrac{13}{2}$

Find an equation for each of the following graphs.

21.

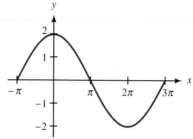

22.

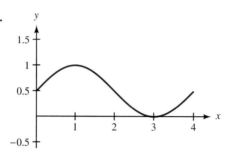

Sketch the following between $x = 0$ and $x = 4\pi$.

23. $y = \dfrac{1}{2}x - \sin x$

24. $y = \sin x + \cos 2x$

25. Graph $y = \cos^{-1} x$.

26. Graph $y = \arcsin x$.

Evaluate each expression without using a calculator and write your answer in radians.

27. $\sin^{-1}\left(\dfrac{1}{2}\right)$

28. $\cos^{-1}\left(-\dfrac{\sqrt{3}}{2}\right)$

29. $\arctan(-1)$

30. $\arcsin(1)$

Use a calculator to evaluate each expression to the nearest tenth of a degree.

31. $\arcsin(0.5934)$

32. $\arctan(-0.8302)$

33. $\arccos(-0.6981)$

34. $\arcsin(-0.2164)$

Evaluate without using a calculator.

35. $\tan\left(\cos^{-1}\dfrac{2}{3}\right)$

36. $\cos\left(\tan^{-1}\dfrac{2}{3}\right)$

For each expression below, write an equivalent expression that involves x only. (Assume x is positive.)

37. $\sin(\cos^{-1} x)$

38. $\tan(\sin^{-1} x)$

5

Identities and Formulas

I I I I I I I I I I

▶ **To the Student**

Recall that an identity in mathematics is a statement that two quantities are equal for all replacements of the variable for which they are defined. For instance, the statement

$$\tan \theta = \frac{\sin \theta}{\cos \theta}$$

is a trigonometric identity. We will begin this chapter by reviewing the list of eight basic trigonometric identities and their more common equivalent forms.

The most important information needed to be successful in this chapter is the material on identities that we developed from the definition of the six trigonometric functions in Chapter 1. You need to have the basic identities and their common equivalent forms memorized. Also important is your knowledge of the exact values of the trigonometric functions of 30°, 45°, and 60°. These exact values appear in a number of problems throughout the chapter, and the more quickly you can recall them, the easier it will be for you to work the problems involving them.

SECTION 5.1 **Proving Identities**

We began proving identities in Chapter 1. In this section, we will extend the work we did in Chapter 1 to include proving more complicated identities. For review, here are the basic identities and some of their more important equivalent forms.

Table 1

	Basic identities	**Common equivalent forms**
Reciprocal	$\csc \theta = \dfrac{1}{\sin \theta}$	$\sin \theta = \dfrac{1}{\csc \theta}$
	$\sec \theta = \dfrac{1}{\cos \theta}$	$\cos \theta = \dfrac{1}{\sec \theta}$
	$\cot \theta = \dfrac{1}{\tan \theta}$	$\tan \theta = \dfrac{1}{\cot \theta}$
Ratio	$\tan \theta = \dfrac{\sin \theta}{\cos \theta}$	
	$\cot \theta = \dfrac{\cos \theta}{\sin \theta}$	
Pythagorean	$\cos^2 \theta + \sin^2 \theta = 1$	$\sin^2 \theta = 1 - \cos^2 \theta$
		$\sin \theta = \pm\sqrt{1 - \cos^2 \theta}$
		$\cos^2 \theta = 1 - \sin^2 \theta$
		$\cos \theta = \pm\sqrt{1 - \sin^2 \theta}$
	$1 + \tan^2 \theta = \sec^2 \theta$	
	$1 + \cot^2 \theta = \csc^2 \theta$	

Note The last two Pythagorean identities can be derived from $\cos^2 \theta + \sin^2 \theta = 1$ by dividing each side by $\cos^2 \theta$ and $\sin^2 \theta$, respectively. For example, if we divide each side of $\cos^2 \theta + \sin^2 \theta$ by $\cos^2 \theta$, we have

$$\cos^2 \theta + \sin^2 \theta = 1$$

$$\frac{\cos^2 \theta + \sin^2 \theta}{\cos^2 \theta} = \frac{1}{\cos^2 \theta}$$

$$\frac{\cos^2 \theta}{\cos^2 \theta} + \frac{\sin^2 \theta}{\cos^2 \theta} = \frac{1}{\cos^2 \theta}$$

$$1 + \tan^2 \theta = \sec^2 \theta$$

To derive the last Pythagorean identity, we would need to divide both sides of $\cos^2 \theta + \sin^2 \theta = 1$ by $\sin^2 \theta$ to obtain $1 + \cot^2 \theta = \csc^2 \theta$.

The rest of this section is concerned with using the basic identities (or their equivalent forms) listed above, along with our knowledge of algebra, to prove other identities.

Recall that an identity in trigonometry is a statement that two expressions are equal for all replacements of the variable for which each expression is defined. To prove (or verify) a trigonometric identity, we use trigonometric substitutions and algebraic manipulations to either

1. Transform the right side of the identity into the left side, or
2. Transform the left side of the identity into the right side.

The main thing to remember in proving identities is to work on each side of the identity separately. We do not want to use properties from algebra that involve both sides of the identity—like the addition property of equality. We prove identities in order to develop the ability to transform one trigonometric expression into another. When we encounter problems in other courses that require the use of the techniques used to verify identities, we usually find that the solution to these problems hinges on transforming an expression containing trigonometric functions into less complicated expressions. In these cases, we do not usually have an equal sign to work with.

▶ **EXAMPLE 1**   Prove $\sin \theta \cot \theta = \cos \theta$.

Proof   To prove this identity, we transform the left side into the right side.

$$\sin \theta \cot \theta = \sin \theta \cdot \frac{\cos \theta}{\sin \theta} \qquad \text{Ratio identity}$$

$$= \frac{\sin \theta \cos \theta}{\sin \theta} \qquad \text{Multiply}$$

$$= \cos \theta \qquad \text{Divide out common factor } \sin \theta$$

In this example, we have transformed the left side into the right side. Remember, we verify identities by transforming one expression into another.   ◀

▶ **EXAMPLE 2**   Prove $\tan x + \cos x = \sin x (\sec x + \cot x)$.

Proof   We can begin by applying the distributive property to the right side to multiply through by $\sin x$. Then we can change each expression on the right side to an equivalent expression involving only $\sin x$ and $\cos x$.

$$\sin x (\sec x + \cot x) = \sin x \sec x + \sin x \cot x \qquad \text{Multiply}$$

$$= \sin x \cdot \frac{1}{\cos x} + \sin x \cdot \frac{\cos x}{\sin x} \qquad \begin{array}{l}\text{Reciprocal and} \\ \text{ratio identities}\end{array}$$

$$= \frac{\sin x}{\cos x} + \cos x \qquad \text{Multiply}$$

$$= \tan x + \cos x \qquad \text{Ratio identity}$$

In this case, we transformed the right side into the left side.   ◀

Before we go on to the next example, let's list some guidelines that may be useful in learning how to prove identities.

Guidelines for Proving Identities

1. It is usually best to work on the more complicated side first.
2. Look for trigonometric substitutions involving the basic identities that may help simplify things.
3. Look for algebraic operations, such as adding fractions, the distributive property, or factoring, that may simplify the side you are working with or that will at least lead to an expression that will be easier to simplify.
4. If you cannot think of anything else to do, change everything to sines and cosines and see if that helps.
5. Always keep an eye on the side you are not working with to be sure you are working toward it. There is a certain sense of direction that accompanies a successful proof.

Probably the best advice is to remember that these are simply guidelines. The best way to become proficient at proving trigonometric identities is to practice. The more identities you prove, the more you will able to prove and the more confident you will become. *Don't be afraid to stop and start over if you don't seem to be getting anywhere.* With most identities, there are a number of different proofs that will lead to the same result. Some of the proofs will be longer than others.

▶ **EXAMPLE 3** Prove $\dfrac{\cos^4 t - \sin^4 t}{\cos^2 t} = 1 - \tan^2 t$.

Proof In this example, factoring the numerator on the left side will reduce the exponents there from 4 to 2.

$$\frac{\cos^4 t - \sin^4 t}{\cos^2 t} = \frac{(\cos^2 t + \sin^2 t)(\cos^2 t - \sin^2 t)}{\cos^2 t} \qquad \text{Factor}$$

$$= \frac{1\,(\cos^2 t - \sin^2 t)}{\cos^2 t} \qquad \text{Pythagorean identity}$$

$$= \frac{\cos^2 t}{\cos^2 t} - \frac{\sin^2 t}{\cos^2 t} \qquad \text{Separate into two fractions}$$

$$= 1 - \tan^2 t \qquad \text{Ratio identity} \qquad ◀$$

▶ **EXAMPLE 4** Prove $1 + \cos \theta = \dfrac{\sin^2 \theta}{1 - \cos \theta}$.

Proof We begin this proof by applying an alternate form of the Pythagorean identity to the right side to write $\sin^2 \theta$ as $1 - \cos^2 \theta$. Then we factor $1 - \cos^2 \theta$ as the difference of two squares and reduce to lowest terms.

$$\frac{\sin^2 \theta}{1 - \cos \theta} = \frac{1 - \cos^2 \theta}{1 - \cos \theta} \qquad \text{Pythagorean identity}$$

$$= \frac{(1 - \cos \theta)(1 + \cos \theta)}{1 - \cos \theta} \qquad \text{Factor}$$

$$= 1 + \cos \theta \qquad \text{Reduce} \qquad ◀$$

▶ **EXAMPLE 5** Prove $\tan x + \cot x = \sec x \csc x$.

Proof We begin this proof by writing the left side in terms of $\sin x$ and $\cos x$. Then we simplify the left side by finding a common denominator in order to add the resulting fractions.

$$\tan x + \cot x = \frac{\sin x}{\cos x} + \frac{\cos x}{\sin x}$$ Change to sines and cosines

$$= \frac{\sin x}{\cos x} \cdot \frac{\sin x}{\sin x} + \frac{\cos x}{\sin x} \cdot \frac{\cos x}{\cos x}$$ LCD

$$= \frac{\sin^2 x + \cos^2 x}{\cos x \sin x}$$ Add fractions

$$= \frac{1}{\cos x \sin x}$$ Pythagorean identity

$$= \frac{1}{\cos x} \cdot \frac{1}{\sin x}$$ Write as separate fractions

$$= \sec x \csc x$$ Reciprocal identities ◀

▶ **EXAMPLE 6** Prove $\dfrac{\sin \alpha}{1 + \cos \alpha} + \dfrac{1 + \cos \alpha}{\sin \alpha} = 2 \csc \alpha$.

Proof The common denominator for the left side of the equation is $\sin \alpha (1 + \cos \alpha)$. We multiply the first fraction by $(\sin \alpha)/(\sin \alpha)$ and the second fraction by $(1 + \cos \alpha)/(1 + \cos \alpha)$ to produce two equivalent fractions with the same denominator.

$$\frac{\sin \alpha}{1 + \cos \alpha} + \frac{1 + \cos \alpha}{\sin \alpha}$$

$$= \frac{\sin \alpha}{\sin \alpha} \cdot \frac{\sin \alpha}{1 + \cos \alpha} + \frac{1 + \cos \alpha}{\sin \alpha} \cdot \frac{1 + \cos \alpha}{1 + \cos \alpha}$$ LCD

$$= \frac{\sin^2 \alpha + (1 + \cos \alpha)^2}{\sin \alpha (1 + \cos \alpha)}$$ Add numerators

$$= \frac{\sin^2 \alpha + 1 + 2 \cos \alpha + \cos^2 \alpha}{\sin \alpha (1 + \cos \alpha)}$$ Expand $(1 + \cos \alpha)^2$

$$= \frac{2 + 2 \cos \alpha}{\sin \alpha (1 + \cos \alpha)}$$ Pythagorean identity

$$= \frac{2(1 + \cos \alpha)}{\sin \alpha (1 + \cos \alpha)}$$ Factor out a 2

$$= \frac{2}{\sin \alpha}$$ Reduce

$$= 2 \csc \alpha$$ Reciprocal identity ◀

▶ **EXAMPLE 7** Prove $\dfrac{1 + \sin t}{\cos t} = \dfrac{\cos t}{1 - \sin t}$.

Proof The trick to proving this identity requires that we multiply the numerator and denominator on the right side by $1 + \sin t$. (This is similar to rationalizing the denominator.)

$$\frac{\cos t}{1 - \sin t} = \frac{\cos t}{1 - \sin t} \cdot \frac{1 + \sin t}{1 + \sin t} \qquad \text{Multiply numerator and denominator by } 1 + \sin t$$

$$= \frac{\cos t \,(1 + \sin t)}{1 - \sin^2 t} \qquad \text{Multiply out the denominator}$$

$$= \frac{\cos t \,(1 + \sin t)}{\cos^2 t} \qquad \text{Pythagorean identity}$$

$$= \frac{1 + \sin t}{\cos t} \qquad \text{Reduce}$$

Note that it would have been just as easy for us to verify this identity by multiplying the numerator and denominator on the left side by $1 - \sin t$. ◀

PROBLEM SET 5.1

Prove that each of the following identities is true:

1. $\cos \theta \tan \theta = \sin \theta$

2. $\sec \theta \cot \theta = \csc \theta$

3. $\csc \theta \tan \theta = \sec \theta$

4. $\tan \theta \cot \theta = 1$

5. $\dfrac{\tan A}{\sec A} = \sin A$

6. $\dfrac{\cot A}{\csc A} = \cos A$

7. $\sec \theta \cot \theta \sin \theta = 1$

8. $\tan \theta \csc \theta \cos \theta = 1$

9. $\cos x \,(\csc x + \tan x) = \cot x + \sin x$

10. $\sin x \,(\sec x + \csc x) = \tan x + 1$

11. $\cot x - 1 = \cos x \,(\csc x - \sec x)$

12. $\tan x \,(\cos x + \cot x) = \sin x + 1$

13. $\cos^2 x \,(1 + \tan^2 x) = 1$

14. $\sin^2 x \,(\cot^2 x + 1) = 1$

15. $(1 - \sin x)(1 + \sin x) = \cos^2 x$

16. $(1 - \cos x)(1 + \cos x) = \sin^2 x$

17. $\dfrac{\cos^4 t - \sin^4 t}{\sin^2 t} = \cot^2 t - 1$

18. $\dfrac{\sin^4 t - \cos^4 t}{\sin^2 t \cos^2 t} = \sec^2 t - \csc^2 t$

19. $1 + \sin \theta = \dfrac{\cos^2 \theta}{1 - \sin \theta}$

20. $1 - \sin \theta = \dfrac{\cos^2 \theta}{1 + \sin \theta}$

21. $\dfrac{1 - \sin^4 \theta}{1 + \sin^2 \theta} = \cos^2 \theta$

22. $\dfrac{1 - \cos^4 \theta}{1 + \cos^2 \theta} = \sin^2 \theta$

23. $\sec^2 \theta - \tan^2 \theta = 1$

24. $\csc^2 \theta - \cot^2 \theta = 1$

25. $\sec^4 \theta - \tan^4 \theta = \dfrac{1 + \sin^2 \theta}{\cos^2 \theta}$

26. $\csc^4 \theta - \cot^4 \theta = \dfrac{1 + \cos^2 \theta}{\sin^2 \theta}$

27. $\tan \theta - \cot \theta = \dfrac{\sin^2 \theta - \cos^2 \theta}{\sin \theta \cos \theta}$

28. $\sec \theta - \csc \theta = \dfrac{\sin \theta - \cos \theta}{\sin \theta \cos \theta}$

29. $\csc B - \sin B = \cot B \cos B$

30. $\sec B - \cos B = \tan B \sin B$

31. $\cot \theta \cos \theta + \sin \theta = \csc \theta$

32. $\tan \theta \sin \theta + \cos \theta = \sec \theta$

33. $\dfrac{\cos x}{1 + \sin x} + \dfrac{1 + \sin x}{\cos x} = 2 \sec x$

34. $\dfrac{\cos x}{1 + \sin x} - \dfrac{1 - \sin x}{\cos x} = 0$

35. $\dfrac{1}{1 + \cos x} + \dfrac{1}{1 - \cos x} = 2 \csc^2 x$

36. $\dfrac{1}{1 - \sin x} + \dfrac{1}{1 + \sin x} = 2 \sec^2 x$

37. $\dfrac{1 - \sec x}{1 + \sec x} = \dfrac{\cos x - 1}{\cos x + 1}$

38. $\dfrac{\csc x - 1}{\csc x + 1} = \dfrac{1 - \sin x}{1 + \sin x}$

39. $\dfrac{\cos t}{1 + \sin t} = \dfrac{1 - \sin t}{\cos t}$

40. $\dfrac{\sin t}{1 + \cos t} = \dfrac{1 - \cos t}{\sin t}$

41. $\dfrac{(1 - \sin t)^2}{\cos^2 t} = \dfrac{1 - \sin t}{1 + \sin t}$

42. $\dfrac{\sin^2 t}{(1 - \cos t)^2} = \dfrac{1 + \cos t}{1 - \cos t}$

43. $\dfrac{\sec \theta + 1}{\tan \theta} = \dfrac{\tan \theta}{\sec \theta - 1}$

44. $\dfrac{\csc \theta - 1}{\cot \theta} = \dfrac{\cot \theta}{\csc \theta + 1}$

45. $\dfrac{1 - \sin x}{1 + \sin x} = (\sec x - \tan x)^2$

46. $\dfrac{1 + \cos x}{1 - \cos x} = (\csc x + \cot x)^2$

47. $\sec x + \tan x = \dfrac{1}{\sec x - \tan x}$

48. $\dfrac{1}{\csc x - \cot x} = \csc x + \cot x$

49. $\dfrac{\sin x + 1}{\cos x + \cot x} = \tan x$

50. $\dfrac{\cos x + 1}{\cot x} = \sin x + \tan x$

51. $\sin^4 A - \cos^4 A = 1 - 2 \cos^2 A$

52. $\cos^4 A - \sin^4 A = 1 - 2 \sin^2 A$

53. $\dfrac{\sin^2 B - \tan^2 B}{1 - \sec^2 B} = \sin^2 B$

54. $\dfrac{\cot^2 B - \cos^2 B}{\csc^2 B - 1} = \cos^2 B$

55. $\dfrac{\sec^4 y - \tan^4 y}{\sec^2 y + \tan^2 y} = 1$

56. $\dfrac{\csc^2 y + \cot^2 y}{\csc^4 y - \cot^4 y} = 1$

57. $\dfrac{\sin^3 A - 8}{\sin A - 2} = \sin^2 A + 2 \sin A + 4$

58. $\dfrac{1 - \cos^3 A}{1 - \cos A} = \cos^2 A + \cos A + 1$

59. $\dfrac{1 - \tan^3 t}{1 - \tan t} = \sec^2 t + \tan t$

60. $\dfrac{1 + \cot^3 t}{1 + \cot t} = \csc^2 t - \cot t$

61. $\dfrac{\sec B}{\sin B + 1} = \dfrac{1 - \sin B}{\cos^3 B}$

62. $\dfrac{1 - \cos B}{\csc B} = \dfrac{\sin^3 B}{1 + \cos B}$

63. $\dfrac{\tan x}{\sin x - \cos x} = \dfrac{\sin^2 x + \sin x \cos x}{\cos x - 2 \cos^3 x}$

64. $\dfrac{\cot^2 x}{\sin x + \cos x} = \dfrac{\cos^2 x \sin x - \cos^3 x}{2 \sin^4 x - \sin^2 x}$

The following two identities appeared in the book *Practical Mathematics* published in 1945 by the National Educational Alliance. Verify each identity.

65. $\dfrac{1}{\tan \theta + \cot \theta} = \sin \theta \cos \theta$

66. $\dfrac{\sin \theta + \tan \theta}{1 + \cos \theta} = \tan \theta$

The identities below are from the book *Plane and Spherical Trigonometry* written by Leonard M. Passano and published by The Macmillan Company in 1918. Verify each identity.

67. $\dfrac{\sec^2 \alpha}{\sin^2 \alpha} = \csc^2 \alpha + \sec^2 \alpha$

68. $\sec^2 x \csc^2 x = \sec^2 x + \csc^2 x$

69. $(1 + \sec x)(1 - \cos x) = \tan^2 x \cos x$

70. $\cot^2 \theta - \cos^2 \theta = \cot^2 \theta \cos^2 \theta$

The following identities are from the book *Plane and Spherical Trigonometry with Tables* by Rosenbach, Whitman, and Moskovitz, and published by Ginn and Company in 1937. Verify each identity.

71. $(\tan \theta + \cot \theta)^2 = \sec^2 \theta + \csc^2 \theta$

72. $\dfrac{\tan^2 \psi + 2}{1 + \tan^2 \psi} = 1 + \cos^2 \psi$

73. $\dfrac{1 + \sin \phi}{1 - \sin \phi} - \dfrac{1 - \sin \phi}{1 + \sin \phi} = 4 \tan \phi \sec \phi$

74. $\dfrac{\cos \beta}{1 - \tan \beta} + \dfrac{\sin \beta}{1 - \cot \beta} = \sin \beta + \cos \beta$

Show that each of the following statements is not an identity by finding a value of θ that makes the statement false.

75. $\sin \theta = \sqrt{1 - \cos^2 \theta}$

76. $\sin \theta + \cos \theta = 1$

77. $\sin \theta = \dfrac{1}{\cos \theta}$

78. $\tan^2 \theta + \cot^2 \theta = 1$

79. $\sqrt{\sin^2 \theta + \cos^2 \theta} = \sin \theta + \cos \theta$

80. $\sin \theta \cos \theta = 1$

81. Show that $\sin (A + B)$ is not, in general, equal to $\sin A + \sin B$ by substituting 30° for A and 60° for B in both expressions and simplifying. (An example like this, which shows that a statement is not true for certain values of the variables, is called a *counterexample* for the statement.)

82. Show that $\sin 2x \neq 2 \sin x$ by substituting 30° for x and then simplifying both sides.

Review Problems

The problems that follow review material we covered in Sections 1.4 and 3.2. Reviewing these problems will help you with some of the material in the next section.

83. If $\sin A = 3/5$ and A terminates in quadrant I, find $\cos A$ and $\tan A$.

84. If $\cos B = -5/13$ with B in quadrant III, find $\sin B$ and $\tan B$.

Give the exact value of each of the following:

85. $\sin \dfrac{\pi}{3}$ **86.** $\cos \dfrac{\pi}{3}$ **87.** $\cos \dfrac{\pi}{6}$ **88.** $\sin \dfrac{\pi}{6}$

Convert to degrees.

89. $\pi/12$ **90.** $5\pi/12$ **91.** $7\pi/12$ **92.** $11\pi/12$

SECTION 5.2   Sum and Difference Formulas

The expressions $\sin (A + B)$ and $\cos (A + B)$ occur frequently enough in mathematics that it is necessary to find expressions equivalent to them that involve sines and cosines of single angles. The most obvious question to begin with is

$$\text{Is } \sin (A + B) = \sin A + \sin B?$$

The answer is no. Substituting almost any pair of numbers for A and B in the formula will yield a false statement. As a counterexample, we can let $A = 30°$ and $B = 60°$ in the above formula and then simplify each side. (A counterexample is an example that shows that a statement is not, in general, true.)

$$\sin (30° + 60°) \stackrel{?}{=} \sin 30° + \sin 60°$$

$$\sin 90° \stackrel{?}{=} \frac{1}{2} + \frac{\sqrt{3}}{2}$$

$$1 \neq \frac{1 + \sqrt{3}}{2}$$

The formula just doesn't work. The next question is, what are the formulas for $\sin (A + B)$ and $\cos (A + B)$? The answer to that question is what this section is all about. Let's start by deriving the formula for $\cos (A + B)$.

We begin by drawing angle A in standard position and then adding B and $-B$ to it. Figure 1 shows these angles in relation to the unit circle. Note that the points on the unit circle through which the terminal sides of the angles A, $A + B$, and $-B$ pass have been labeled with the sines and cosines of those angles.

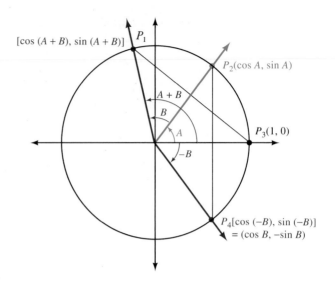

Figure 1

To derive the formula for cos $(A + B)$, we simply have to see that line segment P_1P_3 is equal to line segment P_2P_4. (From geometry, they are chords cut off by equal central angles.)

$$P_1P_3 = P_2P_4$$

Squaring both sides gives us

$$(P_1P_3)^2 = (P_2P_4)^2$$

Now, applying the distance formula, we have

$$[\cos (A + B) - 1]^2 + [\sin (A + B) - 0]^2 = (\cos A - \cos B)^2 + (\sin A + \sin B)^2$$

Let's call this Equation 1. Taking the left side of Equation 1, expanding it, and then simplifying by using the first Pythagorean identity gives us

Left Side of Equation 1

$$\cos^2 (A + B) - 2 \cos (A + B) + 1 + \sin^2 (A + B) \qquad \text{Expand squares}$$
$$= -2 \cos (A + B) + 2 \qquad \text{Pythagorean identity}$$

Applying the same two steps to the right side of Equation 1 gives us

Right Side of Equation 1

$$\cos^2 A - 2 \cos A \cos B + \cos^2 B + \sin^2 A + 2 \sin A \sin B + \sin^2 B$$
$$= -2 \cos A \cos B + 2 \sin A \sin B + 2$$

Equating the simplified versions of the left and right sides of Equation 1 we have

$$-2 \cos (A + B) + 2 = -2 \cos A \cos B + 2 \sin A \sin B + 2$$

Adding -2 to both sides and then dividing both sides by -2 gives us the formula we are after:

$$\cos (A + B) = \cos A \cos B - \sin A \sin B$$

This is the first formula in a series of formulas for trigonometric functions of the sum or difference of two angles. It must be memorized. Before we derive the others, let's look at some of the ways we can use our first formula.

▶ **EXAMPLE 1** Find the exact value for cos 75°.

Solution We write 75° as 45° + 30° and then apply the formula for $\cos (A + B)$.

$$
\begin{aligned}
\cos 75° &= \cos (45° + 30°) \\
&= \cos 45° \cos 30° - \sin 45° \sin 30° \\
&= \frac{\sqrt{2}}{2} \cdot \frac{\sqrt{3}}{2} - \frac{\sqrt{2}}{2} \cdot \frac{1}{2} \\
&= \frac{\sqrt{6} - \sqrt{2}}{4}
\end{aligned}
$$

◀

▶ **EXAMPLE 2** Show that $\cos (x + 2\pi) = \cos x$.

Solution Applying the formula for $\cos (A + B)$, we have

$$
\begin{aligned}
\cos (x + 2\pi) &= \cos x \cos 2\pi - \sin x \sin 2\pi \\
&= \cos x \cdot 1 - \sin x \cdot 0 \\
&= \cos x
\end{aligned}
$$

Notice that this is not a new relationship. We already know that if two angles are coterminal, then their cosines are equal—and $x + 2\pi$ and x are coterminal. What we have done here is shown this to be true with a formula instead of the definition of cosine. ◀

▶ **EXAMPLE 3** Write $\cos 3x \cos 2x - \sin 3x \sin 2x$ as a single cosine.

Solution We apply the formula for $\cos (A + B)$ in the reverse direction from the way we applied it in the first two examples.

$$
\begin{aligned}
\cos 3x \cos 2x - \sin 3x \sin 2x &= \cos (3x + 2x) \\
&= \cos 5x
\end{aligned}
$$

◀

Here is the derivation of the formula for $\cos (A - B)$. It involves the formula for $\cos (A + B)$ and the formulas for even and odd functions.

$$
\begin{aligned}
\cos (A - B) &= \cos [A + (-B)] && \text{Write } A - B \text{ as a sum} \\
&= \cos A \cos (-B) - \sin A \sin (-B) && \text{Sum formula} \\
&= \cos A \cos B - \sin A (-\sin B) && \text{Cosine is an even} \\
& && \text{function, sine is odd} \\
&= \cos A \cos B + \sin A \sin B
\end{aligned}
$$

The only difference in the formulas for the expansion of $\cos (A + B)$ and $\cos (A - B)$ is the sign between the two terms. Here are both formulas again.

$$\cos (A + B) = \cos A \cos B - \sin A \sin B$$
$$\cos (A - B) = \cos A \cos B + \sin A \sin B$$

Again, both formulas are important and should be memorized.

▶ **EXAMPLE 4** Show that $\cos (90° - A) = \sin A$.

Solution We will need this formula when we derive the formula for $\sin (A + B)$.

$$\cos (90° - A) = \cos 90° \cos A + \sin 90° \sin A$$
$$= 0 \cdot \cos A + 1 \cdot \sin A$$
$$= \sin A$$
◀

Note that the formula we just derived is not a new formula. The angles $90° - A$ and A are complementary angles and we already know the sine of an angle is always equal to the cosine of its complement. We could also state it this way:

$$\sin (90° - A) = \cos A$$

We can use this information to derive the formula for $\sin (A + B)$. To understand this derivation, you must recognize that $A + B$ and $90° - (A + B)$ are complementary angles.

$$\sin (A + B) = \cos [90° - (A + B)] \qquad \text{The sine of an angle is the cosine of its complement}$$
$$= \cos [90° - A - B] \qquad \text{Remove parentheses}$$
$$= \cos [(90° - A) - B] \qquad \text{Regroup within brackets}$$

Now we expand using the formula for the cosine of a difference

$$= \cos (90° - A) \cos B + \sin (90° - A) \sin B$$
$$= \sin A \cos B + \cos A \sin B$$

This gives us an expansion formula for $\sin (A + B)$.

$$\sin(A + B) = \sin A \cos B + \cos A \sin B$$

This is the formula for the sine of a sum. To find the formula for $\sin (A - B)$, we write $A - B$ as $A + (-B)$ and proceed as follows:

$$\sin (A - B) = \sin [A + (-B)]$$
$$= \sin A \cos (-B) + \cos A \sin (-B)$$
$$= \sin A \cos B - \cos A \sin B$$

This gives us the formula for the sine of a difference.

$$\sin (A - B) = \sin A \cos B - \cos A \sin B$$

▶ **EXAMPLE 5**   Graph $y = 4 \sin 5x \cos 3x - 4 \cos 5x \sin 3x$ from $x = 0$ to $x = 2\pi$.

Solution   To write the equation in the form $y = A \sin Bx$, we factor 4 from each term on the right and then apply the formula for $\sin (A - B)$ to the remaining expression to write it as a single trigonometric function.

$$
\begin{aligned}
y &= 4 \sin 5x \cos 3x - 4 \cos 5x \sin 3x \\
&= 4 (\sin 5x \cos 3x - \cos 5x \sin 3x) \\
&= 4 \sin (5x - 3x) \\
&= 4 \sin 2x
\end{aligned}
$$

The graph of $y = 4 \sin 2x$ will have an amplitude of 4 and a period of $2\pi/2 = \pi$. The graph is shown in Figure 2.

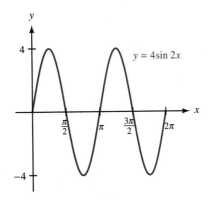

Figure 2   ◀

▶ **EXAMPLE 6**   Find the exact value of $\sin \dfrac{\pi}{12}$.

Solution   We have to write $\pi/12$ in terms of two numbers the exact values of which are known. The numbers $\pi/3$ and $\pi/4$ will work since their difference is $\pi/12$.

$$
\begin{aligned}
\sin \frac{\pi}{12} &= \sin \left(\frac{\pi}{3} - \frac{\pi}{4} \right) \\
&= \sin \frac{\pi}{3} \cos \frac{\pi}{4} - \cos \frac{\pi}{3} \sin \frac{\pi}{4} \\
&= \frac{\sqrt{3}}{2} \cdot \frac{\sqrt{2}}{2} - \frac{1}{2} \cdot \frac{\sqrt{2}}{2} \\
&= \frac{\sqrt{6} - \sqrt{2}}{4}
\end{aligned}
$$

This is the same answer we obtained in Example 1 when we found the exact value of cos 75°. It should be, though, because $\pi/12 = 15°$, which is the complement of 75°, and the cosine of an angle is equal to the sine of its complement. ◀

▶ **EXAMPLE 7**   If $\sin A = \frac{3}{5}$ with A in QI and $\cos B = -\frac{5}{13}$ with B in QIII, find $\sin (A + B)$, $\cos (A + B)$, and $\tan (A + B)$.

Solution   We have $\sin A$ and $\cos B$. We need to find $\cos A$ and $\sin B$ before we can apply any of our formulas. Some equivalent forms of our first Pythagorean identity will help here.

If $\sin A = \frac{3}{5}$ with A in QI, then

$$\cos A = \sqrt{1 - \sin^2 A} = \sqrt{1 - \left(\frac{3}{5}\right)^2} = \frac{4}{5}$$

If $\cos B = -\frac{5}{13}$ with B in QIII, then

$$\sin B = -\sqrt{1 - \left(-\frac{5}{13}\right)^2} = -\frac{12}{13}$$

We have

$$\sin A = \frac{3}{5} \qquad \sin B = -\frac{12}{13}$$

$$\cos A = \frac{4}{5} \qquad \cos B = -\frac{5}{13}$$

Therefore,

$$\sin (A + B) = \sin A \cos B + \cos A \sin B$$
$$= \frac{3}{5}\left(-\frac{5}{13}\right) + \frac{4}{5}\left(-\frac{12}{13}\right)$$
$$= -\frac{63}{65}$$

$$\cos (A + B) = \cos A \cos B - \sin A \sin B$$
$$= \frac{4}{5}\left(-\frac{5}{13}\right) - \frac{3}{5}\left(-\frac{12}{13}\right)$$
$$= \frac{16}{65}$$

$$\tan (A + B) = \frac{\sin (A + B)}{\cos (A + B)}$$
$$= \frac{-63/65}{16/65}$$
$$= -\frac{63}{16}$$

◀

Notice also that $A + B$ must terminate in quadrant IV because

$$\sin (A + B) < 0 \quad \text{and} \quad \cos (A + B) > 0$$

While working through the last part of Example 7, you may have wondered if there is a separate formula for $\tan (A + B)$. (More likely, you are hoping there isn't.) There is, and it is derived from the formulas we already have.

$$\tan (A + B) = \frac{\sin (A + B)}{\cos (A + B)}$$

$$= \frac{\sin A \cos B + \cos A \sin B}{\cos A \cos B - \sin A \sin B}$$

To be able to write this last line in terms of tangents only, we must divide numerator and denominator by $\cos A \cos B$.

$$= \frac{\dfrac{\sin A \cos B}{\cos A \cos B} + \dfrac{\cos A \sin B}{\cos A \cos B}}{\dfrac{\cos A \cos B}{\cos A \cos B} - \dfrac{\sin A \sin B}{\cos A \cos B}}$$

$$= \frac{\tan A + \tan B}{1 - \tan A \tan B}$$

The formula for $\tan (A + B)$ is

$$\boxed{\tan (A + B) = \frac{\tan A + \tan B}{1 - \tan A \tan B}}$$

Since tangent is an odd function, the formula for $\tan (A - B)$ will look like this:

$$\boxed{\tan (A - B) = \frac{\tan A - \tan B}{1 + \tan A \tan B}}$$

▶ **EXAMPLE 8**   If $\sin A = \frac{3}{5}$ with A in QI and $\cos B = -\frac{5}{13}$ with B in QIII, find $\tan (A + B)$ by using the formula

$$\tan (A + B) = \frac{\tan A + \tan B}{1 - \tan A \tan B}$$

Solution   The angles A and B as given here are the same ones used previously in Example 7. Looking over Example 7 again, we find that

$$\tan A = \frac{3}{4} \quad \text{and} \quad \tan B = \frac{12}{5}$$

Therefore,

$$\tan (A + B) = \frac{\tan A + \tan B}{1 - \tan A \tan B}$$

$$= \frac{\dfrac{3}{4} + \dfrac{12}{5}}{1 - \dfrac{3}{4} \cdot \dfrac{12}{5}}$$

$$= \frac{\dfrac{15}{20} + \dfrac{48}{20}}{1 - \dfrac{9}{5}}$$

$$= \frac{\dfrac{63}{20}}{-\dfrac{4}{5}}$$

$$= -\frac{63}{16}$$

which is the same result we obtained previously. ◀

PROBLEM SET 5.2

Find exact values for each of the following:

1. $\sin 15°$ | **2.** $\sin 75°$
3. $\tan 15°$ | **4.** $\tan 75°$

5. $\sin \dfrac{7\pi}{12}$ | **6.** $\cos \dfrac{7\pi}{12}$

7. $\cos 105°$ | **8.** $\sin 105°$

Show that each of the following is true:

9. $\sin (x + 2\pi) = \sin x$ | **10.** $\cos (x - 2\pi) = \cos x$

11. $\cos \left(x - \dfrac{\pi}{2} \right) = \sin x$ | **12.** $\sin \left(x - \dfrac{\pi}{2} \right) = -\cos x$

13. $\cos (180° - \theta) = -\cos \theta$ | **14.** $\sin (180° - \theta) = \sin \theta$
15. $\sin (90° + \theta) = \cos \theta$ | **16.** $\cos (90° + \theta) = -\sin \theta$

17. $\tan \left(x + \dfrac{\pi}{4} \right) = \dfrac{1 + \tan x}{1 - \tan x}$ | **18.** $\tan \left(x - \dfrac{\pi}{4} \right) = \dfrac{\tan x - 1}{\tan x + 1}$

19. $\sin \left(\dfrac{3\pi}{2} - x \right) = -\cos x$ | **20.** $\cos \left(x - \dfrac{3\pi}{2} \right) = -\sin x$

Write each expression as a single trigonometric function.

21. $\sin 3x \cos 2x + \cos 3x \sin 2x$
22. $\cos 3x \cos 2x + \sin 3x \sin 2x$
23. $\cos 5x \cos x - \sin 5x \sin x$
24. $\sin 8x \cos x - \cos 8x \sin x$
25. $\cos 15° \cos 75° - \sin 15° \sin 75°$
26. $\cos 15° \cos 75° + \sin 15° \sin 75°$

Graph each of the following from $x = 0$ to $x = 2\pi$.

27. $y = \sin 5x \cos 3x - \cos 5x \sin 3x$
28. $y = \sin x \cos 2x + \cos x \sin 2x$
29. $y = 3 \cos 7x \cos 5x + 3 \sin 7x \sin 5x$
30. $y = 2 \cos 4x \cos x + 2 \sin 4x \sin x$

31. Graph one complete cycle of $y = \sin x \cos \pi/4 + \cos x \sin \pi/4$ by first rewriting the right side in the form $\sin (A + B)$.

32. Graph one complete cycle of $y = \sin x \cos \pi/6 - \cos x \sin \pi/6$ by first rewriting the right side in the form $\sin (A - B)$.

33. Graph one complete cycle of $y = 2 (\sin x \cos \pi/3 + \cos x \sin \pi/3)$ by first rewriting the right side in the form $2 \sin (A + B)$.

34. Graph one complete cycle of $y = 2 (\sin x \cos \pi/3 - \cos x \sin \pi/3)$ by first rewriting the right side in the form $2 \sin (A - B)$.

35. Let $\sin A = \frac{3}{5}$ with A in QII and $\sin B = -\frac{5}{13}$ with B in QIII. Find $\sin (A + B)$, $\cos (A + B)$, and $\tan (A + B)$. In what quadrant does $A + B$ terminate?

36. Let $\cos A = -\frac{5}{13}$ with A in QII and $\sin B = \frac{3}{5}$ with B in QI. Find $\sin (A - B)$, $\cos (A - B)$, and $\tan (A - B)$. In what quadrant does $A - B$ terminate?

37. If $\sin A = 1/\sqrt{5}$ with A in QI and $\tan B = \frac{3}{4}$ with B in QI, find $\tan (A + B)$ and $\cot (A + B)$. In what quadrant does $A + B$ terminate?

38. If $\sec A = \sqrt{5}$ with A in QI and $\sec B = \sqrt{10}$ with B in QI, find $\sec (A + B)$. [First find $\cos (A + B)$.]

39. If $\tan (A + B) = 3$ and $\tan B = \frac{1}{2}$, find $\tan A$.

40. If $\tan (A + B) = 2$ and $\tan B = \frac{1}{3}$, find $\tan A$.

41. Write a formula for $\sin 2x$ by writing $\sin 2x$ as $\sin (x + x)$ and using the formula for the sine of a sum.

42. Write a formula for $\cos 2x$ by writing $\cos 2x$ as $\cos (x + x)$ and using the formula for the cosine of a sum.

Prove each identity.

43. $\sin (90° + x) + \sin (90° - x) = 2 \cos x$
44. $\sin (90° + x) - \sin (90° - x) = 0$
45. $\cos (x - 90°) - \cos (x + 90°) = 2 \sin x$
46. $\cos (x + 90°) + \cos (x - 90°) = 0$

47. $\sin \left(\dfrac{\pi}{6} + x \right) + \sin \left(\dfrac{\pi}{6} - x \right) = \cos x$

48. $\cos \left(\dfrac{\pi}{3} + x \right) + \cos \left(\dfrac{\pi}{3} - x \right) = \cos x$

49. $\cos \left(x + \dfrac{\pi}{4} \right) + \cos \left(x - \dfrac{\pi}{4} \right) = \sqrt{2} \cos x$

50. $\sin\left(\dfrac{\pi}{4} + x\right) + \sin\left(\dfrac{\pi}{4} - x\right) = \sqrt{2}\cos x$

51. $\sin\left(\dfrac{3\pi}{2} + x\right) + \sin\left(\dfrac{3\pi}{2} - x\right) = -2\cos x$

52. $\cos\left(x + \dfrac{3\pi}{2}\right) + \cos\left(x - \dfrac{3\pi}{2}\right) = 0$

53. $\sin(A + B) + \sin(A - B) = 2\sin A \cos B$

54. $\cos(A + B) + \cos(A - B) = 2\cos A \cos B$

55. $\dfrac{\sin(A - B)}{\cos A \cos B} = \tan A - \tan B$

56. $\dfrac{\cos(A + B)}{\sin A \cos B} = \cot A - \tan B$

57. $\sec(A + B) = \dfrac{\cos(A - B)}{\cos^2 A - \sin^2 B}$

58. $\sec(A - B) = \dfrac{\cos(A + B)}{\cos^2 A - \sin^2 B}$

Another method of verifying identities is with graphing. If the graphs of $y = f(x)$ and $y = g(x)$ coincide for all values of x, then $f(x)$ and $g(x)$ are equal.

59. Graph $y = \sin x$ and $y = \cos\left(\dfrac{\pi}{2} - x\right)$ on the same coordinate system to verify the identity $\sin x = \cos\left(\dfrac{\pi}{2} - x\right)$.

60. Graph $y = \cos x$ and $y = \sin\left(\dfrac{\pi}{2} - x\right)$ on the same coordinate system to verify the identity $\cos x = \sin\left(\dfrac{\pi}{2} - x\right)$.

61. Graph $y = -\cos x$ and $y = \cos(\pi - x)$ on the same coordinate system to verify the identity $-\cos x = \cos(\pi - x)$.

62. Graph $y = -\sin x$ and $y = \cos\left(\dfrac{\pi}{2} + x\right)$ on the same coordinate system to verify the identity $-\sin x = \cos\left(\dfrac{\pi}{2} + x\right)$.

Review Problems

The problems that follow review material we covered in Section 4.2.

Graph one complete cycle of each of the following:

63. $y = 4\sin 2x$

64. $y = 2\sin 4x$

65. $y = 3\sin\dfrac{1}{2}x$

66. $y = 5\sin\dfrac{1}{3}x$

67. $y = 2\cos \pi x$

68. $y = \cos 2\pi x$

69. $y = \csc 3x$

70. $y = \sec 3x$

71. $y = \dfrac{1}{2} \cos 3x$

72. $y = \dfrac{1}{2} \sin 3x$

73. $y = \dfrac{1}{2} \sin \dfrac{\pi}{2} x$

74. $y = 2 \sin \dfrac{\pi}{2} x$

SECTION 5.3

Double-Angle Formulas

We will begin this section by deriving the formulas for sin 2A and cos 2A using the formulas for sin (A + B) and cos (A + B). The formulas we derive for sin 2A and cos 2A are called *double-angle* formulas. Here is the derivation of the formula for sin 2A.

$$\sin 2A = \sin (A + A) \qquad \text{Write } 2A \text{ as } A + A$$
$$= \sin A \cos A + \cos A \sin A \qquad \text{Sum formula}$$
$$= \sin A \cos A + \sin A \cos A \qquad \text{Commutative property}$$
$$= 2 \sin A \cos A$$

The last line gives us our first double-angle formula.

$$\boxed{\sin 2A = 2 \sin A \cos A}$$

The first thing to notice about this formula is that it indicates the 2 in sin 2A *cannot* be factored out and written as a coefficient. That is,

$$\sin 2A \neq 2 \sin A$$

For example, if $A = 30°$, $\sin 2 \cdot 30° = \sin 60° = \sqrt{3}/2$, which is not the same as $2 \sin 30° = 2(\frac{1}{2}) = 1$.

▶ **EXAMPLE 1**   If $\sin A = \frac{3}{5}$ with A in QII, find sin 2A.

Solution   In order to apply the formula for sin 2A, we must first find cos A. Since A terminates in QII, cos A is negative.

$$\cos A = -\sqrt{1 - \sin^2 A} = -\sqrt{1 - \left(\frac{3}{5}\right)^2} = -\sqrt{\frac{16}{25}} = -\frac{4}{5}$$

Now we can apply the formula for sin 2A.

$$\sin 2A = 2 \sin A \cos A$$
$$= 2 \left(\frac{3}{5}\right)\left(-\frac{4}{5}\right)$$
$$= -\frac{24}{25} \qquad \blacktriangleleft$$

We can also use our new formula to expand the work we did previously with identities.

▶ **EXAMPLE 2**   Prove $(\sin \theta + \cos \theta)^2 = 1 + \sin 2\theta$.

Proof

$$\begin{aligned}
(\sin \theta + \cos \theta)^2 &= \sin^2 \theta + 2 \sin \theta \cos \theta + \cos^2 \theta &&\text{Expand} \\
&= 1 + 2 \sin \theta \cos \theta &&\text{Pythagorean} \\
& &&\text{identity} \\
&= 1 + \sin 2\theta &&\text{Double-angle} \\
& &&\text{identity} \qquad \blacktriangleleft
\end{aligned}$$

▶ **EXAMPLE 3**   Prove $\sin 2x = \dfrac{2 \cot x}{1 + \cot^2 x}$.

Proof

$$\begin{aligned}
\frac{2 \cot x}{1 + \cot^2 x} &= \frac{2 \cdot \dfrac{\cos x}{\sin x}}{1 + \dfrac{\cos^2 x}{\sin^2 x}} &&\text{Ratio identity} \\[2em]
&= \frac{2 \sin x \cos x}{\sin^2 x + \cos^2 x} &&\text{Multiply numerator and} \\
& &&\text{denominator by } \sin^2 x \\
&= 2 \sin x \cos x &&\text{Pythagorean identity} \\
&= \sin 2x &&\text{Double-angle identity} \qquad \blacktriangleleft
\end{aligned}$$

There are three forms of the double-angle formula for $\cos 2A$. The first involves both $\sin A$ and $\cos A$, the second involves only $\cos A$, and the third involves only $\sin A$. Here is how we obtain the three formulas.

$$\begin{aligned}
\cos 2A &= \cos (A + A) &&\text{Write } 2A \text{ as } A + A \\
&= \cos A \cos A - \sin A \sin A &&\text{Sum formula} \\
&= \cos^2 A - \sin^2 A
\end{aligned}$$

To write this last formula in terms of $\cos A$ only, we substitute $1 - \cos^2 A$ for $\sin^2 A$.

$$\begin{aligned}
\cos 2A &= \cos^2 A - (1 - \cos^2 A) \\
&= \cos^2 A - 1 + \cos^2 A \\
&= 2 \cos^2 A - 1
\end{aligned}$$

To write the formula in terms of $\sin A$ only, we substitute $1 - \sin^2 A$ for $\cos^2 A$ in the last line above.

$$\begin{aligned}
\cos 2A &= 2 \cos^2 A - 1 \\
&= 2(1 - \sin^2 A) - 1 \\
&= 2 - 2 \sin^2 A - 1 \\
&= 1 - 2 \sin^2 A
\end{aligned}$$

Here are the three forms of the double-angle formula for $\cos 2A$.

$$
\begin{array}{ll}
\cos 2A = \cos^2 A - \sin^2 A & \text{First form} \\
\qquad\; = 2 \cos^2 A - 1 & \text{Second form} \\
\qquad\; = 1 - 2 \sin^2 A & \text{Third form}
\end{array}
$$

▶ **EXAMPLE 4** If $\sin A = 1/\sqrt{5}$, find $\cos 2A$.

Solution In this case, since we are given $\sin A$, applying the third form of the formula for $\cos 2A$ will give us the answer more quickly than applying either of the other two forms.

$$
\begin{aligned}
\cos 2A &= 1 - 2 \sin^2 A \\
&= 1 - 2 \left(\frac{1}{\sqrt{5}} \right)^2 \\
&= 1 - \frac{2}{5} \\
&= \frac{3}{5}
\end{aligned}
$$
◀

▶ **EXAMPLE 5** Prove $\cos 4x = 8 \cos^4 x - 8 \cos^2 x + 1$.

Proof We can write $\cos 4x$ as $\cos 2 \cdot 2x$ and apply our double-angle formula. Since the right side is written in terms of $\cos x$ only, we will choose the second form of our double-angle formula for $\cos 2A$.

$$
\begin{aligned}
\cos 4x &= \cos 2 \cdot 2x & \\
&= 2 \cos^2 2x - 1 & \text{Double-angle formula} \\
&= 2 (2 \cos^2 x - 1)^2 - 1 & \text{Double-angle formula} \\
&= 2 (4 \cos^4 x - 4 \cos^2 x + 1) - 1 & \text{Square} \\
&= 8 \cos^4 x - 8 \cos^2 x + 2 - 1 & \text{Distribute} \\
&= 8 \cos^4 x - 8 \cos^2 x + 1 & \text{Simplify}
\end{aligned}
$$
◀

▶ **EXAMPLE 6** Graph $y = 3 - 6 \sin^2 x$ from $x = 0$ to $x = 2\pi$.

Solution To write the equation in the form $y = A \cos Bx$, we factor 3 from each term on the right side and then apply the formula for $\cos 2A$ to the remaining expression to write it as a single trigonometric function.

$$
\begin{aligned}
y &= 3 - 6 \sin^2 x & \\
&= 3(1 - 2 \sin^2 x) & \text{Factor 3 from each term} \\
&= 3 \cos 2x & \text{Double-angle formula}
\end{aligned}
$$

The graph of $y = 3 \cos 2x$ will have an amplitude of 3 and a period of $2\pi/2 = \pi$. The graph is shown in Figure 1.

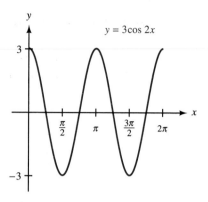

Figure 1

▶ **EXAMPLE 7** Prove $\tan \theta = \dfrac{1 - \cos 2\theta}{\sin 2\theta}$.

Proof

$$\dfrac{1 - \cos 2\theta}{\sin 2\theta} = \dfrac{1 - (1 - 2\sin^2 \theta)}{2 \sin \theta \cos \theta} \qquad \text{Double-angle formula}$$

$$= \dfrac{2 \sin^2 \theta}{2 \sin \theta \cos \theta} \qquad \text{Simplify numerator}$$

$$= \dfrac{\sin \theta}{\cos \theta} \qquad \begin{array}{l}\text{Divide out common} \\ \text{factor } 2 \sin \theta\end{array}$$

$$= \tan \theta \qquad \text{Ratio identity}$$

We end this section by deriving the formula for $\tan 2A$.

$$\tan 2A = \tan (A + A)$$

$$= \dfrac{\tan A + \tan A}{1 - \tan A \tan A}$$

$$= \dfrac{2 \tan A}{1 - \tan^2 A}$$

Our double-angle formula for $\tan 2A$ is

$$\boxed{\tan 2A = \dfrac{2 \tan A}{1 - \tan^2 A}}$$

▶ **EXAMPLE 8** Simplify $\dfrac{2 \tan 15°}{1 - \tan^2 15°}$.

Solution The expression has the same form as the right side of our double-angle formula for tan 2A. Therefore,

$$\frac{2 \tan 15°}{1 - \tan^2 15°} = \tan 2 \cdot 15°$$

$$= \tan 30°$$

$$= \frac{1}{\sqrt{3}}$$ ◀

▶ **EXAMPLE 9** If $x = 3 \tan \theta$, write the expression below in terms of just x.

$$\frac{\theta}{2} + \frac{\sin 2\theta}{4}$$

Solution To substitute for the first term above, we need to write θ in terms of x. To do so, we solve the equation $x = 3 \tan \theta$ for θ.

$$\text{If} \qquad 3 \tan \theta = x$$

$$\text{then} \qquad \tan \theta = \frac{x}{3}$$

$$\text{and} \qquad \theta = \tan^{-1} \frac{x}{3}$$

Next, since the inverse tangent function can take on values only between $-\pi/2$ and $\pi/2$, we can visualize θ by drawing a right triangle in which θ is one of the acute angles. Since $\tan \theta = x/3$, we label the side opposite θ with x and the side adjacent to θ with 3.

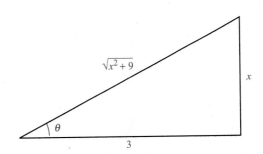

Figure 2

By the Pythagorean Theorem, the hypotenuse of the triangle in Figure 2 must be $\sqrt{x^2 + 9}$, which means $\sin \theta = x/\sqrt{x^2 + 9}$ and $\cos \theta = 3/\sqrt{x^2 + 9}$. Now we are ready to simplify and substitute to solve our problem.

$$\frac{\theta}{2} + \frac{\sin 2\theta}{4} = \frac{\theta}{2} + \frac{2\sin\theta\cos\theta}{4}$$

$$= \frac{\theta}{2} + \frac{\sin\theta\cos\theta}{2}$$

$$= \frac{1}{2}(\theta + \sin\theta\cos\theta)$$

$$= \frac{1}{2}\left(\tan^{-1}\frac{x}{3} + \frac{x}{\sqrt{x^2+9}} \cdot \frac{3}{\sqrt{x^2+9}}\right)$$

$$= \frac{1}{2}\left(\tan^{-1}\frac{x}{3} + \frac{3x}{x^2+9}\right)$$

Note that we do not have to use absolute value symbols when we multiply and simplify the square roots in the second to the last line above because we know that $x^2 + 9$ is always positive.   ◄

PROBLEM SET 5.3

Let $\sin A = -\frac{3}{5}$ with A in QIII and find

1. $\sin 2A$ **2.** $\cos 2A$
3. $\tan 2A$ **4.** $\cot 2A$

Let $\cos x = 1/\sqrt{10}$ with x in QIV and find

5. $\cos 2x$ **6.** $\sin 2x$
7. $\cot 2x$ **8.** $\tan 2x$

Let $\tan \theta = \frac{5}{12}$ with θ in QI and find

9. $\sin 2\theta$ **10.** $\cos 2\theta$
11. $\csc 2\theta$ **12.** $\sec 2\theta$

Let $\csc t = \sqrt{5}$ with t in QII and find

13. $\cos 2t$ **14.** $\sin 2t$
15. $\sec 2t$ **16.** $\csc 2t$

Graph each of the following from $x = 0$ to $x = 2\pi$.

17. $y = 4 - 8\sin^2 x$ **18.** $y = 2 - 4\sin^2 x$
19. $y = 6\cos^2 x - 3$ **20.** $y = 4\cos^2 x - 2$
21. $y = 1 - 2\sin^2 2x$ **22.** $y = 2\cos^2 2x - 1$

Use exact values to show that each of the following is true.

23. $\sin 60° = 2\sin 30°\cos 30°$ **24.** $\cos 60° = 1 - 2\sin^2 30°$
25. $\cos 120° = \cos^2 60° - \sin^2 60°$ **26.** $\sin 90° = 2\sin 45°\cos 45°$

27. If $\tan A = \frac{3}{4}$, find $\tan 2A$.   **28.** If $\tan A = -\sqrt{3}$, find $\tan 2A$.

Simplify each of the following.

29. $2 \sin 15° \cos 15°$

30. $\cos^2 15° - \sin^2 15°$

31. $1 - 2 \sin^2 75°$

32. $2 \cos^2 105° - 1$

33. $\sin \dfrac{\pi}{12} \cos \dfrac{\pi}{12}$

34. $\sin \dfrac{\pi}{8} \cos \dfrac{\pi}{8}$

35. $\dfrac{\tan 22.5°}{1 - \tan^2 22.5°}$

36. $\dfrac{\tan \dfrac{3\pi}{8}}{1 - \tan^2 \dfrac{3\pi}{8}}$

Prove each of the following identities.

37. $(\sin x - \cos x)^2 = 1 - \sin 2x$

38. $(\cos x - \sin x)(\cos x + \sin x) = \cos 2x$

39. $\cos^2 \theta = \dfrac{1 + \cos 2\theta}{2}$

40. $\sin^2 \theta = \dfrac{1 - \cos 2\theta}{2}$

41. $\cot \theta = \dfrac{\sin 2\theta}{1 - \cos 2\theta}$

42. $\cos 2\theta = \dfrac{1 - \tan^2 \theta}{1 + \tan^2 \theta}$

43. $2 \csc 2x = \tan x + \cot x$

44. $2 \cot 2x = \cot x - \tan x$

45. $\sin 3\theta = 3 \sin \theta - 4 \sin^3 \theta$

46. $\cos 3\theta = 4 \cos^3 \theta - 3 \cos \theta$

47. $\cos^4 x - \sin^4 x = \cos 2x$

48. $2 \sin^4 x + 2 \sin^2 x \cos^2 x = 1 - \cos 2x$

49. $\cot \theta - \tan \theta = \dfrac{\cos 2\theta}{\sin \theta \cos \theta}$

50. $\csc \theta - 2 \sin \theta = \dfrac{\cos 2\theta}{\sin \theta}$

51. $\sin 4A = 4 \sin A \cos^3 A - 4 \sin^3 A \cos A$

52. $\cos 4A = \cos^4 A - 6 \cos^2 A \sin^2 A + \sin^4 A$

53. $\cot 2B = \dfrac{\cos B - \sin B \tan B}{\sec B \sin 2B}$

54. $\sec 2B = \dfrac{\sec^2 B \csc^2 B}{\csc^2 B - \sec^2 B}$

55. $\csc 2t = \dfrac{\sec t + \csc t}{2 \sin t + 2 \cos t}$

56. $\tan 2t = \dfrac{2 \cot t}{\csc^2 t - 2}$

57. $\dfrac{1 - \tan x}{1 + \tan x} = \dfrac{1 - \sin 2x}{\cos 2x}$

58. $\dfrac{2 - 2 \cos 2x}{\sin 2x} = \sec x \csc x - \cot x + \tan x$

59. If $x = 5 \tan \theta$, write the expression $\dfrac{\theta}{2} - \dfrac{\sin 2\theta}{4}$ in terms of just x.

60. If $x = 4 \sin \theta$, write the expression $\dfrac{\theta}{2} + \dfrac{\sin 2\theta}{4}$ in terms of just x.

61. If $x = 3 \sin \theta$, write the expression $\dfrac{\theta}{2} - \dfrac{\sin 2\theta}{4}$ in terms of just x.

62. If $x = 2 \sin \theta$, write the expression $2\theta - \tan 2\theta$ in terms of just x.

Review Problems

The problems that follow review material we covered in Section 4.5.

Graph each of the following from $x = 0$ to $x = 4\pi$.

63. $y = 2 - 2 \cos x$ **64.** $y = 2 + 2 \cos x$

65. $y = 3 - 3 \cos x$ **66.** $y = 3 + 3 \cos x$

67. $y = \cos x + \dfrac{1}{2} \sin 2x$ **68.** $y = \sin x + \dfrac{1}{2} \cos 2x$

Graph each of the following from $x = 0$ to $x = 8$.

69. $y = \dfrac{1}{2} x + \sin \pi x$ **70.** $y = x + \sin \dfrac{\pi}{2} x$

SECTION 5.4   # Half-Angle Formulas

In this section, we will derive formulas for $\sin \dfrac{A}{2}$ and $\cos \dfrac{A}{2}$. These formulas are called half-angle formulas and are derived from the double-angle formulas for $\cos 2A$.

In Section 5.3, we developed three ways to write the formula for $\cos 2A$, two of which were

$$\cos 2A = 1 - 2 \sin^2 A \qquad \text{and} \qquad \cos 2A = 2 \cos^2 A - 1$$

Since the choice of the letter we use to denote the angles in these formulas is arbitrary, we can use an x instead of A.

$$\cos 2x = 1 - 2 \sin^2 x \qquad \text{and} \qquad \cos 2x = 2 \cos^2 x - 1$$

Let us exchange sides in the first formula and solve for $\sin x$.

$$1 - 2 \sin^2 x = \cos 2x \qquad \qquad \text{Exchange sides}$$

$$-2 \sin^2 x = -1 + \cos 2x \qquad \qquad \text{Add } -1 \text{ to both sides}$$

$$\sin^2 x = \frac{1 - \cos 2x}{2} \qquad \qquad \text{Divide both sides by } -2$$

$$\sin x = \pm \sqrt{\frac{1 - \cos 2x}{2}} \qquad \qquad \begin{array}{l} \text{Take the square root} \\ \text{of both sides} \end{array}$$

Since every value of x can be written as $\frac{1}{2}$ of some other number A, we can replace x with $A/2$. This is equivalent to saying $2x = A$.

$$\sin \frac{A}{2} = \pm \sqrt{\frac{1 - \cos A}{2}}$$

This last expression is the half-angle formula for $\sin \dfrac{A}{2}$. To find the half-angle formula for $\cos \dfrac{A}{2}$, we solve $\cos 2x = 2 \cos^2 x - 1$ for $\cos x$ and then replace x with $A/2$ (and $2x$ with A). Without showing the steps involved in this process, here is the result:

$$\cos \frac{A}{2} = \pm \sqrt{\frac{1 + \cos A}{2}}$$

In both half-angle formulas, the sign in front of the radical, $+$ or $-$, is determined by the quadrant in which $A/2$ terminates.

▶ **EXAMPLE 1**   If $\cos A = \frac{3}{5}$ with $270° < A < 360°$, find $\sin \dfrac{A}{2}$, $\cos \dfrac{A}{2}$, and $\tan \dfrac{A}{2}$.

Solution   First of all, we determine the quadrant in which $A/2$ terminates.

$$270° < A < 360° \Rightarrow \frac{270°}{2} < \frac{A}{2} < \frac{360°}{2}$$

$$\text{or } 135° < \frac{A}{2} < 180° \Rightarrow \frac{A}{2} \in \text{QII}$$

In quadrant II, sine is positive, and cosine and tangent are negative.

$$\sin \frac{A}{2} = \sqrt{\frac{1 - \cos A}{2}} \qquad \cos \frac{A}{2} = -\sqrt{\frac{1 + \cos A}{2}}$$

$$= \sqrt{\frac{1 - 3/5}{2}} \qquad\qquad = -\sqrt{\frac{1 + 3/5}{2}}$$

$$= \sqrt{\frac{1}{5}} \qquad\qquad = -\sqrt{\frac{4}{5}}$$

$$= \frac{1}{\sqrt{5}} \qquad\qquad = -\frac{2}{\sqrt{5}}$$

$$\tan \frac{A}{2} = \frac{\sin \dfrac{A}{2}}{\cos \dfrac{A}{2}} = \frac{\dfrac{1}{\sqrt{5}}}{-\dfrac{2}{\sqrt{5}}} = -\frac{1}{2}$$

◀

▶ **EXAMPLE 2** If $\sin A = -\frac{12}{13}$ with $180° < A < 270°$, find the six trigonometric functions of $A/2$.

Solution To use the half-angle formulas, we need to find $\cos A$.

because $A \in$ QIII

$$\cos A = \overset{\downarrow}{-}\sqrt{1 - \sin^2 A} = -\sqrt{1 - \left(-\frac{12}{13}\right)^2} = -\sqrt{\frac{25}{169}} = -\frac{5}{13}$$

Also, $A/2$ terminates in QII because

$$180° < A < 270°$$

$$\frac{180°}{2} < \frac{A}{2} < \frac{270°}{2}$$

$$90° < \frac{A}{2} < 135° \Rightarrow \frac{A}{2} \in \text{QII}$$

In quadrant II, sine is positive and cosine is negative.

$$\sin\frac{A}{2} = \sqrt{\frac{1 - (-5/13)}{2}} \qquad \cos\frac{A}{2} = -\sqrt{\frac{1 + (-5/13)}{2}}$$

$$= \sqrt{\frac{9}{13}} \qquad\qquad = -\sqrt{\frac{4}{13}}$$

$$= \frac{3}{\sqrt{13}} \qquad\qquad = -\frac{2}{\sqrt{13}}$$

Now that we have sine and cosine of $A/2$, we can apply the ratio identity for tangent to find $\tan\dfrac{A}{2}$.

$$\tan\frac{A}{2} = \frac{\sin\dfrac{A}{2}}{\cos\dfrac{A}{2}} = \frac{\dfrac{3}{\sqrt{13}}}{-\dfrac{2}{\sqrt{13}}} = -\frac{3}{2}$$

Next we apply our reciprocal identities to find cosecant, secant, and cotangent of $A/2$.

$$\csc\frac{A}{2} = \frac{1}{\sin\dfrac{A}{2}} = \frac{\sqrt{13}}{3} \qquad \sec\frac{A}{2} = \frac{1}{\cos\dfrac{A}{2}} = -\frac{\sqrt{13}}{2}$$

$$\cot\frac{A}{2} = \frac{1}{\tan\dfrac{A}{2}} = -\frac{2}{3}$$

◀

In the previous two examples, we found $\tan \dfrac{A}{2}$ by using the ratio of $\sin \dfrac{A}{2}$ to $\cos \dfrac{A}{2}$.

There are formulas that allow us to find $\tan \dfrac{A}{2}$ directly from $\sin A$ and $\cos A$. In Example 7 of Section 5.3, we proved the following identity:

$$\tan \theta = \frac{1 - \cos 2\theta}{\sin 2\theta}$$

If we let $\theta = \dfrac{A}{2}$ in this identity, we obtain a formula for $\tan \dfrac{A}{2}$ that involves only $\sin A$ and $\cos A$. Here it is.

$$\boxed{\tan \frac{A}{2} = \frac{1 - \cos A}{\sin A}}$$

If we multiply the numerator and denominator of the right side of this formula by $1 + \cos A$ and simplify the result, we have a second formula for $\tan \dfrac{A}{2}$.

$$\boxed{\tan \frac{A}{2} = \frac{\sin A}{1 + \cos A}}$$

▶ **EXAMPLE 3**   Find tan 15°.

Solution   Since $15° = 30°/2$, we can use a half-angle formula to find $\tan 15°$.

$$\tan 15° = \tan \frac{30°}{2}$$

$$= \frac{1 - \cos 30°}{\sin 30°}$$

$$= \frac{1 - \dfrac{\sqrt{3}}{2}}{\dfrac{1}{2}}$$

$$= 2 - \sqrt{3} \qquad \blacktriangleleft$$

▶ **Example 4** Graph $y = 4 \cos^2 \dfrac{x}{2}$ from $x = 0$ to $x = 4\pi$.

Solution Applying our half-angle formula for $\cos \dfrac{x}{2}$ to the right side, we have

$$y = 4 \cos^2 \frac{x}{2}$$

$$= 4 \left(\pm \sqrt{\frac{1 + \cos x}{2}} \right)^2$$

$$= 4 \left(\frac{1 + \cos x}{2} \right)$$

$$= 2 + 2 \cos x$$

We graph $y = 2 + 2 \cos x$ using the method developed in Section 4.5. The graph is shown in Figure 1.

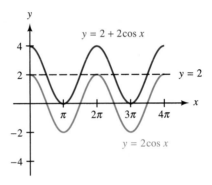

Figure 1 ◀

▶ **EXAMPLE 5** Prove $\sin^2 \dfrac{x}{2} = \dfrac{\tan x - \sin x}{2 \tan x}$.

Proof We can use a half-angle formula on the left side. In this case, since we have $\sin^2 (x/2)$, we write the half-angle formula without the square root sign. After that, we multiply the numerator and denominator on the left side by $\tan x$ because the right side has $\tan x$ in both the numerator and the denominator.

$$\sin^2 \frac{x}{2} = \frac{1 - \cos x}{2} \qquad \text{Square of half-angle formula}$$

$$= \frac{\tan x}{\tan x} \cdot \frac{1 - \cos x}{2} \qquad \begin{array}{l}\text{Multiply numerator and}\\ \text{denominator by } \tan x\end{array}$$

$$= \frac{\tan x - \tan x \cos x}{2 \tan x} \qquad \text{Distributive property}$$

$$= \frac{\tan x - \sin x}{2 \tan x} \qquad \tan x \cos x \text{ is } \sin x$$

◀

PROBLEM SET 5.4

Note For the following problems, assume that all the given angles are in simplest form so that if A is in QIV you may assume that $270° < A < 360°$.

If $\cos A = \frac{1}{2}$ with A in QIV, find

1. $\sin \dfrac{A}{2}$

2. $\cos \dfrac{A}{2}$

3. $\csc \dfrac{A}{2}$

4. $\sec \dfrac{A}{2}$

If $\sin A = -3/5$ with A in QIII, find

5. $\cos \dfrac{A}{2}$

6. $\sin \dfrac{A}{2}$

7. $\sec \dfrac{A}{2}$

8. $\csc \dfrac{A}{2}$

If $\sin B = -\frac{1}{3}$ with B in QIII, find

9. $\sin \dfrac{B}{2}$

10. $\csc \dfrac{B}{2}$

11. $\cos \dfrac{B}{2}$

12. $\sec \dfrac{B}{2}$

13. $\tan \dfrac{B}{2}$

14. $\cot \dfrac{B}{2}$

If $\sin A = \frac{4}{5}$ with A in QII, and $\sin B = \frac{3}{5}$ with B in QI, find

15. $\sin \dfrac{A}{2}$

16. $\cos \dfrac{A}{2}$

17. $\cos 2A$

18. $\sin 2A$

19. $\sec 2A$

20. $\csc 2A$

21. $\cos \dfrac{B}{2}$

22. $\sin \dfrac{B}{2}$

23. $\sin (A + B)$

24. $\cos (A + B)$

25. $\cos (A - B)$

26. $\sin (A - B)$

Graph each of the following from $x = 0$ to $x = 4\pi$.

27. $y = 4 \sin^2 \dfrac{x}{2}$

28. $y = 6 \cos^2 \dfrac{x}{2}$

29. $y = 2 \cos^2 \dfrac{x}{2}$

30. $y = 2 \sin^2 \dfrac{x}{2}$

Use half-angle formulas to find exact values for each of the following:

31. $\cos 15°$ **32.** $\tan 15°$

33. $\sin 75°$ **34.** $\cos 75°$

35. $\cos 105°$ **36.** $\sin 105°$

Prove the following identities.

37. $\sin^2 \dfrac{\theta}{2} = \dfrac{\csc \theta - \cot \theta}{2 \csc \theta}$ **38.** $2 \cos^2 \dfrac{\theta}{2} = \dfrac{\sin^2 \theta}{1 - \cos \theta}$

39. $\sec^2 \dfrac{A}{2} = \dfrac{2 \sec A}{\sec A + 1}$ **40.** $\csc^2 \dfrac{A}{2} = \dfrac{2 \sec A}{\sec A - 1}$

41. $\tan \dfrac{B}{2} = \csc B - \cot B$ **42.** $\tan \dfrac{B}{2} = \dfrac{\sec B}{\sec B \csc B + \csc B}$

43. $\tan \dfrac{x}{2} + \cot \dfrac{x}{2} = 2 \csc x$ **44.** $\tan \dfrac{x}{2} - \cot \dfrac{x}{2} = -2 \cot x$

45. $\cos^2 \dfrac{\theta}{2} = \dfrac{\tan \theta + \sin \theta}{2 \tan \theta}$ **46.** $2 \sin^2 \dfrac{\theta}{2} = \dfrac{\sin^2 \theta}{1 + \cos \theta}$

47. $\cos^4 \theta = \dfrac{1}{4} + \dfrac{\cos 2\theta}{2} + \dfrac{\cos^2 2\theta}{4}$ **48.** $4 \sin^4 \theta = 1 - 2 \cos 2\theta + \cos^2 2\theta$

Review Problems

The following problems review material we covered in Sections 4.5 and 4.6. Reviewing these problems will help you with the next section.

Evaluate without using a calculator or tables.

49. $\sin \left(\arcsin \dfrac{3}{5} \right)$ **50.** $\cos \left(\arcsin \dfrac{3}{5} \right)$

51. $\cos (\arctan 2)$ **52.** $\sin (\arctan 2)$

Write an equivalent expression that involves x only. (Assume x is positive.)

53. $\sin (\tan^{-1} x)$ **54.** $\cos (\tan^{-1} x)$

55. $\tan (\sin^{-1} x)$ **56.** $\tan (\cos^{-1} x)$

57. Graph $y = \sin^{-1} x$ **58.** Graph $y = \cos^{-1} x$

SECTION 5.5   ## Additional Identities

There are two main parts to this section, both of which rely on the work we have done previously with identities and formulas. In the first part of this section, we will extend our work on identities to include problems that involve inverse trigonometric functions. In the second part, we will use the formulas we obtained for the sine and cosine of a sum or difference to write some new formulas involving sums and products.

Identities and Formulas Involving Inverse Functions

The solution to our first example combines our knowledge of inverse trigonometric functions with our formula for sin $(A + B)$.

▶ **EXAMPLE 1**   Evaluate sin (arcsin $\frac{3}{5}$ + arctan 2) without using a calculator.

Solution   We can simplify things somewhat if we let $\alpha = \arcsin \frac{3}{5}$ and $\beta = \arctan 2$.

$$\sin \left(\arcsin \frac{3}{5} + \arctan 2 \right) = \sin (\alpha + \beta)$$

$$= \sin \alpha \cos \beta + \cos \alpha \sin \beta$$

Drawing and labeling a triangle for α and another for β, we have

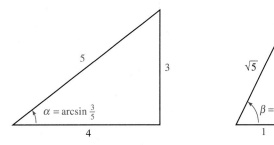

Figure 1

From the triangles in Figure 1, we have

$$\sin \alpha = \frac{3}{5} \qquad \sin \beta = \frac{2}{\sqrt{5}}$$

$$\cos \alpha = \frac{4}{5} \qquad \cos \beta = \frac{1}{\sqrt{5}}$$

Substituting these numbers into

$$\sin \alpha \cos \beta + \cos \alpha \sin \beta$$

gives us

$$\frac{3}{5} \cdot \frac{1}{\sqrt{5}} + \frac{4}{5} \cdot \frac{2}{\sqrt{5}} = \frac{11}{5\sqrt{5}}$$

◀

Calculator Note To work this problem on a calculator, we would use the following sequence:

$$0.6 \ \boxed{\sin^{-1}} \ \boxed{+} \ 2 \ \boxed{\tan^{-1}} \ \boxed{=} \ \boxed{\sin}$$

The display would show 0.9839 to four decimal places, which is the decimal approximation of $\frac{11}{5\sqrt{5}}$. It is appropriate to check your work on problems like this by using your calculator. The concepts are best understood, however, by working through the problems without using a calculator.

Here is a similar example involving inverse trigonometric functions and a double-angle identity.

▶ **EXAMPLE 2** Write $\sin (2 \tan^{-1} x)$ as an equivalent expression involving only x. (Assume x is positive.)

Solution We begin by letting $\theta = \tan^{-1} x$ and then draw a right triangle with an acute angle of θ. If we label the opposite side with x and the adjacent side with 1, the ratio of the side opposite θ to the side adjacent to θ is $x/1 = x$.

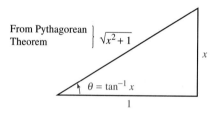

From Pythagorean Theorem $\sqrt{x^2 + 1}$

x

$\theta = \tan^{-1} x$

1

Figure 2

From Figure 2, we have $\sin \theta = x/\sqrt{x^2 + 1}$ and $\cos \theta = 1/\sqrt{x^2 + 1}$. Therefore,

$$\begin{aligned}
\sin (2 \tan^{-1} x) &= \sin 2\theta && \text{Substitute } \theta \text{ for } \tan^{-1} x \\
&= 2 \sin \theta \cos \theta && \text{Double-angle identity} \\
&= 2 \cdot \frac{x}{\sqrt{x^2 + 1}} \cdot \frac{1}{\sqrt{x^2 + 1}} && \text{From Figure 2} \\
&= \frac{2x}{x^2 + 1} && \text{Multiplication}
\end{aligned}$$

◀

To conclude our work with identities in this chapter, we will derive some additional formulas that contain sums and products of sines and cosines.

Product to Sum Formulas

If we add the formula for $\sin (A - B)$ to the formula for $\sin (A + B)$, we will eventually arrive at a formula for the product $\sin A \cos B$.

$$\sin A \cos B + \cos A \sin B = \sin (A + B)$$
$$\underline{\sin A \cos B - \cos A \sin B = \sin (A - B)}$$
$$2 \sin A \cos B \qquad\qquad = \sin (A + B) + \sin (A - B)$$

Dividing both sides of this result by 2 gives us

$$\sin A \cos B = \frac{1}{2} [\sin (A + B) + \sin (A - B)] \tag{1}$$

By similar methods, we can derive the formulas that follow.

$$\cos A \sin B = \frac{1}{2} [\sin (A + B) - \sin (A - B)] \tag{2}$$

$$\cos A \cos B = \frac{1}{2} [\cos (A + B) + \cos (A - B)] \tag{3}$$

$$\sin A \sin B = \frac{1}{2} [\cos (A - B) - \cos (A + B)] \tag{4}$$

These four product formulas are of use in calculus. The reason they are useful is that they indicate how we can convert a product into a sum. In calculus, it is sometimes much easier to work with sums of trigonometric functions than it is to work with products.

▶ **EXAMPLE 3**   Verify product formula (3) for $A = 30°$ and $B = 120°$.

Solution   Substituting $A = 30°$ and $B = 120°$ into

$$\cos A \cos B = \frac{1}{2} [\cos (A + B) + \cos (A - B)]$$

we have

$$\cos 30° \cos 120° = \frac{1}{2} [\cos 150° + \cos (-90°)]$$

$$\frac{\sqrt{3}}{2} \cdot \left(-\frac{1}{2}\right) = \frac{1}{2}\left(-\frac{\sqrt{3}}{2} + 0\right)$$

$$-\frac{\sqrt{3}}{4} = -\frac{\sqrt{3}}{4} \qquad \text{A true statement}$$

◀

▶ **EXAMPLE 4**   Write $10 \cos 5x \sin 3x$ as a sum or difference.

Solution   Product formula (2) is appropriate for an expression of the form $\cos A \sin B$.

$$10 \cos 5x \sin 3x = 10 \cdot \frac{1}{2} [\sin (5x + 3x) - \sin (5x - 3x)]$$

$$= 5 (\sin 8x - \sin 2x) \qquad \blacktriangleleft$$

Sum to Product Formulas

By some simple manipulations we can change our product formulas into sum formulas. If we take the formula for $\sin A \cos B$, exchange sides, and then multiply through by 2 we have

$$\sin (A + B) + \sin (A - B) = 2 \sin A \cos B$$

If we let $\alpha = A + B$ and $\beta = A - B$, then we can solve for A by adding the left sides and the right sides.

$$A + B = \alpha$$
$$\underline{A - B = \beta}$$
$$2A \qquad = \alpha + \beta$$
$$A = \frac{\alpha + \beta}{2}$$

By subtracting the expression for β from the expression for α, we have

$$B = \frac{\alpha - \beta}{2}$$

Writing the equation $\sin (A + B) + \sin (A - B) = 2 \sin A \cos B$ in terms of α and β gives us our first sum formula:

$$\boxed{\sin \alpha + \sin \beta = 2 \sin \frac{\alpha + \beta}{2} \cos \frac{\alpha - \beta}{2}} \qquad (5)$$

Similarly, the following sum formulas can be derived from the other product formulas:

$$\boxed{\sin \alpha - \sin \beta = 2 \cos \frac{\alpha + \beta}{2} \sin \frac{\alpha - \beta}{2}} \qquad (6)$$

$$\boxed{\cos \alpha + \cos \beta = 2 \cos \frac{\alpha + \beta}{2} \cos \frac{\alpha - \beta}{2}} \qquad (7)$$

$$\boxed{\cos \alpha - \cos \beta = -2 \sin \frac{\alpha + \beta}{2} \sin \frac{\alpha - \beta}{2}} \qquad (8)$$

▶ **EXAMPLE 5**   Verify sum formula (7) for $\alpha = 30°$ and $\beta = 90°$.

Solution   We substitute $\alpha = 30°$ and $\beta = 90°$ into sum formula (7) and simplify each side of the resulting equation.

$$\cos 30° + \cos 90° = 2 \cos \frac{30° + 90°}{2} \cos \frac{30° - 90°}{2}$$

$$\cos 30° + \cos 90° = 2 \cos 60° \cos (-30°)$$

$$\frac{\sqrt{3}}{2} + 0 = 2 \left(\frac{1}{2}\right)\left(\frac{\sqrt{3}}{2}\right)$$

$$\frac{\sqrt{3}}{2} = \frac{\sqrt{3}}{2} \qquad \text{A true statement}$$

◀

▶ **EXAMPLE 6**   Verify the identity.

$$-\tan x = \frac{\cos 3x - \cos x}{\sin 3x + \sin x}$$

Proof   Applying the formulas for $\cos \alpha - \cos \beta$ and $\sin \alpha + \sin \beta$ to the right side and then simplifying, we arrive at $-\tan x$.

$$\frac{\cos 3x - \cos x}{\sin 3x + \sin x} = \frac{-2 \sin \dfrac{3x + x}{2} \sin \dfrac{3x - x}{2}}{2 \sin \dfrac{3x + x}{2} \cos \dfrac{3x - x}{2}} \qquad \begin{array}{l}\text{Sum to}\\ \text{product formulas}\end{array}$$

$$= \frac{-2 \sin 2x \sin x}{2 \sin 2x \cos x} \qquad \text{Simplify}$$

$$= -\frac{\sin x}{\cos x} \qquad \begin{array}{l}\text{Divide out}\\ \text{common factors}\end{array}$$

$$= -\tan x \qquad \text{Ratio identity}$$

◀

PROBLEM SET 5.5

Evaluate each expression below without using a calculator. (Assume any variables represent positive numbers.)

1. $\sin \left(\arcsin \dfrac{3}{5} - \arctan 2 \right)$

2. $\cos \left(\arcsin \dfrac{3}{5} - \arctan 2 \right)$

3. $\cos \left(\tan^{-1} \dfrac{1}{2} + \sin^{-1} \dfrac{1}{2} \right)$

4. $\sin \left(\tan^{-1} \dfrac{1}{2} - \sin^{-1} \dfrac{1}{2} \right)$

5. $\sin \left(2 \cos^{-1} \dfrac{1}{\sqrt{5}} \right)$

6. $\sin \left(2 \tan^{-1} \dfrac{3}{4} \right)$

7. $\tan (\sin^{-1} x)$

8. $\tan (\cos^{-1} x)$

9. $\sin (2 \sin^{-1} x)$

10. $\sin (2 \cos^{-1} x)$

11. $\cos (2 \cos^{-1} x)$

12. $\cos (2 \sin^{-1} x)$

13. Verify product formula (4) for $A = 30°$ and $B = 120°$.

14. Verify product formula (1) for $A = 120°$ and $B = 30°$.

Rewrite each expression as a sum or difference, then simplify if possible.

15. $10 \sin 5x \cos 3x$

16. $10 \sin 5x \sin 3x$

17. $\cos 8x \cos 2x$

18. $\cos 2x \sin 8x$

19. $\sin 60° \cos 30°$

20. $\cos 90° \cos 180°$

21. $\sin 4\pi \sin 2\pi$

22. $\cos 3\pi \sin \pi$

23. Verify sum formula (6) for $\alpha = 30°$ and $\beta = 90°$.

24. Verify sum formula (8) for $\alpha = 90°$ and $\beta = 30°$.

Rewrite each expression as a product. Simplify if possible.

25. $\sin 7x + \sin 3x$

26. $\cos 5x - \cos 3x$

27. $\cos 45° + \cos 15°$

28. $\sin 75° - \sin 15°$

29. $\sin \dfrac{7\pi}{12} - \sin \dfrac{\pi}{12}$

30. $\cos \dfrac{\pi}{12} + \cos \dfrac{7\pi}{12}$

Verify each identity.

31. $-\cot x = \dfrac{\sin 3x + \sin x}{\cos 3x - \cos x}$

32. $\cot x = \dfrac{\cos 3x + \cos x}{\sin 3x - \sin x}$

33. $\cot x = \dfrac{\sin 4x + \sin 6x}{\cos 4x - \cos 6x}$

34. $-\tan 4x = \dfrac{\cos 3x - \cos 5x}{\sin 3x - \sin 5x}$

35. $\tan 4x = \dfrac{\sin 5x + \sin 3x}{\cos 3x + \cos 5x}$

36. $\cot 2x = \dfrac{\sin 3x - \sin x}{\cos x - \cos 3x}$

Review Problems

The problems that follow review material we covered in Section 4.3.

Graph one complete cycle.

37. $y = \sin \left(x + \dfrac{\pi}{4} \right)$

38. $y = \sin \left(x - \dfrac{\pi}{4} \right)$

39. $y = \cos \left(x - \dfrac{\pi}{3} \right)$

40. $y = \cos \left(x + \dfrac{\pi}{3} \right)$

41. $y = \sin \left(2x - \dfrac{\pi}{2} \right)$

42. $y = \sin \left(2x - \dfrac{\pi}{3} \right)$

43. $y = \dfrac{1}{2} \cos \left(3x - \dfrac{\pi}{2} \right)$

44. $y = \dfrac{4}{3} \cos \left(3x - \dfrac{\pi}{2} \right)$

45. $y = 3 \sin \left(\pi x - \dfrac{\pi}{2} \right)$

46. $y = 4 \sin \left(2\pi x - \dfrac{\pi}{2} \right)$

CHAPTER 5 SUMMARY

Examples

Basic Identities [5.1]

	Basic identities	Common equivalent forms
Reciprocal	$\csc \theta = \dfrac{1}{\sin \theta}$	$\sin \theta = \dfrac{1}{\csc \theta}$
	$\sec \theta = \dfrac{1}{\cos \theta}$	$\cos \theta = \dfrac{1}{\sec \theta}$
	$\cot \theta = \dfrac{1}{\tan \theta}$	$\tan \theta = \dfrac{1}{\cot \theta}$
Ratio	$\tan \theta = \dfrac{\sin \theta}{\cos \theta}$	
	$\cot \theta = \dfrac{\cos \theta}{\sin \theta}$	
Pythagorean	$\cos^2 \theta + \sin^2 \theta = 1$	$\sin^2 \theta = 1 - \cos^2 \theta$
		$\sin \theta = \pm\sqrt{1 - \cos^2 \theta}$
		$\cos^2 \theta = 1 - \sin^2 \theta$
		$\cos \theta = \pm\sqrt{1 - \sin^2 \theta}$
	$1 + \tan^2 \theta = \sec^2 \theta$	
	$1 + \cot^2 \theta = \csc^2 \theta$	

1. To prove

$\tan x + \cos x$
$\qquad = \sin x \,(\sec x + \cot x),$

we can multiply through by $\sin x$ on the right side and then change to sines and cosines.

$\sin x \,(\sec x + \cot x)$
$\quad = \sin x \sec x + \sin x \cot x$

$\quad = \sin x \cdot \dfrac{1}{\cos x} + \sin x \cdot \dfrac{\cos x}{\sin x}$

$\quad = \dfrac{\sin x}{\cos x} + \cos x$

$\quad = \tan x + \cos x$

Proving Identities [5.1]

An identity in trigonometry is a statement that two expressions are equal for all replacements of the variable for which each expression is defined. To prove a trigonometric identity, we use trigonometric substitutions and algebraic manipulations to either

1. Transform the right side into the left side, or
2. Transform the left side into the right side.

Remember to work on each side separately. We do not want to use properties from algebra that involve both sides of the identity—like the addition property of equality.

2. To find the exact value for cos 75°, we write 75° as 45° + 30° and then apply the formula for cos $(A + B)$.

cos 75°
$= \cos(45° + 30°)$
$= \cos 45° \cos 30° - \sin 45° \sin 30°$
$= \dfrac{\sqrt{2}}{2} \cdot \dfrac{\sqrt{3}}{2} - \dfrac{\sqrt{2}}{2} \cdot \dfrac{1}{2}$
$= \dfrac{\sqrt{6} - \sqrt{2}}{4}$

Sum and Difference Formulas [5.2]

$$\sin(A + B) = \sin A \cos B + \cos A \sin B$$
$$\sin(A - B) = \sin A \cos B - \cos A \sin B$$
$$\cos(A + B) = \cos A \cos B - \sin A \sin B$$
$$\cos(A - B) = \cos A \cos B + \sin A \sin B$$
$$\tan(A + B) = \frac{\tan A + \tan B}{1 - \tan A \tan B}$$
$$\tan(A - B) = \frac{\tan A - \tan B}{1 + \tan A \tan B}$$

3. If $\sin A = \frac{3}{5}$ with A in QII, then

$\cos 2A = 1 - 2 \sin^2 A$
$= 1 - 2 \left(\dfrac{3}{5}\right)^2$
$= \dfrac{7}{25}$

Double-Angle Formulas [5.3]

$$\sin 2A = 2 \sin A \cos A$$

$\cos 2A = \cos^2 A - \sin^2 A$	First form
$= 2 \cos^2 A - 1$	Second form
$= 1 - 2 \sin^2 A$	Third form

$$\tan 2A = \frac{2 \tan A}{1 - \tan^2 A}$$

4. We can use a half-angle formula to find the exact value of sin 15° by writing 15° as 30°/2.

$\sin 15° = \sin \dfrac{30°}{2}$
$= \sqrt{\dfrac{1 - \cos 30°}{2}}$
$= \sqrt{\dfrac{1 - \sqrt{3}/2}{2}}$
$= \sqrt{\dfrac{2 - \sqrt{3}}{4}}$
$= \dfrac{\sqrt{2 - \sqrt{3}}}{2}$

Half-Angle Formulas [5.4]

$$\sin \frac{A}{2} = \pm\sqrt{\frac{1 - \cos A}{2}}$$
$$\cos \frac{A}{2} = \pm\sqrt{\frac{1 + \cos A}{2}}$$
$$\tan \frac{A}{2} = \frac{1 - \cos A}{\sin A} = \frac{\sin A}{1 + \cos A}$$

5. We can write the product

$$10 \cos 5x \sin 3x$$

as a difference by applying the second product to sum formula:

$10 \cos 5x \sin 3x$
$= 10 \cdot \frac{1}{2} [\sin(5x + 3x) - \sin(5x - 3x)]$
$= 5 (\sin 8x - \sin 2x)$

Product to Sum Formulas [5.5]

$$\sin A \cos B = \frac{1}{2} [\sin(A + B) + \sin(A - B)]$$
$$\cos A \sin B = \frac{1}{2} [\sin(A + B) - \sin(A - B)]$$
$$\cos A \cos B = \frac{1}{2} [\cos(A + B) + \cos(A - B)]$$
$$\sin A \sin B = \frac{1}{2} [\cos(A - B) - \cos(A + B)]$$

6. Prove

$$-\tan x = \frac{\cos 3x - \cos x}{\sin 3x + \sin x}$$

Proof

$$
\begin{aligned}
&\frac{\cos 3x - \cos x}{\sin 3x + \sin x} \\[2mm]
&= \frac{-2 \sin \dfrac{3x + x}{2} \sin \dfrac{3x - x}{2}}{2 \sin \dfrac{3x + x}{2} \cos \dfrac{3x - x}{2}} \\[2mm]
&= \frac{-2 \sin 2x \sin x}{2 \sin 2x \cos x} \\[2mm]
&= -\frac{\sin x}{\cos x} \\[2mm]
&= -\tan x
\end{aligned}
$$

Sum to Product Formulas [5.5]

$$\sin \alpha + \sin \beta = 2 \sin \frac{\alpha + \beta}{2} \cos \frac{\alpha - \beta}{2}$$

$$\sin \alpha - \sin \beta = 2 \cos \frac{\alpha + \beta}{2} \sin \frac{\alpha - \beta}{2}$$

$$\cos \alpha + \cos \beta = 2 \cos \frac{\alpha + \beta}{2} \cos \frac{\alpha - \beta}{2}$$

$$\cos \alpha - \cos \beta = -2 \sin \frac{\alpha + \beta}{2} \sin \frac{\alpha - \beta}{2}$$

CHAPTER 5 TEST

Prove each identity.

1. $\tan \theta = \sin \theta \sec \theta$

2. $\dfrac{\cot \theta}{\csc \theta} = \cos \theta$

3. $(\sec x - 1)(\sec x + 1) = \tan^2 x$

4. $\sec \theta - \cos \theta = \tan \theta \sin \theta$

5. $\dfrac{\cos t}{1 - \sin t} = \dfrac{1 + \sin t}{\cos t}$

6. $\dfrac{1}{1 - \sin t} + \dfrac{1}{1 + \sin t} = 2 \sec^2 t$

7. $\sin (\theta - 90°) = -\cos \theta$

8. $\cos \left(\dfrac{\pi}{2} + \theta \right) = -\sin \theta$

9. $\cos^4 A - \sin^4 A = \cos 2A$

10. $\cot A = \dfrac{\sin 2A}{1 - \cos 2A}$

11. $\cot x - \tan x = \dfrac{\cos 2x}{\sin x \cos x}$

12. $\tan \dfrac{x}{2} = \dfrac{\tan x}{\sec x + 1}$

Let $\sin A = -\frac{3}{5}$ with $270° \le A \le 360°$ and $\sin B = \frac{12}{13}$ with $90° \le B \le 180°$ and find

13. $\sin (A + B)$

14. $\cos (A - B)$

15. $\cos 2B$

16. $\sin 2B$

17. $\sin \dfrac{A}{2}$

18. $\cos \dfrac{A}{2}$

Find exact values for each of the following:

19. $\sin 75°$

20. $\cos 15°$

21. $\tan \dfrac{\pi}{12}$

22. $\cot \dfrac{\pi}{12}$

Write each expression as a single trigonometric function.

23. $\cos 4x \cos 5x - \sin 4x \sin 5x$

24. $\sin 15° \cos 75° + \cos 15° \sin 75°$

25. If $\sin A = -1/\sqrt{5}$ with $180° \le A \le 270°$, find $\cos 2A$ and $\cos A/2$.

26. If $\sec A = \sqrt{10}$ with $0° \le A \le 90°$, find $\sin 2A$ and $\sin A/2$.

27. Find $\tan A$ if $\tan B = \dfrac{1}{2}$ and $\tan (A + B) = 3$.

28. Find $\cos x$ if $\cos 2x = \dfrac{1}{2}$.

Evaluate each expression below without using a calculator. (Assume any variables represent positive numbers.)

29. $\cos \left(\arcsin \dfrac{4}{5} - \arctan 2 \right)$

30. $\sin \left(\arccos \dfrac{4}{5} + \arctan 2 \right)$

31. $\cos (2 \sin^{-1} x)$

32. $\sin (2 \cos^{-1} x)$

33. Rewrite the product $\sin 6x \sin 4x$ as a sum or difference.

34. Rewrite the sum $\cos 15° + \cos 75°$ as a product and simplify.

6

Equations

I I I I I I I I I I

▶ **To the Student**

In this chapter, we will solve equations that contain trigonometric functions. We begin by solving simple equations that have the form of the linear and quadratic equations you solved in algebra. In Section 6.2, we include equations whose solutions depend on the use of the basic trigonometric identities. In Section 6.3, we also include equations that contain multiples of angles. We end the chapter with a look at parametric equations and some additional techniques of graphing.

To be successful in this chapter, you need a good working knowledge of the trigonometric identities and formulas presented in Chapter 5. It is also very important that you know the exact values for trigonometric functions of 30°, 45°, and 60°.

Solving Trigonometric Equations

The solution set for an equation is the set of all numbers which, when used in place of the variable, make the equation a true statement. For example, the solution set for the equation $4x^2 - 9 = 0$ is $\{-\frac{3}{2}, \frac{3}{2}\}$ since these are the only two numbers that, when used in place of x, turn the equation into a true statement.

In algebra, the first kind of equations you learned to solve were linear (or first-degree) equations in one variable. Solving these equations was accomplished by applying two important properties: the *addition property of equality* and the *multiplication property of equality*. These two properties were stated as follows:

Addition Property of Equality

For any three algebraic expressions A, B, and C

$$\text{If } A = B$$
$$\text{then } A + C = B + C$$

In Words: Adding the same quantity to both sides of an equation will not change the solution set.

Multiplication Property of Equality

For any three algebraic expressions A, B, and C, with $C \neq 0$,

$$\text{If } A = B$$
$$\text{then } AC = BC$$

In Words: Multiplying both sides of an equation by the same nonzero quantity will not change the solution set.

Here is an example that shows how we use these two properties to solve a linear equation in one variable.

▶ **EXAMPLE 1**   Solve for x: $5x + 7 = 2x - 5$.

Solution

$$5x + 7 = 2x - 5$$
$$3x + 7 = -5 \qquad \text{Add } -2x \text{ to each side}$$
$$3x = -12 \qquad \text{Add } -7 \text{ to each side}$$
$$x = -4 \qquad \text{Multiply each side by } \frac{1}{3}$$

Notice in the last step we could just as easily have divided both sides by 3 instead of multiplying both sides by $\frac{1}{3}$. Division by a number and multiplication by its reciprocal are equivalent operations.   ◀

The process of solving trigonometric equations is very similar to the process of solving algebraic equations. With trigonometric equations, we look for values of an *angle* that will make the equation into a true statement. We usually begin by solving for a specific trigonometric function of that angle and then use the concepts we have developed earlier to find the angle. Here are some examples that illustrate this procedure.

▶ **EXAMPLE 2**   Solve for x: $2 \sin x - 1 = 0$.

Solution   We can solve for $\sin x$ using our methods from algebra. We then use our knowledge of trigonometry to find x.

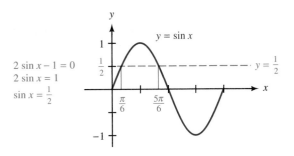

$2 \sin x - 1 = 0$

$2 \sin x = 1$

$\sin x = \dfrac{1}{2}$

Figure 1

From Figure 1 we can see that if we are looking for radian solutions between 0 and 2π, then x is either $\pi/6$ or $5\pi/6$. On the other hand, if we want degree solutions between $0°$ and $360°$, then our solutions will be $30°$ and $150°$. Without the aid of Figure 1, we would reason that, since $\sin x = \frac{1}{2}$, the reference angle for x is $30°$. Then, since $\frac{1}{2}$ is a positive number and the sine function is positive in quadrants I and II, x must be $30°$ or $150°$.

Solutions Between $0°$ and $360°$ or 0 and 2π

In degrees	In radians
$x = 30°$ or $x = 150°$	$x = \dfrac{\pi}{6}$ or $x = \dfrac{5\pi}{6}$

Since the sine function is periodic with period 2π (or $360°$), adding multiples of 2π (or $360°$) will give us all solutions.

All Solutions (k is an integer)

In degrees	In radians
$x = 30° + 360°k$	$x = \dfrac{\pi}{6} + 2k\pi$
or $x = 150° + 360°k$	or $x = \dfrac{5\pi}{6} + 2k\pi$

◀

▶ **EXAMPLE 3** Solve $2 \sin \theta - 3 = 0$, if $0° \le \theta < 360°$.

Solution We begin by solving for $\sin \theta$.

$$2 \sin \theta - 3 = 0$$

$$2 \sin \theta = 3 \qquad \text{Add 3 to both sides}$$

$$\sin \theta = \dfrac{3}{2} \qquad \text{Divide both sides by 2}$$

Since $\sin \theta$ is between -1 and 1 for all values of θ, $\sin \theta$ can never be $\frac{3}{2}$. Therefore, there is no solution to our equation. To justify our conclusion further, we can graph $y = 2 \sin x - 3$. The graph is a sine curve with amplitude 2 that has been shifted down 3 units vertically. The graph is shown in Figure 2.

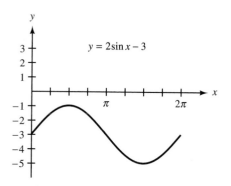

Figure 2

Since the graph in Figure 2 does not cross the x-axis, there is no solution to the equation $2 \sin x - 3 = 0$.   ◀

▶ **EXAMPLE 4**   Find all degree solutions to $\sin (2A - 50°) = \dfrac{\sqrt{3}}{2}$.

Solution   The expression $2A - 50°$ must be coterminal with $60°$ or $120°$, since $\sin 60° = \sqrt{3}/2$ and $\sin 120° = \sqrt{3}/2$. Therefore,

$$2A - 50° = 60° + 360°k \qquad \text{or} \qquad 2A - 50° = 120° + 360°k$$

where k is an integer. To solve each of these equations for A, we first add $50°$ to each side of the equations and then divide each side by 2. Here are the steps involved in doing so:

$$2A - 50° = 60° + 360°k \quad \text{or} \quad 2A - 50° = 120° + 360°k$$
$$2A = 110° + 360°k \qquad\qquad 2A = 170° + 360°k \quad \text{Add } 50° \text{ to each side}$$

$$A = \frac{110° + 360°k}{2} \qquad\qquad A = \frac{170° + 360°k}{2} \quad \text{Divide each side by 2}$$

$$A = \frac{110°}{2} + \frac{360°k}{2} \qquad\qquad A = \frac{170°}{2} + \frac{360°k}{2}$$

$$A = 55° + 180°k \qquad\qquad\quad A = 85° + 180°k$$

Note   Unless directed otherwise, let's agree to write all degree solutions to our equations in decimal degrees, to the nearest tenth of a degree.   ◀

▶ **EXAMPLE 5**   Solve $3 \sin \theta - 2 = 7 \sin \theta - 1$, if $0° \le \theta < 360°$.

Solution   We can solve for $\sin \theta$ by collecting all the variable terms on the left side and all the constant terms on the right side.

$$3 \sin \theta - 2 = 7 \sin \theta - 1$$
$$-4 \sin \theta - 2 = -1 \qquad \text{Add } -7 \sin \theta \text{ to each side}$$
$$-4 \sin \theta = 1 \qquad \text{Add 2 to each side}$$
$$\sin \theta = -\frac{1}{4} \qquad \text{Divide each side by } -4$$

Since we have not memorized the angle whose sine is $-\frac{1}{4}$, we must convert $-\frac{1}{4}$ to a decimal and use a calculator to find the reference angle.

$$\sin \theta = -\frac{1}{4} = -0.2500$$

We find that the angle whose sine is nearest to 0.2500 is 14.5°. Therefore, the reference angle is 14.5°. Since $\sin \theta$ is negative, θ will terminate in quadrant III or IV.

In quadrant III we have
$$\theta = 180° + 14.5°$$
$$= 194.5°$$

In quadrant IV we have
$$\theta = 360° - 14.5°$$
$$= 345.5°$$

Calculator Note   Remember, because of the restricted values on your calculator, if you enter -0.2500 and press the $\sin^{-1}$ key, your calculator will display approximately $-14.5°$, which is incorrect. The best way to proceed is to use your calculator to find the reference angle by entering 0.2500 and pressing the $\sin^{-1}$ key. Then do the rest of the calculations as we have here.   ◀

The next kind of trigonometric equation we will solve is quadratic in form. In algebra, the two most common methods of solving quadratic equations are factoring and applying the quadratic formula. Here is an example that reviews the factoring method.

▶ **EXAMPLE 6**   Solve $2x^2 - 9x = 5$ for x.

Solution   We begin by writing the equation in standard form (0 on one side — decreasing powers of the variable on the other). We then factor the left side and set each factor equal to 0.

$$2x^2 - 9x = 5$$
$$2x^2 - 9x - 5 = 0 \qquad \text{Standard form}$$
$$(2x + 1)(x - 5) = 0 \qquad \text{Factor}$$
$$2x + 1 = 0 \quad \text{or} \quad x - 5 = 0 \qquad \text{Set each factor to 0}$$
$$x = -\frac{1}{2} \quad \text{or} \quad x = 5 \qquad \text{Solve resulting equations}$$

The two solutions, $x = -\frac{1}{2}$ and $x = 5$, are the only two numbers that satisfy the original equation.   ◀

▶ **EXAMPLE 7**   Solve $2 \cos^2 t - 9 \cos t = 5$, if $0 \leq t < 2\pi$.

Solution   This equation is the equation from Example 6 with $\cos t$ in place of x. The fact that $0 \leq t < 2\pi$ indicates we are to write our solutions in radians.

$$2 \cos^2 t - 9 \cos t = 5$$

$$2 \cos^2 t - 9 \cos t - 5 = 0 \qquad \text{Standard form}$$

$$(2 \cos t + 1)(\cos t - 5) = 0 \qquad \text{Factor}$$

$$2 \cos t + 1 = 0 \quad \text{or} \quad \cos t - 5 = 0 \qquad \text{Set each factor to 0}$$

$$\cos t = -\frac{1}{2} \quad \text{or} \quad \cos t = 5$$

The first result, $\cos t = -\frac{1}{2}$, gives us $t = 2\pi/3$ or $t = 4\pi/3$. The second result, $\cos t = 5$, has no solution. For any value of t, $\cos t$ must be between -1 and 1. It can never be 5.   ◀

▶ **EXAMPLE 8**   Solve $2 \sin^2 \theta + 2 \sin \theta - 1 = 0$, if $0° \leq \theta < 360°$.

Solution   The equation is already in standard form. If we try to factor the left side, however, we find it does not factor. We must use the quadratic formula. The quadratic formula states that the solutions to the equation

$$ax^2 + bx + c = 0$$

will be

$$x = \frac{-b \pm \sqrt{b^2 - 4ac}}{2a}$$

In our case, the coefficients a, b, and c are

$$a = 2, \qquad b = 2, \qquad c = -1$$

Using these numbers, we can solve for $\sin \theta$ as follows:

$$\sin \theta = \frac{-2 \pm \sqrt{4 - 4(2)(-1)}}{2(2)}$$

$$= \frac{-2 \pm \sqrt{12}}{4}$$

$$= \frac{-2 \pm 2\sqrt{3}}{4}$$

$$= \frac{-1 \pm \sqrt{3}}{2}$$

Using the approximation $\sqrt{3} = 1.7321$, we arrive at the following decimal approximations for $\sin\theta$:

$$\sin\theta = \frac{-1 + 1.7321}{2} \quad \text{or} \quad \sin\theta = \frac{-1 - 1.7321}{2}$$

$$\sin\theta = 0.3661 \quad \text{or} \quad \sin\theta = -1.3661$$

We will not obtain any solutions from the second expression, $\sin\theta = -1.3661$, since $\sin\theta$ must be between -1 and 1. For $\sin\theta = 0.3661$, we use a calculator to find the angle whose sine is nearest to 0.3661. That angle is $21.5°$, and it is the reference angle for θ. Since $\sin\theta$ is positive, θ must terminate in quadrant I or II. Therefore,

$$\theta = 21.5° \quad \text{or} \quad \theta = 180° - 21.5° = 158.5° \quad \blacktriangleleft$$

PROBLEM SET 6.1

Solve each equation for θ if $0° \leq \theta < 360°$. Do not use a calculator.

1. $2\sin\theta = 1$
2. $2\cos\theta = 1$
3. $2\cos\theta - \sqrt{3} = 0$
4. $2\cos\theta + \sqrt{3} = 0$
5. $2\tan\theta + 2 = 0$
6. $\sqrt{3}\cot\theta - 1 = 0$

Solve each equation for t if $0 \leq t < 2\pi$. Give all answers as exact values in radians. Do not use a calculator.

7. $4\sin t - \sqrt{3} = 2\sin t$
8. $\sqrt{3} + 5\sin t = 3\sin t$
9. $2\cos t = 6\cos t - \sqrt{12}$
10. $5\cos t + \sqrt{12} = \cos t$
11. $3\sin t + 5 = -2\sin t$
12. $3\sin t + 4 = 4$

Find all solutions in the interval $0° \leq \theta < 360°$. Use a calculator on the last step and write all answers to the nearest tenth of a degree.

13. $4\sin\theta - 3 = 0$
14. $4\sin\theta + 3 = 0$
15. $2\cos\theta - 5 = 3\cos\theta - 2$
16. $4\cos\theta - 1 = 3\cos\theta + 4$
17. $\sin\theta - 3 = 5\sin\theta$
18. $\sin\theta - 4 = -2\sin\theta$

Solve for x, if $0 \leq x < 2\pi$. Write your answers in exact values only.

19. $(\sin x - 1)(2\sin x - 1) = 0$
20. $(\cos x - 1)(2\cos x - 1) = 0$
21. $\tan x(\tan x - 1) = 0$
22. $\tan x(\tan x + 1) = 0$
23. $\sin x + 2\sin x \cos x = 0$
24. $\cos x - 2\sin x \cos x = 0$
25. $2\sin^2 x - \sin x - 1 = 0$
26. $2\cos^2 x + \cos x - 1 = 0$

Solve for θ, if $0° \leq \theta < 360°$.

27. $(2\cos\theta + \sqrt{3})(2\cos\theta + 1) = 0$
28. $(2\sin\theta - \sqrt{3})(2\sin\theta - 1) = 0$
29. $\sqrt{3}\tan\theta - 2\sin\theta\tan\theta = 0$
30. $\tan\theta - 2\cos\theta\tan\theta = 0$
31. $2\cos^2\theta + 11\cos\theta = -5$
32. $2\sin^2\theta - 7\sin\theta = -3$

Use the quadratic formula to find all solutions in the interval $0° \leq \theta < 360°$ to the nearest tenth of a degree.

33. $2 \sin^2 \theta - 2 \sin \theta - 1 = 0$ **34.** $2 \cos^2 \theta + 2 \cos \theta - 1 = 0$
35. $\cos^2 \theta + \cos \theta - 1 = 0$ **36.** $\sin^2 \theta - \sin \theta - 1 = 0$
37. $2 \sin^2 \theta + 1 = 4 \sin \theta$ **38.** $1 - 4 \cos \theta = -2 \cos^2 \theta$

Write expressions representing all solutions to the equations you solved in the problems below.

39. Problem 1   **40.** Problem 2   **41.** Problem 7   **42.** Problem 8
43. Problem 11   **44.** Problem 12   **45.** Problem 13   **46.** Problem 14

Find all degree solutions to the following equations.

47. $\cos (2A - 50°) = \dfrac{\sqrt{3}}{2}$

48. $\sin (2A + 50°) = \dfrac{\sqrt{3}}{2}$

49. $\sin (3A + 30°) = \dfrac{1}{2}$

50. $\cos (3A + 30°) = \dfrac{1}{2}$

51. $\cos (4A - 20°) = -\dfrac{1}{2}$

52. $\sin (4A - 20°) = -\dfrac{1}{2}$

53. $\sin (5A + 15°) = -\dfrac{1}{\sqrt{2}}$

54. $\cos (5A + 15°) = -\dfrac{1}{\sqrt{2}}$

If a projectile (such as a bullet) is fired into the air with an initial velocity of v at an angle of elevation θ, then the height h of the projectile at time t is given by

$$h = -16t^2 + vt \sin \theta$$

55. Give the equation for the height, if v is 1,500 ft/sec and θ is 30°.
56. Give the equation for h, if v is 600 ft/sec and θ is 45°. (Leave your answer in exact value form.)
57. Use the equation found in Problem 55 to find the height of the object after 2 seconds.
58. Use the equation found in Problem 56 to find the height of the object after $\sqrt{3}$ seconds to the nearest tenth.
59. Find the angle of elevation θ of a rifle barrel, if a bullet fired at 1,500 ft/sec takes 2 seconds to reach a height of 750 ft to the nearest tenth.
60. Find the angle of elevation of a rifle, if a bullet fired at 1,500 ft/sec takes 3 seconds to reach a height of 750 ft. Give your answer to the nearest tenth of a degree.

Review Problems

The problems that follow review material we covered in Sections 5.2 and 5.3. Reviewing these problems will help you with the next section.

61. Write the double-angle formula for $\sin 2A$.
62. Write $\cos 2A$ in terms of $\sin A$ only.

63. Write cos $2A$ in terms of cos A only.
64. Write cos $2A$ in terms of sin A and cos A.
65. Expand sin $(\theta + 45°)$ and then simplify.
66. Expand sin $(\theta + 30°)$ and then simplify.
67. Find the exact value of sin $75°$.
68. Find the exact value of cos $105°$.

69. Prove the identity $\cos 2x = \dfrac{1 - \tan^2 x}{1 + \tan^2 x}$.

70. Prove the identity $\sin 2x = \dfrac{2}{\tan x + \cot x}$.

SECTION 6.2   # More on Trigonometric Equations

In this section, we will use our knowledge of identities to replace some parts of the equations we are solving with equivalent expressions that will make the equations easier to solve. Here are some examples.

▶ **EXAMPLE 1**   Solve $2 \cos x - 1 = \sec x$, if $0 \le x < 2\pi$.

Solution   To solve this equation as we have solved the equations in the previous section, we must write each term using the same trigonometric function. To do so, we can use a reciprocal identity to write sec x in terms of cos x.

$$2 \cos x - 1 = \frac{1}{\cos x}$$

To clear the equation of fractions, we multiply both sides by cos x. (Note that we must assume cos $x \ne 0$ in order to multiply both sides by it. If we obtain solutions for which cos $x = 0$, we will have to discard them.)

$$\cos x(2 \cos x - 1) = \frac{1}{\cos x} \cdot \cos x$$

$$2 \cos^2 x - \cos x = 1$$

We are left with a quadratic equation that we write in standard form and then solve.

$$2 \cos^2 x - \cos x - 1 = 0 \qquad \text{Standard form}$$

$$(2 \cos x + 1)(\cos x - 1) = 0 \qquad \text{Factor}$$

$$2 \cos x + 1 = 0 \quad \text{or} \quad \cos x - 1 = 0 \qquad \text{Set each factor to 0}$$

$$\cos x = -\frac{1}{2} \quad \text{or} \quad \cos x = 1$$

$$x = \frac{2\pi}{3}, \frac{4\pi}{3} \quad \text{or} \quad x = 0$$

The solutions are 0, $2\pi/3$, and $4\pi/3$.   ◀

▶ **EXAMPLE 2**   Solve $\sin 2\theta + \sqrt{2} \cos \theta = 0, 0° \leq \theta < 360°$.

Solution   In order to solve this equation, both trigonometric functions must be functions of the same angle. As the equation stands now, one angle is 2θ, while the other is θ. We can write everything as a function of θ by using the double-angle identity $\sin 2\theta = 2 \sin \theta \cos \theta$.

$$\sin 2\theta + \sqrt{2} \cos \theta = 0$$

$$2 \sin \theta \cos \theta + \sqrt{2} \cos \theta = 0 \qquad \sin 2\theta = 2 \sin \theta \cos \theta$$

$$\cos \theta \, (2 \sin \theta + \sqrt{2}) = 0 \qquad \text{Factor out } \cos \theta$$

$$\cos \theta = 0 \quad \text{or} \quad 2 \sin \theta + \sqrt{2} = 0 \qquad \text{Set each factor to 0}$$

$$\sin \theta = -\frac{\sqrt{2}}{2}$$

$$\theta = 90°, 270° \qquad \text{or} \qquad \theta = 225°, 315° \qquad \blacktriangleleft$$

▶ **EXAMPLE 3**   Solve $\cos 2\theta + 3 \sin \theta - 2 = 0$, if $0° \leq \theta < 360°$.

Solution   We have the same problem with this equation that we did with the equation in Example 2. We must rewrite $\cos 2\theta$ in terms of functions of just θ. Recall that there are three forms of the double-angle identity for $\cos 2\theta$. We choose the double-angle identity that involves $\sin \theta$ only, since the middle term of our equation involves $\sin \theta$, and it is best to have all terms involve the same trigonometric function.

$$\cos 2\theta + 3 \sin \theta - 2 = 0$$

$$1 - 2 \sin^2 \theta + 3 \sin \theta - 2 = 0 \qquad \cos 2\theta = 1 - 2 \sin^2 \theta$$

$$-2 \sin^2 \theta + 3 \sin \theta - 1 = 0 \qquad \text{Simplify}$$

$$2 \sin^2 \theta - 3 \sin \theta + 1 = 0 \qquad \text{Multiply each side by } -1$$

$$(2 \sin \theta - 1)(\sin \theta - 1) = 0 \qquad \text{Factor}$$

$$2 \sin \theta - 1 = 0 \quad \text{or} \quad \sin \theta - 1 = 0 \qquad \text{Set factors to 0}$$

$$\sin \theta = \frac{1}{2} \qquad\qquad \sin \theta = 1$$

$$\theta = 30°, 150° \quad \text{or} \quad \theta = 90° \qquad \blacktriangleleft$$

▶ **EXAMPLE 4**   Solve $4 \cos^2 x + 4 \sin x - 5 = 0, 0 \leq x < 2\pi$.

Solution   We cannot factor and solve this quadratic equation until each term involves the same trigonometric function. If we change the $\cos^2 x$ in the first term to $1 - \sin^2 x$, we will obtain an equation that involves the sine function only.

$$4 \cos^2 x + 4 \sin x - 5 = 0$$

$4(1 - \sin^2 x) + 4 \sin x - 5 = 0$	$\cos^2 x = 1 - \sin^2 x$
$4 - 4 \sin^2 x + 4 \sin x - 5 = 0$	Distributive property
$-4 \sin^2 x + 4 \sin x - 1 = 0$	Add 4 and -5
$4 \sin^2 x - 4 \sin x + 1 = 0$	Multiply each side by -1
$(2 \sin x - 1)^2 = 0$	Factor
$2 \sin x - 1 = 0$	Set factor to 0

$$\sin x = \frac{1}{2}$$

$$x = \frac{\pi}{6}, \frac{5\pi}{6} \qquad \blacktriangleleft$$

▶ **EXAMPLE 5**   Solve $\sin \theta - \cos \theta = 1$, if $0° \le \theta < 360°$.

Solution   If we separate $\sin \theta$ and $\cos \theta$ on opposite sides of the equal sign, and then square both sides of the equation, we will be able to use an identity to write the equation in terms of one trigonometric function only.

$\sin \theta - \cos \theta = 1$	
$\sin \theta = 1 + \cos \theta$	Add $\cos \theta$ to each side
$\sin^2 \theta = (1 + \cos \theta)^2$	Square each side
$\sin^2 \theta = 1 + 2 \cos \theta + \cos^2 \theta$	Expand $(1 + \cos \theta)^2$
$1 - \cos^2 \theta = 1 + 2 \cos \theta + \cos^2 \theta$	$\sin^2 \theta = 1 - \cos^2 \theta$
$0 = 2 \cos \theta + 2 \cos^2 \theta$	Standard form
$0 = 2 \cos \theta(1 + \cos \theta)$	Factor

$2 \cos \theta = 0$	or $\quad 1 + \cos \theta = 0$	Set factors to 0
$\cos \theta = 0$	$\cos \theta = -1$	
$\theta = 90°, 270°$	or $\quad \theta = 180°$	

We have three possible solutions, some of which may be extraneous since we squared both sides of the equation in Step 2. Any time we raise both sides of an equation to an even power, we have the possibility of introducing extraneous solutions. We must check each possible solution in our original equation.

Checking $\qquad \theta = 90°$	*Checking* $\qquad \theta = 180°$
$\sin 90° - \cos 90° \overset{?}{=} 1$	$\sin 180° - \cos 180° \overset{?}{=} 1$
$1 - 0 \overset{?}{=} 1$	$0 - (-1) \overset{?}{=} 1$
$1 = 1$	$1 = 1$
$\theta = 90°$ is a solution	$\theta = 180°$ is a solution

Checking $\qquad \theta \overset{?}{=} 270°$

$\sin 270° - \cos 270° \overset{?}{=} 1$

$-1 - 0 \overset{?}{=} 1$

$-1 \neq 1$

$\theta = 270°$ is not a solution

All possible solutions, except $\theta = 270°$, produce true statements when used in place of the variable in the original equation. $\theta = 270°$ is an extraneous solution produced by squaring both sides of the equation. Our solution set is $\{90°, 180°\}$.

◀

PROBLEM SET 6.2

Solve each equation for θ if $0° \leq \theta < 360°$. Give your answers in degrees.

1. $\sqrt{3} \sec \theta = 2$ **2.** $\sqrt{2} \csc \theta = 2$

3. $\sqrt{2} \csc \theta + 5 = 3$ **4.** $2\sqrt{3} \sec \theta + 7 = 3$

5. $4 \sin \theta - 2 \csc \theta = 0$ **6.** $4 \cos \theta - 3 \sec \theta = 0$

7. $\sec \theta - 2 \tan \theta = 0$ **8.** $\csc \theta + 2 \cot \theta = 0$

9. $\sin 2\theta - \cos \theta = 0$ **10.** $2 \sin \theta + \sin 2\theta = 0$

11. $2 \cos \theta + 1 = \sec \theta$ **12.** $2 \sin \theta - 1 = \csc \theta$

Solve each equation for x if $0 \leq x < 2\pi$. Give your answers in radians using exact values only.

13. $\cos 2x - 3 \sin x - 2 = 0$ **14.** $\cos 2x - \cos x - 2 = 0$

15. $\cos x - \cos 2x = 0$ **16.** $\sin x = -\cos 2x$

17. $2 \cos^2 x + \sin x - 1 = 0$ **18.** $2 \sin^2 x - \cos x - 1 = 0$

19. $4 \sin^2 x + 4 \cos x - 5 = 0$ **20.** $4 \cos^2 x - 4 \sin x - 5 = 0$

21. $2 \sin x + \cot x - \csc x = 0$ **22.** $2 \cos x + \tan x = \sec x$

23. $\sin x + \cos x = \sqrt{2}$ **24.** $\sin x - \cos x = \sqrt{2}$

Solve for θ if $0° \leq \theta < 360°$.

25. $\sqrt{3} \sin \theta + \cos \theta = \sqrt{3}$ **26.** $\sin \theta - \sqrt{3} \cos \theta = \sqrt{3}$

27. $\sqrt{3} \sin \theta - \cos \theta = 1$ **28.** $\sin \theta - \sqrt{3} \cos \theta = 1$

29. $\sin \dfrac{\theta}{2} - \cos \theta = 0$ **30.** $\sin \dfrac{\theta}{2} + \cos \theta = 1$

31. $\cos \dfrac{\theta}{2} - \cos \theta = 1$ **32.** $\cos \dfrac{\theta}{2} - \cos \theta = 0$

For each equation, find all degree solutions in the interval $0° \leq \theta < 360°$. If rounding is necessary, round to the nearest tenth of a degree.

33. $6 \cos \theta + 7 \tan \theta = \sec \theta$ **34.** $13 \cot \theta + 11 \csc \theta = 6 \sin \theta$

35. $23 \csc^2 \theta - 22 \cot \theta \csc \theta - 15 = 0$ **36.** $18 \sec^2 \theta - 17 \tan \theta \sec \theta - 12 = 0$

37. $7 \sin^2 \theta - 9 \cos 2\theta = 0$ **38.** $16 \cos 2\theta - 18 \sin^2 \theta = 0$

Write expressions that give all solutions to the equations you solved in the problems given below.

39. Problem 3
40. Problem 4
41. Problem 23
42. Problem 24
43. Problem 31
44. Problem 32
45. In the human body, the value of θ that makes the following expression 0 is the angle at which an artery of radius r will branch off from a larger artery of radius R in order to minimize the energy loss due to friction. Show that the following expression is 0 when $\cos \theta = r^4/R^4$.

$$r^4 \csc^2 \theta - R^4 \csc \theta \cot \theta$$

46. Find the value of θ that makes the expression in Problem 45 zero, if $r = 2$ mm and $R = 4$ mm. (Give your answer to the nearest tenth of a degree.)

Solving the following equations will require you to use the quadratic formula. Solve each equation for θ between $0°$ and $360°$, and round your answers to the nearest tenth of a degree.

47. $2 \sin^2 \theta - 2 \cos \theta - 1 = 0$
48. $2 \cos^2 \theta + 2 \sin \theta - 1 = 0$
49. $\cos^2 \theta + \sin \theta = 0$
50. $\sin^2 \theta = \cos \theta$
51. $2 \sin^2 \theta = 3 - 4 \cos \theta$
52. $4 \sin \theta = 3 - 2 \cos^2 \theta$

Review Problems

The problems that follow review material we covered in Section 5.4.

If $\sin A = \dfrac{2}{3}$ with A in the interval $0° \le A \le 90°$, find

53. $\sin \dfrac{A}{2}$ **54.** $\cos \dfrac{A}{2}$

55. $\csc \dfrac{A}{2}$ **56.** $\sec \dfrac{A}{2}$

57. $\tan \dfrac{A}{2}$ **58.** $\cot \dfrac{A}{2}$

59. Graph $y = 4 \sin^2 \dfrac{x}{2}$.   **60.** Graph $y = 6 \cos^2 \dfrac{x}{2}$.

61. Use a half-angle formula to find $\sin 22.5°$.
62. Use a half-angle formula to find $\cos 15°$.

Trigonometric Equations Involving Multiple Angles

In this section, we will consider equations that contain multiple angles. We will use most of the same techniques to solve these equations that we have used in the past. We have to be careful at the last step, however, when our equations contain multiple angles. Here is an example.

▶ **EXAMPLE 1**   Solve $\cos 2\theta = \sqrt{3}/2$, if $0° \le \theta < 360°$.

Solution   The equation cannot be simplified further. Since we are looking for θ in the interval $0° \le \theta < 360°$, we must first find all values of 2θ in the interval $0° \le 2\theta < 720°$ that satisfy the equation. Here's why:

$$\text{If } 0° \le \theta < 360°, \quad \text{then} \quad 2(0°) \le 2\theta < 2(360°)$$
$$\text{or} \quad 0° \le 2\theta < 720°$$

To find all values of 2θ between $0°$ and $720°$ that satisfy our original equation, we first find all solutions to $\cos 2\theta = \sqrt{3}/2$ between $0°$ and $360°$. Then we add $360°$ to each of these solutions to obtain all solutions between $0°$ and $720°$. That is,

$$\text{If } \cos 2\theta = \frac{\sqrt{3}}{2}$$

$$\overbrace{30° + 360°} \qquad \overbrace{330° + 360°}$$
$$\text{then} \quad 2\theta = 30° \quad \text{or} \quad 2\theta = 330° \quad \text{or} \quad 2\theta = 390° \quad \text{or} \quad 2\theta = 690°$$

Dividing both sides of each equation by 2 gives us all θ between $0°$ and $360°$ that satisfy $\cos 2\theta = \sqrt{3}/2$.

$$\theta = 15° \quad \text{or} \quad \theta = 165° \quad \text{or} \quad \theta = 195° \quad \text{or} \quad \theta = 345° \quad ◀$$

If we had originally been asked to find *all solutions* to the equation in Example 1, instead of only those between $0°$ and $360°$, our work would have been a little less complicated. To find all solutions, we would first find all values of 2θ between $0°$ and $360°$ that satisfy the equation, and then we would add on multiples of $360°$, since the period of the cosine function is $360°$. After that, it is just a matter of dividing everything by 2.

$$\text{If} \quad \cos 2\theta = \frac{\sqrt{3}}{2}$$

$$\text{then} \quad 2\theta = 30° + 360°k \quad \text{or} \quad 2\theta = 330° + 360°k \qquad k \text{ is an integer}$$
$$\theta = 15° + 180°k \quad \text{or} \quad \theta = 165° + 180°k \qquad \text{Divide by 2}$$

The last line gives us *all values* of θ that satisfy $\cos 2\theta = \sqrt{3}/2$. Note that $k = 0$ and $k = 1$ will give us the four solutions between $0°$ and $360°$ that we found in Example 1.

▶ **Example 2**   Find all solutions to $\tan 3x = 1$, if x is measured in radians with exact values.

Solution   First we find all values of $3x$ in the interval $0 \le 3x < \pi$ that satisfy $\tan 3x = 1$, and then we add on multiples of π because the period of the tangent function is π. After that, we simply divide by 3 to solve for x.

$$\text{If } \tan 3x = 1$$

$$\text{then } 3x = \frac{\pi}{4} + k\pi \qquad k \text{ is an integer}$$

$$x = \frac{\pi}{12} + \frac{k\pi}{3} \qquad \text{Divide by 3}$$

Note that $k = 0$, 1, and 2 will give us all values of x between 0 and π that satisfy $\tan 3x = 1$.   ◀

▶ **EXAMPLE 3**   Solve: $\sin 2x \cos x + \cos 2x \sin x = 1/\sqrt{2}$, if $0 \le x < 2\pi$.

Solution   We can simplify the left side by using the formula for $\sin (A + B)$.

$$\sin 2x \cos x + \cos 2x \sin x = \frac{1}{\sqrt{2}}$$

$$\sin (2x + x) = \frac{1}{\sqrt{2}}$$

$$\sin 3x = \frac{1}{\sqrt{2}}$$

First we find *all* possible solutions for x:

$$3x = \frac{\pi}{4} + 2k\pi \qquad \text{or} \qquad 3x = \frac{3\pi}{4} + 2k\pi \qquad k \text{ is an integer}$$

$$x = \frac{\pi}{12} + \frac{2k\pi}{3} \qquad \text{or} \qquad x = \frac{\pi}{4} + \frac{2k\pi}{3} \qquad \text{Divide by 3}$$

To find those solutions that lie in the interval $0 \le x < 2\pi$, we let k take on values of 0, 1, and 2. Doing so results in the following solutions:

$$x = \frac{\pi}{12}, \ \frac{\pi}{4}, \ \frac{3\pi}{4}, \ \frac{11\pi}{12}, \ \frac{17\pi}{12}, \ \text{and} \ \frac{19\pi}{12}$$ ◀

▶ **EXAMPLE 4**   Find all solutions to $2 \sin^2 3\theta - \sin 3\theta - 1 = 0$, if θ is measured in degrees.

Solution   We have an equation that is quadratic in $\sin 3\theta$. We factor and solve as usual.

$$2 \sin^2 3\theta - \sin 3\theta - 1 = 0 \qquad \text{Standard form}$$
$$(2 \sin 3\theta + 1)(\sin 3\theta - 1) = 0 \qquad \text{Factor}$$
$$2 \sin 3\theta + 1 = 0 \quad \text{or} \quad \sin 3\theta - 1 = 0 \qquad \text{Set factors to 0}$$
$$\sin 3\theta = -\frac{1}{2} \quad \text{or} \quad \sin 3\theta = 1$$
$$3\theta = 210° + 360°k \quad \text{or} \quad 3\theta = 330° + 360°k \quad \text{or} \quad 3\theta = 90° + 360°k$$
$$\theta = 70° + 120°k \quad \text{or} \quad \theta = 110° + 120°k \quad \text{or} \quad \theta = 30° + 120°k \quad \blacktriangleleft$$

▶ **EXAMPLE 5** Find all solutions, in radians, for $\tan^2 3x = 1$.

Solution Taking the square root of both sides we have

$$\tan^2 3x = 1$$
$$\tan 3x = \pm 1 \qquad \text{Square root of both sides}$$

Since the period of the tangent function is π, we find all solutions to $\tan 3x = 1$ and $\tan 3x = -1$ between 0 and π, and then we add multiples of π to these solutions. Finally, we divide each side of the resulting equations by 3.

$$3x = \frac{\pi}{4} + k\pi \qquad \text{or} \qquad \frac{3\pi}{4} + k\pi$$

$$x = \frac{\pi}{12} + \frac{k\pi}{3} \qquad \text{or} \qquad \frac{\pi}{4} + \frac{k\pi}{3} \qquad\qquad \blacktriangleleft$$

▶ **EXAMPLE 6** Solve $\sin \theta - \cos \theta = 1$ if $0° \le \theta < 360°$.

Solution We have solved this equation before—in Section 6.2. Here is a second solution.

$$\sin \theta - \cos \theta = 1$$
$$(\sin \theta - \cos \theta)^2 = 1^2 \qquad \text{Square both sides}$$
$$\sin^2 \theta - 2 \sin \theta \cos \theta + \cos^2 \theta = 1 \qquad \text{Expand left side}$$
$$-2 \sin \theta \cos \theta + 1 = 1 \qquad \sin^2 \theta + \cos^2 \theta = 1$$
$$-2 \sin \theta \cos \theta = 0 \qquad \text{Add } -1 \text{ to both sides}$$
$$-\sin 2\theta = 0 \qquad \text{Double-angle identity}$$
$$2\theta = 0°, 180°, 360°, 540°$$
$$\theta = 0°, 90°, 180°, 270°$$

Since we squared both sides of the equation in Step 2, we must check all the possible solutions to see if they satisfy the original equation. Doing so gives us solutions $\theta = 90°$ and $180°$. The other two are extraneous. ◀

PROBLEM SET 6.3

Find all solutions if $0° \leq \theta < 360°$.

1. $\sin 2\theta = \sqrt{3}/2$
2. $\sin 2\theta = -\sqrt{3}/2$
3. $\tan 2\theta = -1$
4. $\cot 2\theta = 1$
5. $\cos 3\theta = -1$
6. $\sin 3\theta = -1$

Find all solutions if $0 \leq x < 2\pi$. Use exact values only.

7. $\sin 2x = 1/\sqrt{2}$
8. $\cos 2x = 1/\sqrt{2}$
9. $\sec 3x = -1$
10. $\csc 3x = 1$
11. $\tan 2x = \sqrt{3}$
12. $\tan 2x = -\sqrt{3}$

Find all degree solutions for each of the following:

13. $\sin 2\theta = 1/2$
14. $\sin 2\theta = -\sqrt{3}/2$
15. $\cos 3\theta = 0$
16. $\cos 3\theta = -1$
17. $\sin 10\theta = \sqrt{3}/2$
18. $\cos 8\theta = 1/2$

Find all solutions if $0 \leq x < 2\pi$. Use exact values only.

19. $\sin 2x \cos x + \cos 2x \sin x = 1/2$
20. $\sin 2x \cos x + \cos 2x \sin x = -1/2$
21. $\cos 2x \cos x - \sin 2x \sin x = -\sqrt{3}/2$
22. $\cos 2x \cos x - \sin 2x \sin x = 1/\sqrt{2}$

Find all solutions in radians using exact values only.

23. $\sin 3x \cos 2x + \cos 3x \sin 2x = 1$
24. $\sin 2x \cos 3x + \cos 2x \sin 3x = -1$
25. $\sin^2 4x = 1$
26. $\cos^2 4x = 1$
27. $\cos^3 5x = -1$
28. $\sin^3 5x = -1$

Find all degree solutions.

29. $2 \sin^2 3\theta + \sin 3\theta - 1 = 0$
30. $2 \sin^2 3\theta + 3 \sin 3\theta + 1 = 0$
31. $2 \cos^2 2\theta + 3 \cos 2\theta + 1 = 0$
32. $2 \cos^2 2\theta - \cos 2\theta - 1 = 0$
33. $\tan^2 3\theta = 3$
34. $\cot^2 3\theta = 1$

Find all solutions if $0° \leq \theta < 360°$.

35. $\cos \theta - \sin \theta = 1$
36. $\sin \theta - \cos \theta = 1$
37. $\sin \theta + \cos \theta = -1$
38. $\cos \theta - \sin \theta = -1$
39. $\sin^2 2\theta - 4 \sin 2\theta - 1 = 0$
40. $\cos^2 3\theta - 6 \cos 3\theta + 4 = 0$
41. $4 \cos^2 3\theta - 8 \cos 3\theta + 1 = 0$
42. $2 \sin^2 2\theta - 6 \sin 2\theta + 3 = 0$
43. $2 \cos^2 4\theta + 2 \sin 4\theta = 1$
44. $2 \sin^2 4\theta - 2 \cos 4\theta = 1$

45. In Example 7 of Section 4.4, we found the equation that gives the height h of a passenger on a Ferris wheel at any time t during the ride to be

$$h = 139 - 125 \cos \frac{\pi}{10} t$$

where h is given in feet and t is given in minutes. Use this equation to find the time at which a passenger will be 100 ft above the ground. Round your answer to the nearest tenth of a minute.

46. In Problem 31 of Problem Set 4.4, you found the equation that gives the height h of a passenger on a Ferris wheel at any time t during the ride to be

$$h = 110.5 - 98.5 \cos \frac{2\pi}{15} t$$

where the units for h are feet and the units for t are minutes. Use this equation to find the time at which a passenger will be 100 ft above the ground. Round your answer to the nearest tenth of a minute.

47. The formula below gives the relationship between the number of sides n, the radius r, and the length of each side l in a regular polygon. Find n, if $l = r$.

$$l = 2r \sin \frac{180°}{n}$$

48. If central angle θ cuts off a chord of length c in a circle of radius r, then the relationship between θ, c, and r is given by

$$2r \sin \frac{\theta}{2} = c$$

Find θ, if $c = \sqrt{3}r$.

49. In Example 4 of Section 3.5, we found the equation that gives d in terms of t in Figure 1 to be $d = 10 \tan \pi t$. If a person is standing against the wall, 10 ft from point A, how long after the light is at point A will the person see the light? (*Hint:* You must find t when d is 10.)

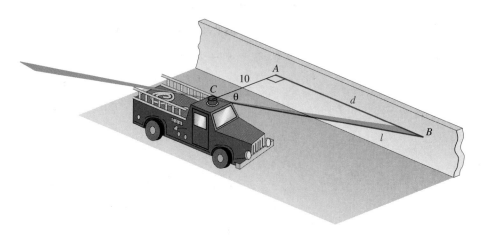

Figure 1

50. In Problem 21 of Problem Set 3.5, you found the equation that gives d in terms of t in Figure 2 to be $d = 100 \tan \frac{1}{2} \pi t$. Two people are sitting on the wall. One of them is directly opposite the lighthouse, while the other person is 100 ft further down the wall. How long after one of them sees the light does the other one see the light? (*Hint:* There are two solutions depending on who sees the light first.)

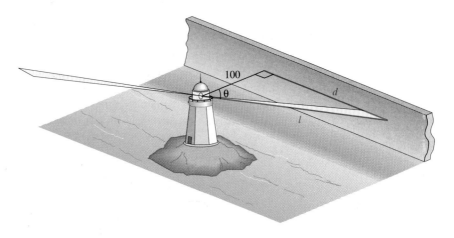

Figure 2

51. Find the smallest positive value of t for which $\sin 2\pi t = \frac{1}{2}$.

52. Find the smallest positive value of t for which $\sin 2\pi t = 1/\sqrt{2}$.

Review Problems

The problems that follow review material we covered in Sections 5.1 and 5.4.

Prove each identity.

53. $\dfrac{\sin x}{1 + \cos x} = \dfrac{1 - \cos x}{\sin x}$

54. $\dfrac{\sin^2 x}{(1 - \cos x)^2} = \dfrac{1 + \cos x}{1 - \cos x}$

55. $\dfrac{1}{1 + \cos t} + \dfrac{1}{1 - \cos t} = 2\csc^2 t$

56. $\dfrac{1}{1 - \sin t} + \dfrac{1}{1 + \sin t} = 2\sec^2 t$

57. $\tan \dfrac{A}{2} = \dfrac{\sin A}{1 + \cos A}$

58. $\tan \dfrac{A}{2} = \dfrac{1 - \cos A}{\sin A}$

If $\sin A = \dfrac{1}{3}$ with $90° \le A \le 180°$ and $\sin B = \dfrac{3}{5}$ with $0° \le B \le 90°$, find each of the following.

59. $\sin 2A$ **60.** $\cos 2B$ **61.** $\cos \dfrac{A}{2}$ **62.** $\sin \dfrac{B}{2}$

63. $\sin (A + B)$ **64.** $\cos (A - B)$ **65.** $\csc (A + B)$ **66.** $\sec (A - B)$

Parametric Equations and Further Graphing

Up to this point we have had a number of encounters with the Ferris wheel problem. In Section 4.4, we were able to derive an equation that gives the rider's height above the ground at any time t during the ride. We did this by first looking at the graph and then deriving the equation from that graph. Is there a simple way to proceed directly from the diagram of the Ferris wheel problem to an equation that gives height as a function of time? There is, and one of the easiest ways to do so is with *parametric equations,* which we will study in this section.

Many times in mathematics a set of points (x, y) in the plane is described by a pair of equations rather than one equation. For example, $x = \cos t$ and $y = \sin t$, where t is any real number, are a pair of equations that describe a certain curve in the xy-plane. The equations are called *parametric equations* and the variable t is called the *parameter.* The variable t does not appear as part of the graph but rather produces values of x and y that do appear as ordered pairs (x, y) on the graph.

Table 1 shows the values of x and y produced by substituting convenient values of t into each equation.

Plotting each ordered pair (x, y) from the last column of Table 1 on a coordinate system, we see that the set of points form the unit circle, starting at $(1, 0)$ when $t = 0$ and ending at $(1, 0)$ when $t = 2\pi$. (See Figure 1.)

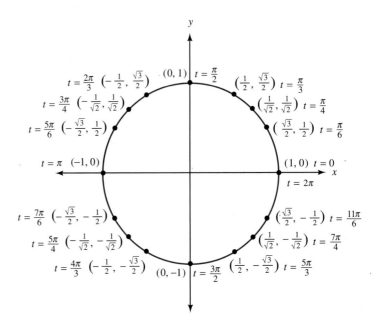

Figure 1

We could have found this graph without going to as much work by using the Pythagorean identity $\cos^2 t + \sin^2 t = 1$. Substituting $x = \cos t$ and $y = \sin t$ into

$$\cos^2 t + \sin^2 t = 1$$

we have

$$x^2 + y^2 = 1$$

which is the equation of the unit circle.

The process of going directly to an equation that contains x and y but not t is called *eliminating the parameter.*

Table 1

t	$x = \cos t$	$y = \sin t$	(x, y)
0	$x = \cos 0 = 1$	$y = \sin 0 = 0$	$(1, 0)$
$\dfrac{\pi}{6}$	$x = \cos \dfrac{\pi}{6} = \dfrac{\sqrt{3}}{2}$	$y = \sin \dfrac{\pi}{6} = \dfrac{1}{2}$	$\left(\dfrac{\sqrt{3}}{2}, \dfrac{1}{2}\right)$
$\dfrac{\pi}{4}$	$x = \cos \dfrac{\pi}{4} = \dfrac{1}{\sqrt{2}}$	$y = \sin \dfrac{\pi}{4} = \dfrac{1}{\sqrt{2}}$	$\left(\dfrac{1}{\sqrt{2}}, \dfrac{1}{\sqrt{2}}\right)$
$\dfrac{\pi}{3}$	$x = \cos \dfrac{\pi}{3} = \dfrac{1}{2}$	$y = \sin \dfrac{\pi}{3} = \dfrac{\sqrt{3}}{2}$	$\left(\dfrac{1}{2}, \dfrac{\sqrt{3}}{2}\right)$
$\dfrac{\pi}{2}$	$x = \cos \dfrac{\pi}{2} = 0$	$y = \sin \dfrac{\pi}{2} = 1$	$(0, 1)$
$\dfrac{2\pi}{3}$	$x = \cos \dfrac{2\pi}{3} = -\dfrac{1}{2}$	$y = \sin \dfrac{2\pi}{3} = \dfrac{\sqrt{3}}{2}$	$\left(-\dfrac{1}{2}, \dfrac{\sqrt{3}}{2}\right)$
$\dfrac{3\pi}{4}$	$x = \cos \dfrac{3\pi}{4} = -\dfrac{1}{\sqrt{2}}$	$y = \sin \dfrac{3\pi}{4} = \dfrac{1}{\sqrt{2}}$	$\left(-\dfrac{1}{\sqrt{2}}, \dfrac{1}{\sqrt{2}}\right)$
$\dfrac{5\pi}{6}$	$x = \cos \dfrac{5\pi}{6} = -\dfrac{\sqrt{3}}{2}$	$y = \sin \dfrac{5\pi}{6} = \dfrac{1}{2}$	$\left(-\dfrac{\sqrt{3}}{2}, \dfrac{1}{2}\right)$
π	$x = \cos \pi = -1$	$y = \sin \pi = 0$	$(-1, 0)$
$\dfrac{7\pi}{6}$	$x = \cos \dfrac{7\pi}{6} = -\dfrac{\sqrt{3}}{2}$	$y = \sin \dfrac{7\pi}{6} = -\dfrac{1}{2}$	$\left(-\dfrac{\sqrt{3}}{2}, -\dfrac{1}{2}\right)$
$\dfrac{5\pi}{4}$	$x = \cos \dfrac{5\pi}{4} = -\dfrac{1}{\sqrt{2}}$	$y = \sin \dfrac{5\pi}{4} = -\dfrac{1}{\sqrt{2}}$	$\left(-\dfrac{1}{\sqrt{2}}, -\dfrac{1}{\sqrt{2}}\right)$
$\dfrac{4\pi}{3}$	$x = \cos \dfrac{4\pi}{3} = -\dfrac{1}{2}$	$y = \sin \dfrac{4\pi}{3} = -\dfrac{\sqrt{3}}{2}$	$\left(-\dfrac{1}{2}, -\dfrac{\sqrt{3}}{2}\right)$
$\dfrac{3\pi}{2}$	$x = \cos \dfrac{3\pi}{2} = 0$	$y = \sin \dfrac{3\pi}{2} = -1$	$(0, -1)$
$\dfrac{5\pi}{3}$	$x = \cos \dfrac{5\pi}{3} = \dfrac{1}{2}$	$y = \sin \dfrac{5\pi}{3} = -\dfrac{\sqrt{3}}{2}$	$\left(\dfrac{1}{2}, -\dfrac{\sqrt{3}}{2}\right)$
$\dfrac{7\pi}{4}$	$x = \cos \dfrac{7\pi}{4} = \dfrac{1}{\sqrt{2}}$	$y = \sin \dfrac{7\pi}{4} = -\dfrac{1}{\sqrt{2}}$	$\left(\dfrac{1}{\sqrt{2}}, -\dfrac{1}{\sqrt{2}}\right)$
$\dfrac{11\pi}{6}$	$x = \cos \dfrac{11\pi}{6} = \dfrac{\sqrt{3}}{2}$	$y = \sin \dfrac{11\pi}{6} = -\dfrac{1}{2}$	$\left(\dfrac{\sqrt{3}}{2}, -\dfrac{1}{2}\right)$
2π	$x = \cos 2\pi = 1$	$y = \sin 2\pi = 0$	$(1, 0)$

▶ **EXAMPLE 1** Eliminate the parameter t from the parametric equations

$$x = 3 \cos t \qquad y = 2 \sin t$$

Solution Again, we will use the identity $\cos^2 t + \sin^2 t = 1$. Before we do so, however, we must solve the first equation for $\cos t$ and the second equation for $\sin t$.

$$x = 3 \cos t \Rightarrow \cos t = \frac{x}{3}$$

$$y = 2 \sin t \Rightarrow \sin t = \frac{y}{2}$$

Substituting $x/3$ and $y/2$ for $\cos t$ and $\sin t$ into the first Pythagorean identity gives us

$$\left(\frac{x}{3}\right)^2 + \left(\frac{y}{2}\right)^2 = 1$$

$$\frac{x^2}{9} + \frac{y^2}{4} = 1$$

which is the equation of an ellipse. The center is at the origin, the x-intercepts are 3 and -3, and the y-intercepts are 2 and -2. Figure 2 shows the graph.

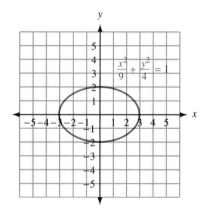

Figure 2

▶ **EXAMPLE 2** Eliminate the parameter t from the equations

$$x = 3 + \sin t \qquad y = \cos t - 2$$

Solution Solving the first equation for $\sin t$ and the second equation for $\cos t$, we have

$$\sin t = x - 3 \qquad \text{and} \qquad \cos t = y + 2$$

Substituting these expressions into the Pythagorean identity for sin t and cos t gives us

$$(x - 3)^2 + (y + 2)^2 = 1$$

which is the equation of a circle with a radius of 1 and center at $(3, -2)$.

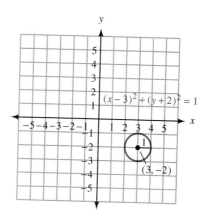

Figure 3   ◀

▶ **EXAMPLE 3**   Eliminate the parameter t.

$$x = 3 + 2 \sec t \qquad y = 2 + 4 \tan t$$

Solution   In this case, we solve for sec t and tan t and then use the identity $1 + \tan^2 t = \sec^2 t$.

$$x = 3 + 2 \sec t \Rightarrow \sec t = \frac{x - 3}{2}$$

$$y = 2 + 4 \tan t \Rightarrow \tan t = \frac{y - 2}{4}$$

$$1 + \left(\frac{y - 2}{4}\right)^2 = \left(\frac{x - 3}{2}\right)^2$$

or

$$\frac{(x - 3)^2}{4} - \frac{(y - 2)^2}{16} = 1$$

This is the equation of a hyperbola.

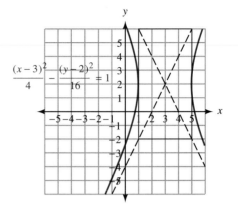

Figure 4 ◀

▶ **EXAMPLE 4** Figure 5 shows a diagram of the Ferris wheel built by George Ferris in 1893. The diameter of this wheel is 250 ft, it rotates through one complete revolution every 20 minutes, and position P_0 is 14 ft above the ground. Find the parametric equations of any point $P(x, y)$ on the wheel. Write the equations in terms of the parameter θ and then in terms of time t.

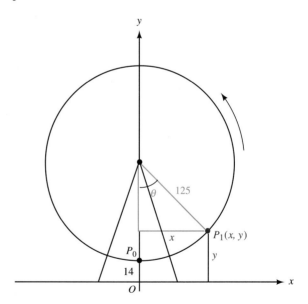

Figure 5

Solution The triangle containing angle θ in Figure 5 is reproduced in Figure 6. Note that the horizontal side of the triangle is simply x. The vertical side is the difference between 139 (the distance from the origin to the center of the wheel) and the y-coordinate of point P_1.

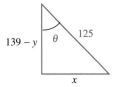

Figure 6

Using right-triangle trigonometry, we write the following relationships:

$$\sin \theta = \frac{x}{125}$$

$$\cos \theta = \frac{139 - y}{125}$$

Solving these equations for x and y, we have

$$x = 125 \sin \theta$$
$$y = 139 - 125 \cos \theta$$

These are parametric equations that give us any point (x, y) on the circumference of the wheel as a function of θ.

To rewrite our equations in terms of time t, we use the fact that a rider will travel once around the wheel every 20 minutes.

$$\frac{2\pi}{20 \text{ minutes}} = \frac{\theta}{t}$$

Solving for θ, we have

$$\theta = \frac{\pi}{10} t$$

We substitute this expression for θ into our parametric equations to obtain

$$\left. \begin{array}{l} x = 125 \sin \dfrac{\pi}{10} t \\[4mm] y = 139 - 125 \cos \dfrac{\pi}{10} t \end{array} \right\} \begin{array}{l} \text{Parametric equations with } x \text{ and } y \\ \text{as functions of } t \end{array}$$

If we graph all the ordered pairs (x, y) using either of our two sets of parametric equations, we produce the Ferris wheel shown earlier in Figure 5. On the other hand, if we graph each of the parametric equations above separately, as if they were two separate functions of t, we have the two curves shown in Figures 7 and 8. Notice that the horizontal axes are the t axes, while the vertical axes are labeled x and y, respectively.

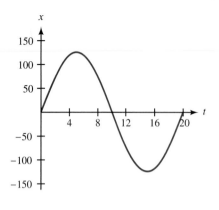

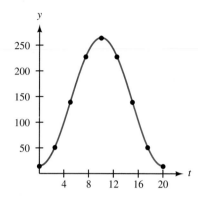

Figure 7   **Figure 8**

Figure 7 shows the distance the rider is from the y-axis at any time t during the ride. (If the sun were directly overhead during the ride, this curve would describe the motion of the shadow of the rider.) Figure 8 gives us the height y the rider is above the ground at time t. If you look back to Example 7 in Section 4.4, you will see that this is the same graph we encountered there.   ◀

PROBLEM SET 6.4

Eliminate the parameter t from each of the following and then sketch the graph of each:

1. $x = \sin t \quad y = \cos t$ **2.** $x = -\sin t \quad y = \cos t$

3. $x = 3 \cos t \quad y = 3 \sin t$ **4.** $x = 2 \cos t \quad y = 2 \sin t$

5. $x = 2 \sin t \quad y = 4 \cos t$ **6.** $x = 3 \sin t \quad y = 4 \cos t$

7. $x = 2 + \sin t \quad y = 3 + \cos t$ **8.** $x = 3 + \sin t \quad y = 2 + \cos t$

9. $x = \sin t - 2 \quad y = \cos t - 3$ **10.** $x = \cos t - 3 \quad y = \sin t + 2$

11. $x = 3 + 2 \sin t \quad y = 1 + 2 \cos t$ **12.** $x = 2 + 3 \sin t \quad y = 1 + 3 \cos t$

13. $x = 3 \cos t - 3 \quad y = 3 \sin t + 1$ **14.** $x = 4 \sin t - 5 \quad y = 4 \cos t - 3$

Eliminate the parameter t in each of the following:

15. $x = \sec t \quad y = \tan t$ **16.** $x = \tan t \quad y = \sec t$

17. $x = 3 \sec t \quad y = 3 \tan t$ **18.** $x = 3 \cot t \quad y = 3 \csc t$

19. $x = 2 + 3 \tan t \quad y = 4 + 3 \sec t$ **20.** $x = 3 + 5 \tan t \quad y = 2 + 5 \sec t$

21. $x = \cos 2t \quad y = \sin t$ **22.** $x = \cos 2t \quad y = \cos t$

23. $x = \sin t \quad y = \sin t$ **24.** $x = \cos t \quad y = \cos t$

25. $x = 3 \sin t \quad y = 2 \sin t$ **26.** $x = 2 \sin t \quad y = 3 \sin t$

27. Figure 9 is a model of the Ferris wheel built in Vienna in 1897. The diameter of the wheel is 197 ft, and one complete revolution takes 15 minutes. The bottom of the wheel is 12 ft above the ground. Use the diagram in Figure 9 to find the parametric equations of any point $P_1(x, y)$ on the wheel. Write the equations in terms of the parameter θ and then in terms of time t.

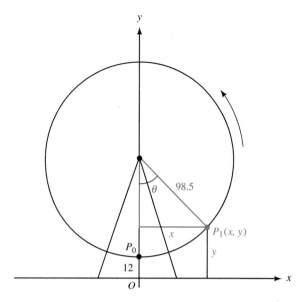

Figure 9

28. Figure 10 is a model of the Ferris wheel called Colossus that was built in St. Louis in 1986. The diameter of the wheel is 165 ft, it rotates at 1.5 rpm, and the bottom of the wheel is 9 ft above the ground. Find the parametric equations of any point $P_1(x, y)$ on the wheel. Write the equations in terms of the parameter θ and then in terms of time t.

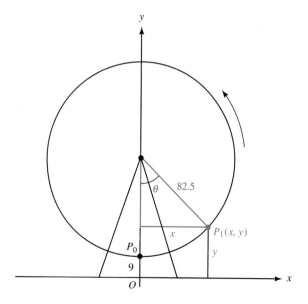

Figure 10

Review Problems

The problems that follow review material we covered in Sections 4.6 and 5.5.

Evaluate each expression. (Assume all variables represent positive numbers.)

29. $\sin\left(\cos^{-1}\dfrac{1}{2}\right)$

30. $\tan\left(\sin^{-1}\dfrac{1}{3}\right)$

31. $\sin\left(\tan^{-1}\dfrac{1}{3} + \sin^{-1}\dfrac{1}{4}\right)$

32. $\cos\left(\tan^{-1}\dfrac{2}{3} + \cos^{-1}\dfrac{1}{3}\right)$

33. $\cos\left(\sin^{-1}x\right)$

34. $\sin\left(\cos^{-1}x\right)$

35. $\cos\left(2\tan^{-1}x\right)$

36. $\sin\left(2\tan^{-1}x\right)$

37. Write $8\sin 3x \cos 2x$ as a sum.

38. Write $\sin 8x + \sin 4x$ as a product.

CHAPTER 6 SUMMARY

Examples

1. Solve for x:

$$2\cos x - \sqrt{3} = 0.$$
$$2\cos x - \sqrt{3} = 0$$
$$2\cos x = \sqrt{3}$$
$$\cos x = \frac{\sqrt{3}}{2}$$

Solutions Between $0°$ and $360°$ or 0 and 2π

In Degrees	In Radians
$x = 30°$	$x = \dfrac{\pi}{6}$
or	or
$x = 330°$	$x = \dfrac{11\pi}{6}$

All Solutions (k is an integer)

In Degrees	In Radians
$x = 30° + 360°k$	$x = \dfrac{\pi}{6} + 2k\pi$
or	or
$x = 330° + 360°k$	$x = \dfrac{11\pi}{6} + 2k\pi$

Solving Simple Trigonometric Equations [6.1]

We solve linear equations in trigonometry by applying the properties of equality developed in algebra. The two most important properties from algebra are stated as follows:

Addition Property of Equality

For any three algebraic expressions A, B, and C,

$$\text{If } A = B$$
$$\text{then } A + C = B + C$$

In Words: Adding the same quantity to both sides of an equation will not change the solution set.

Multiplication Property of Equality

For any three algebraic expressions A, B, and C with $C \neq 0$,

$$\text{If } A = B$$
$$\text{then } AC = BC$$

In Words: Multiplying both sides of an equation by the same nonzero quantity will not change the solution set.

To solve a trigonometric equation that is quadratic in $\sin x$ or $\cos x$, we write it in standard form and then factor it or use the quadratic formula.

2. Solve if $0° \leq \theta < 360°$:

Using Identities in Trigonometric Equations [6.2]

$$\cos 2\theta + 3 \sin \theta - 2 = 0$$
$$1 - 2 \sin^2 \theta + 3 \sin \theta - 2 = 0$$
$$2 \sin^2 \theta - 3 \sin \theta + 1 = 0$$
$$(2 \sin \theta - 1)(\sin \theta - 1) = 0$$
$$2 \sin \theta - 1 = 0 \text{ or } \sin \theta - 1 = 0$$
$$\sin \theta = \frac{1}{2} \qquad \sin \theta = 1$$
$$\theta = 30°, 150°, 90°$$

Sometimes it is necessary to use identities to make trigonometric substitutions when solving equations. Identities are usually required if the equation contains more than one trigonometric function or if there is more than one angle named in the equation. In the example to the left, we begin by replacing $\cos 2\theta$ with $1 - 2 \sin^2 \theta$. Doing so gives us a quadratic equation in $\sin \theta$, which we put in standard form and solve by factoring.

3. $\sin 2x \cos x +$

Equations Involving Multiple Angles [6.3]

$$\cos 2x \sin x = \frac{1}{\sqrt{2}}$$
$$\sin (2x + x) = \frac{1}{\sqrt{2}}$$
$$\sin 3x = \frac{1}{\sqrt{2}}$$

$$3x = \frac{\pi}{4} + 2k\pi \quad \text{or} \quad 3x = \frac{3\pi}{4} + 2k\pi$$

$$x = \frac{\pi}{12} + \frac{2k\pi}{3} \quad \text{or} \quad x = \frac{\pi}{4} + \frac{2k\pi}{3}$$

where k is an integer.

Sometimes the equations we solve in trigonometry reduce to equations that contain multiple angles. When this occurs, we have to be careful in the last step that we do not leave out any solutions. For instance, if we are asked to find all solutions between $x = 0$ and $x = 2\pi$, and our final equation contains $2x$, we must find all values of $2x$ between 0 and 4π in order that x remain between 0 and 2π.

4. Eliminate the parameter t from the equations

$$x = 3 + \sin t \quad y = \cos t - 2$$

Solving for $\sin t$ and $\cos t$ we have

$$\sin t = x - 3 \quad \text{and}$$
$$\cos t = y + 2$$

Substituting these expressions into the Pythagorean identity, we have

$$(x - 3)^2 + (y + 2)^2 = 1$$

which is the equation of a circle with a radius of 1 and center at $(3, -2)$.

Parametric Equations [6.4]

When the coordinates of point (x, y) are described separately by two equations of the form $x = f(t)$ and $y = g(t)$, then the two equations are called *parametric equations* and t is called the *parameter*. One way to graph a set of points (x, y) that are given in terms of the parameter t is to eliminate the parameter and obtain an equation in just x and y that gives the same set of points (x, y).

CHAPTER 6 TEST

Find all solutions in the interval $0° \leq \theta < 360°$. If rounding is necessary, round to the nearest tenth of a degree.

1. $2 \sin \theta - 1 = 0$ **2.** $\sqrt{3} \tan \theta + 1 = 0$

3. $\cos \theta - 2 \sin \theta \cos \theta = 0$ **4.** $\tan \theta - 2 \cos \theta \tan \theta = 0$

5. $4 \cos \theta - 2 \sec \theta = 0$ **6.** $2 \sin \theta - \csc \theta = 1$

7. $\sin \dfrac{\theta}{2} + \cos \theta = 0$ **8.** $\cos \dfrac{\theta}{2} - \cos \theta = 0$

9. $4 \cos 2\theta + 2 \sin \theta = 1$ **10.** $\sin (3\theta - 45°) = -\dfrac{\sqrt{3}}{2}$

11. $\sin \theta + \cos \theta = 1$ **12.** $\sin \theta - \cos \theta = 1$

Find all solutions for the following equations. Write your answers in radians using exact values.

13. $\cos 2x - 3 \cos x = -2$ **14.** $\sqrt{3} \sin x - \cos x = 0$

15. $\sin 2x \cos x + \cos 2x \sin x = -1$ **16.** $\sin^3 4x = 1$

Find all solutions, to the nearest tenth of a degree, in the interval $0° \leq \theta < 360°$.

17. $5 \sin^2 \theta - 3 \sin \theta = 2$ **18.** $4 \cos^2 \theta - 4 \cos \theta = 2$

Eliminate the parameter t from each of the following and then sketch the graph.

19. $x = 3 \cos t \quad y = 3 \sin t$ **20.** $x = \sec t \quad y = \tan t$

21. $x = 3 + 2 \sin t \quad y = 1 + 2 \cos t$

22. $x = 3 \cos t - 3 \quad y = 3 \sin t + 1$

7

Triangles

I   I   I   I   I   I   I   I   I   I

▶ **To the Student**

In Chapter 2, we gave a definition of the six trigonometric functions of an acute angle in a right triangle. Since then, we have used this definition in a number of situations to solve for the different parts of right triangles. In this chapter, we are going to extend our work with triangles to include those that are not necessarily right triangles.

To be successful in this chapter, you should have a good working knowledge of the material we covered in Chapters 2 and 3 and of the Pythagorean identities we derived in Chapter 1. You also need to be proficient at using a calculator to find the value of a trigonometric function or to find an angle when given one of its trigonometric functions.

SECTION 7.1 | **The Law of Sines**

There are many relationships that exist between the sides and angles in a triangle. One such relationship is called the *law of sines,* which states that the ratio of the sine of an angle to the length of the side opposite that angle is constant in any triangle.

Here it is stated in symbols:

Law of Sines

$$\frac{\sin A}{a} = \frac{\sin B}{b} = \frac{\sin C}{c}$$

or, equivalently,

$$\frac{a}{\sin A} = \frac{b}{\sin B} = \frac{c}{\sin C}$$

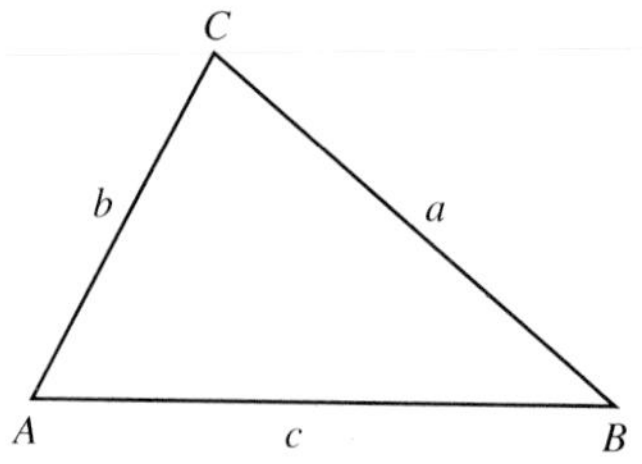

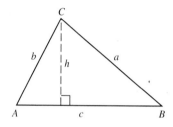

Figure 1

PROOF   The altitude h of the triangle in Figure 2 can be written in terms of $\sin A$ or $\sin B$ depending on which of the two right triangles we are referring to:

$$\sin A = \frac{h}{b} \qquad \sin B = \frac{h}{a}$$

$$h = b \sin A \qquad h = a \sin B$$

Figure 2

Since h is equal to itself, we have

$$h = h$$

$$b \sin A = a \sin B$$

$$\frac{b \sin A}{ab} = \frac{a \sin B}{ab} \qquad \text{Divide both sides by } ab$$

$$\frac{\sin A}{a} = \frac{\sin B}{b} \qquad \text{Divide out common factors}$$

If we do the same kind of thing with the altitude that extends from A, we will have the third ratio in the law of sines, $\dfrac{\sin C}{c}$, equal to the two ratios above.

Note that the derivation of the law of sines will proceed in the same manner if triangle ABC contains an obtuse angle, as in Figure 3.

In triangle BDC we have

$$\sin (180° - B) = \frac{h}{a}$$

but,

$$\sin (180° - B) = \sin 180° \cos B - \cos 180° \sin B$$
$$= (0) \cos B - (-1) \sin B$$
$$= \sin B$$

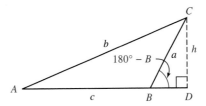

Figure 3

So, $\sin B = h/a$, which is the result we obtained previously. Using triangle ADC, we have $\sin A = h/b$. As you can see, these are the same two expressions we began with when deriving the law of sines for the acute triangle in Figure 2. From this point on, the derivation would match our previous derivation.

We can use the law of sines to find missing parts of triangles for which we are given two angles and a side.

Two Angles and One Side

In our first example, we are given two angles and the side opposite one of them. (You may recall that in geometry these were the parts we needed equal in two triangles in order to prove them congruent using the AAS Theorem.)

▶ **EXAMPLE 1**   In triangle ABC, $A = 30°$, $B = 70°$, and $a = 8.0$ centimeters. Find the length of side c.

Solution   We begin by drawing a picture of triangle ABC (it does not have to be accurate) and labeling it with the information we have been given.

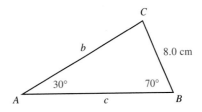

Figure 4

When we use the law of sines, we must have one of the ratios given to us. In this case, since we are given a and A, we have the ratio $\dfrac{a}{\sin A}$. To solve for c, we need first to find angle C. Since the sum of the angles in any triangle is 180°, we have

$$\begin{aligned} C &= 180° - (A + B) \\ &= 180° - (30° + 70°) \\ &= 80° \end{aligned}$$

To find side c, we use the following two ratios given in the law of sines.

$$\frac{c}{\sin C} = \frac{a}{\sin A}$$

To solve for c, we multiply both sides by $\sin C$ and then substitute.

$$\begin{aligned} c &= \frac{a \sin C}{\sin A} & &\text{Multiply both sides by } \sin C \\[2mm] &= \frac{8.0 \sin 80°}{\sin 30°} & &\text{Substitute in given values} \\[2mm] &= \frac{8.0(0.9848)}{0.5000} & &\text{Calculator} \\[2mm] &= 16 \text{ centimeters} & &\text{Rounded to the nearest integer} \end{aligned}$$ ◀

Note The equal sign in the third line above should actually be replaced by the *approximately equal to* symbol, ≈, since the decimal 0.9848 is an approximation to sin 80°. (Remember, most of the trigonometric functions are irrational numbers.) In this chapter, we will use an equal sign in the solutions to all of our examples, even when the ≈ symbol would be more appropriate, in order to make the examples a little easier to follow.

In our next example, we are given two angles and the side included between them (ASA) and are asked to find all the missing parts.

▶ **EXAMPLE 2** Solve triangle ABC if $B = 34°$, $C = 82°$, and $a = 5.6$ centimeters.

Solution We begin by finding angle A so that we have one of the ratios in the law of sines completed.

Angle A

$$\begin{aligned} A &= 180° - (B + C) \\ &= 180° - (34° + 82°) \\ &= 64° \end{aligned}$$

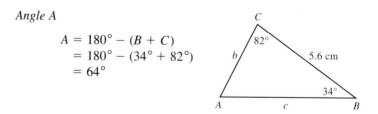

Figure 5

Side b

If $\dfrac{b}{\sin B} = \dfrac{a}{\sin A}$

then $b = \dfrac{a \sin B}{\sin A}$ Multiply both sides by $\sin B$

$= \dfrac{5.6 \sin 34°}{\sin 64°}$ Substitute in given values

$= \dfrac{5.6(0.5592)}{0.8988}$ Calculator

$= 3.5$ centimeters   To the nearest tenth

Side c

If $\dfrac{c}{\sin C} = \dfrac{a}{\sin A}$

then $c = \dfrac{a \sin C}{\sin A}$ Multiply both sides by $\sin C$

$= \dfrac{5.6 \sin 82°}{\sin 64°}$ Substitute in given values

$= \dfrac{5.6(0.9903)}{0.8988}$ Calculator

$= 6.2$ centimeters   To the nearest tenth   ◄

The law of sines, along with some fancy electronic equipment, was used to obtain the results of some of the field events in one of the recent Olympic Games.

Figure 6 is a diagram of a shot-put ring. The shot is tossed (put) from the left and lands at *A*. A small electronic device is then placed at *A* (there is usually a dent in the ground where the shot lands, so it is easy to find where to place the device). The device at *A* sends a signal to a booth in the stands that gives the measures of angles *A* and *B*. The distance *a* is found ahead of time. To find the distance *x*, the law of sines is used.

$$\frac{x}{\sin B} = \frac{a}{\sin A}$$

or $x = \dfrac{a \sin B}{\sin A}$

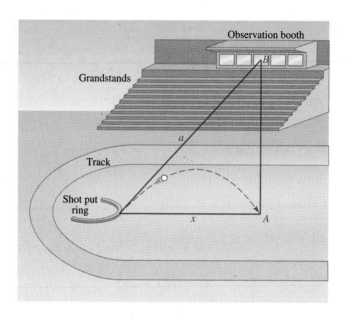

Figure 6

▶ **EXAMPLE 3**   Find x in Figure 6 if $a = 562$ feet, $B = 5.7°$, and $A = 85.3°$.

Solution

$$x = \frac{a \sin B}{\sin A}$$

$$= \frac{562 \sin 5.7°}{\sin 85.3°}$$

$$= \frac{562(0.0993)}{0.9966}$$

$$= 56.0 \text{ feet}$$ ◀

▶ **EXAMPLE 4**   A satellite is circling above the earth as shown in Figure 7. When the satellite is directly above point B, angle A is $75.4°$. If the distance between points B and D on the circumference of the earth is 910 miles and the radius of the earth is 3,960 miles, how far above the earth is the satellite?

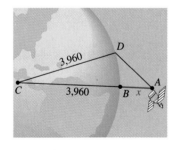

Figure 7

Solution   First we find the radian measure of central angle C by dividing the arc length BD by the radius of the earth. Multiplying this number by $180/\pi$ will give us the degree measure of angle C.

$$C = \overbrace{\frac{910}{3,960}}^{\substack{\text{Angle } C \text{ in} \\ \text{radians}}} \cdot \overbrace{\frac{180}{\pi}}^{\substack{\text{Convert to} \\ \text{degrees}}} = 13.2°$$

Next we find angle *CDA*.

$$\angle CDA = 180° - (75.4° + 13.2°) = 91.4°$$

To find *x*, we use the law of sines.

$$\frac{x + 3{,}960}{\sin 91.4°} = \frac{3{,}960}{\sin 75.4°}$$

$$x + 3{,}960 = \frac{3{,}960 \sin 91.4°}{\sin 75.4°}$$

$$x = \frac{3{,}960 \sin 91.4°}{\sin 75.4°} - 3{,}960$$

$$x = 131 \text{ miles} \qquad \text{To three significant digits} \qquad \blacktriangleleft$$

▶ **EXAMPLE 5**   A hot-air balloon is flying over a dry lake when the wind stops blowing. The balloon comes to a stop 450 feet above the ground at point *D* as shown in Figure 8. A jeep following the balloon runs out of gas at point *A*. The nearest service station is due north of the jeep at point *B*. The bearing of the balloon from the jeep at *A* is N 13° E, while the bearing of the balloon from the service station at *B* is S 19° E. If the angle of elevation of the balloon from *A* is 12°, how far will the people in the jeep have to walk to reach the service station at point *B*?

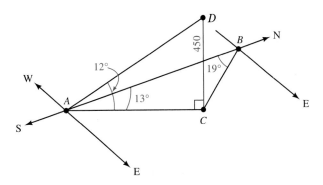

Figure 8

Solution   First we find the distance between *C* and *A* using right triangle trigonometry. Since this is an intermediate calculation, which we will use again, we keep more than two significant digits for *AC*.

$$\tan 12° = \frac{450}{AC}$$

$$AC = \frac{450}{\tan 12°}$$

$$= 2{,}117 \text{ feet}$$

Next we find angle ACB.

$$\angle ACB = 180° - (13° + 19°) = 148°$$

Finally, we find AB using the law of sines.

$$\frac{AB}{\sin 148°} = \frac{2,117}{\sin 19°}$$

$$AB = \frac{2,117 \sin 148°}{\sin 19°}$$

$$= 3,400 \text{ feet} \qquad \text{To two significant digits}$$

Since there are 5,280 feet in a mile, the people at A will walk approximately $3,400/5,280 = 0.6$ miles to get to the service station at B. ◀

Our next example involves vectors. It is taken from the text *College Physics* by Miller and Schroeer, published by Saunders College Publishing.

▶ **EXAMPLE 6** A traffic light weighing 22 pounds is suspended by two wires as shown in Figure 9. Find the magnitude of the tension in wire AB, and the magnitude of the tension in wire AC.

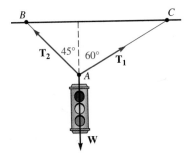

Figure 9

Solution We assume that the traffic light is not moving and is therefore in the state of *static equilibrium*. When an object is in this state, the sum of the forces acting on the object must be 0. It is because of this fact that we can redraw the vectors from Figure 9 and be sure that they form a closed triangle. Figure 10 shows a convenient redrawing of the two tension vectors T_1 and T_2, and the vector W that is due to gravity.

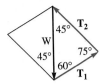

Figure 10

Using the law of sines we have:

$$\frac{|\mathbf{T_1}|}{\sin 45°} = \frac{22}{\sin 75°}$$

$$|\mathbf{T_1}| = \frac{22 \sin 45°}{\sin 75°}$$

$$= 16 \text{ pounds} \quad \text{To two significant figures}$$

$$\frac{|\mathbf{T_2}|}{\sin 60°} = \frac{22}{\sin 75°}$$

$$|\mathbf{T_2}| = \frac{22 \sin 60°}{\sin 75°}$$

$$= 20 \text{ pounds} \quad \text{To two significant figures} \qquad \blacktriangleleft$$

PROBLEM SET 7.1

Each problem that follows refers to triangle ABC.

1. If $A = 40°$, $B = 60°$, and $a = 12$ cm, find b.
2. If $A = 80°$, $B = 30°$, and $b = 14$ cm, find a.
3. If $B = 120°$, $C = 20°$, and $c = 28$ inches, find b.
4. If $B = 110°$, $C = 40°$, and $b = 18$ inches, find c.
5. If $A = 10°$, $C = 100°$, and $a = 24$ yd, find c.
6. If $A = 5°$, $C = 125°$, and $c = 510$ yd, find a.
7. If $A = 50°$, $B = 60°$, and $a = 36$ km, find C and then find c.
8. If $B = 40°$, $C = 70°$, and $c = 42$ km, find A and then find a.
9. If $A = 52°$, $B = 48°$, and $c = 14$ cm, find C and then find a.
10. If $A = 33°$, $C = 82°$, and $b = 18$ cm, find B and then find c.

The information below refers to triangle ABC. In each case, find all the missing parts.

11. $A = 42.5°$, $B = 71.4°$, $a = 215$ inches
12. $A = 110.4°$, $C = 21.8°$, $c = 246$ inches
13. $A = 46°$, $B = 95°$, $c = 6.8$ m

14. $B = 57°, C = 31°, a = 7.3$ m

15. $A = 43° 30', C = 120° 30', a = 3.48$ ft

16. $B = 14° 20', C = 75° 40', b = 2.72$ ft

17. $B = 13.4°, C = 24.8°, a = 315$ cm

18. $A = 105°, B = 45°, c = 630$ cm

19. In triangle ABC, $A = 30°$, $b = 20$ ft, and $a = 2$ ft. Show that it is impossible to solve this triangle by using the law of sines to find $\sin B$.

20. In triangle ABC, $A = 40°$, $b = 20$ ft, and $a = 18$ ft. Use the law of sines to find $\sin B$ and then give two possible values for B.

The circle in Figure 11 has a radius of r and center at C. The distance from A to B is x, the distance from A to D is y, and the length of arc BD is s. For Problems 21 through 24, redraw Figure 11, label it as indicated in each problem, and then solve the problem.

21. If $A = 31°$, $s = 11$, and $r = 12$, find x.

22. If $A = 26°$, $s = 22$, and $r = 20$, find x.

23. If $A = 45°$, $s = 18$, and $r = 15$, find y.

24. If $A = 55°$, $s = 21$, and $r = 22$, find y.

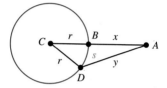

Figure 11

25. A man standing near a radio station antenna observes that the angle of elevation to the top of the antenna is $64°$. He then walks 100 ft further away and observes that the angle of elevation to the top of the antenna is $46°$. Find the height of the antenna to the nearest foot. (*Hint:* Find x first.)

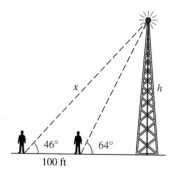

Figure 12

26. A person standing on the street looks up to the top of a building and finds that the angle of elevation is $38°$. She then walks one block further away (440 ft) and finds that the angle of elevation to the top of the building is now $28°$. How far away from the building is she when she makes her second observation? (See Figure 13 on next page.)

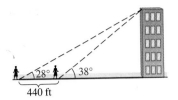

Figure 13

27. A man is flying in a hot-air balloon in a straight line at a constant rate of 5 ft/sec, while keeping it at a constant altitude. As he approaches the parking lot of a market, he notices that the angle of depression from his balloon to a friend's car in the parking lot is 35°. A minute and a half later, after flying directly over his friend's car, he looks back to see his friend getting into the car and observes the angle of depression to be 36°. At that time, what is the distance between him and his friend? (Give your answer to the nearest foot.)

28. A woman entering an outside glass elevator on the ground floor of a hotel glances up to the top of the building across the street and notices that the angle of elevation is 48°. She rides the elevator up three floors (60 ft) and finds that the angle of elevation to the top of the building across the street is 32°. How tall is the building across the street? (Give your answer to the nearest foot.)

29. From a point on the ground, a person notices that a 110-ft antenna on the top of a hill subtends an angle of 0.5°. If the angle of elevation to the bottom of the antenna is 35°, find the height of the hill. (See Figure 14.)

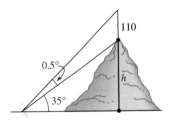

Figure 14

30. A 150-ft antenna is on top of a tall building. From a point on the ground, the angle of elevation to the top of the antenna is 28.5°, while the angle of elevation to the bottom of the antenna from the same point is 23.5°. How tall is the building?

31. Figure 15 is a diagram that shows how Colleen estimates the height of a tree that is on the other side of a stream. She stands at point A facing the tree and finds the angle of elevation from A to the top of the tree to be 51°. Then she turns 105° and walks 25 ft to point B, where she measures the angle between her path and the base of the tree. She finds that angle to be 44°. Use this information to find the height of the tree.

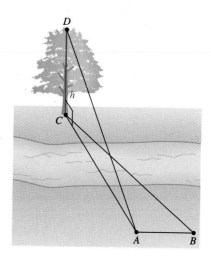

Figure 15

32. A plane makes a forced landing at sea. The last radio signal received at station C gives the bearing of the plane from C as N 55.4° E at an altitude of 1,050 ft. An observer at C sights the plane and gives $\angle DCB$ as 22.5°. How far will a rescue boat at A have to travel to reach any survivors at B, if the bearing of B from A is S 56.4° E? (See Figure 16.)

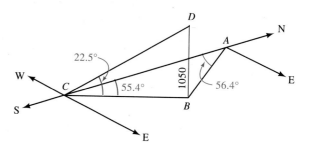

Figure 16

33. A ship is anchored off a long straight shoreline that runs north and south. From two observation points 18 mi apart on shore, the bearings of the ship are N 31° E and S 53° E. What is the distance from the ship to each of the observation points?

34. Tom and Fred are 3.5 mi apart watching a rocket being launched from Vandenberg Air Force Base. Tom estimates the bearing of the rocket from his position to be S 75° W, while Fred estimates that the bearing of the rocket from his position is N 65° W. If Fred is due south of Tom, how far is each of them from the rocket?

35. A tightrope walker is standing still with one foot on the tightrope as shown in Figure 17. If the tightrope walker weighs 125 pounds, find the magnitudes of the tension in the rope toward each of the ends of the rope.

Figure 17

36. A tightrope walker weighing 145 pounds is standing still at the center of a tightrope that is 46.5 feet long. The weight of the walker causes the center of the tightrope to move down 14.5 inches. Find the magnitude of the tension in the tightrope.

37. If you have ever ridden on a chair lift at a ski area and had it stop, you know that the chair will pull down on the cable, dropping you down to a lower height than when the chair is in motion. Figure 18 shows a gondola that is stopped. Find the magnitudes of the tension in the cable toward each end of the cable if the total weight of the gondola and its occupants is 1,850 pounds.

Figure 18

38. A chair lift at a ski resort is stopped halfway between two poles that support the cable to which the chair is attached. The poles are 215 feet apart and the combined weight of the chair and the three people on the chair is 725 pounds. If the weight of the chair and the people riding it causes the chair to move to a position 15.8 feet below the horizontal line that connects the top of the two poles, find the tension in the cable.

Review Problems

The problems that follow review material we covered in Sections 3.1 and 6.1.

Solve each equation for θ if $0° \le \theta < 360°$. If rounding is necessary, round to the nearest tenth of a degree.

39. $2 \sin \theta - \sqrt{2} = 0$ **40.** $5 \cos \theta - 3 = 0$
41. $\sin \theta \cos \theta - 2 \cos \theta = 0$ **42.** $3 \cos \theta - 2 \sin \theta \cos \theta = 0$
43. $2 \sin^2 \theta - 3 \sin \theta = -1$ **44.** $10 \cos^2 \theta + \cos \theta - 3 = 0$
45. $\cos^2 \theta - 4 \cos \theta + 2 = 0$ **46.** $2 \sin^2 \theta - 6 \sin \theta + 3 = 0$

Find all radian solutions to each equation using exact values only.

47. $(\sin x + 1)(2 \sin x - 1) = 0$ **48.** $2 \sin x \cos x - \sqrt{3} \sin x = 0$

Find θ to the nearest tenth of a degree if $0 \le \theta < 360°$, and

49. $\sin \theta = 0.7380$ **50.** $\sin \theta = 0.7965$
51. $\sin \theta = 0.9668$ **52.** $\sin \theta = 0.2351$

SECTION 7.2

The Ambiguous Case

In this section, we will extend the law of sines to solve triangles for which we are given two sides and the angle opposite one of the given sides.

▶ **EXAMPLE 1**   Find angle B in triangle ABC if $a = 2$, $b = 6$, and $A = 30°$.

Solution   Applying the law of sines we have

$$\sin B = \frac{b \sin A}{a}$$

$$= \frac{6 \sin 30°}{2}$$

$$= \frac{6(0.5000)}{2}$$

$$= 1.5$$

Since $\sin B$ can never be larger than 1, no triangle exists for which $a = 2$, $b = 6$, and $A = 30°$. (You may recall from geometry that there was no congruence theorem SSA.) Figure 1 illustrates what went wrong here.

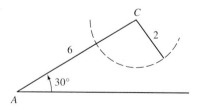

Figure 1

When we are given two sides and an angle opposite one of them (SSA), we have several possibilities for the triangle or triangles that result. As was the case in Example 1, one of the possibilities is that no triangle will fit the given information. If side *a* in Example 1 had been longer than the altitude drawn from vertex *C* but shorter than side *b*, we would have had two triangles that fit the given information, as shown in Figure 2. On the other hand, if side *a* in triangle *ABC* of Example 1 had been longer than side *b*, we would have had only one triangle that fit the given information, as illustrated in Figure 3.

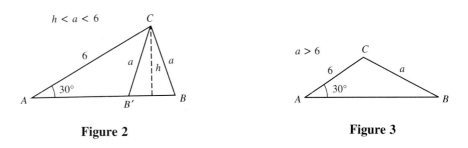

Figure 2   **Figure 3**

Because of the different possibilities that arise in solving a triangle for which we are given two sides and an angle opposite one of the given sides, we call this situation the *ambiguous case.*

▶ **EXAMPLE 2**   Find the missing parts in triangle *ABC* if $a = 54$ centimeters, $b = 62$ centimeters, and $A = 40°$.

Solution   First we solve for sin *B* with the law of sines.

Angle B

$$\sin B = \frac{b \sin A}{a}$$
$$= \frac{62 \sin 40°}{54}$$
$$= 0.7380$$

Now, since sin *B* is positive for any angle in quadrant I or II, we have two possibilities. We will call one of them *B* and the other *B′*.

$$B = 48° \quad \text{or} \quad B' = 180° - 48° = 132°$$

We have two different triangles that can be found with $a = 54$ centimeters, $b = 62$ centimeters, and $A = 40°$. Figure 4 shows both of them. One is labeled *ABC*, while the other is labeled *AB′C*.

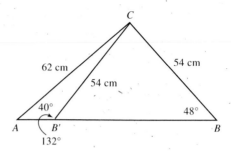

Figure 4

Angles C and C'

Since there are two values for *B*, we have two values for *C*.

$$C = 180 - (A + B) \qquad \text{and} \qquad C' = 180 - (A + B')$$
$$= 180 - (40° + 48°) \qquad\qquad\qquad = 180 - (40° + 132°)$$
$$= 92° \qquad\qquad\qquad\qquad\qquad = 8°$$

Sides c and c'

$$c = \frac{a \sin C}{\sin A} \qquad \text{and} \qquad c' = \frac{a \sin C'}{\sin A}$$

$$= \frac{54 \sin 92°}{\sin 40°} \qquad\qquad\qquad = \frac{54 \sin 8°}{\sin 40°}$$

$$= 84 \text{ centimeters} \qquad\qquad = 12 \text{ centimeters}$$

Figure 5 shows both triangles.

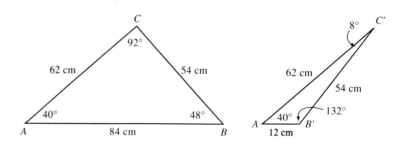

Figure 5

▶ **EXAMPLE 3**   Find the missing parts of triangle ABC if $C = 35.4°$, $a = 205$ feet, and $c = 314$ feet.

Solution   Applying the law of sines, we find $\sin A$.

Angle A

$$\sin A = \frac{a \sin C}{c}$$

$$= \frac{205 \sin 35.4°}{314}$$

$$= 0.3782$$

Since $\sin A$ is positive in quadrants I and II, we have two possible values for A.

$$A = 22.2° \quad \text{and} \quad A' = 180° - 22.2°$$
$$= 157.8°$$

The second possibility, $A' = 157.8°$, will not work, however, since C is $35.4°$ and therefore

$$C + A' = 35.4° + 157.8°$$
$$= 193.2°$$

which is larger than $180°$. This result indicates that there is exactly one triangle that fits the description given in Example 3. In that triangle

$$A = 22.2°$$

Angle B

$$B = 180° - (35.4° + 22.2°)$$
$$= 122.4°$$

Side b

$$b = \frac{c \sin B}{\sin C}$$

$$= \frac{314 \sin 122.4°}{\sin 35.4°}$$

$$= 458 \text{ feet}$$

Figure 6 is a diagram of this triangle.

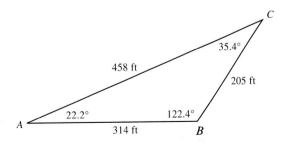

Figure 6

The different cases that can occur when we solve the kinds of triangles given in Examples 1, 2, and 3 become apparent in the process of solving for the missing parts. The following table completes the set of conditions under which we will have 1, 2, or no triangles in the ambiguous case.

In Table 1, we are assuming that we are given angle A and sides a and b in triangle ABC, and that h is the altitude from vertex C.

Table 1

Conditions	Number of Triangles	Diagram
$A < 90°$ and $a < h$	0	
$A > 90°$ and $a < b$	0	
$A < 90°$ and $a = h$	1	
$A < 90°$ and $a \geq b$	1	
$A > 90°$ and $a > b$	1	
$A < 90°$ and $h < a < b$	2	

PROBLEM SET 7.2

For each triangle below, solve for B and use the results to explain why the triangle has the given number of solutions.

1. $A = 30°$, $b = 40$ ft, $a = 10$ ft; no solution
2. $A = 150°$, $b = 30$ ft, $a = 10$ ft; no solution
3. $A = 120°$, $b = 20$ cm, $a = 30$ cm; one solution
4. $A = 30°$, $b = 12$ cm, $a = 6$ cm; one solution
5. $A = 60°$, $b = 18$ m, $a = 16$ m; two solutions
6. $A = 20°$, $b = 40$ m, $a = 30$ m; two solutions

Find all solutions to each of the following triangles:

7. $A = 38°$, $a = 41$ ft, $b = 54$ ft
8. $A = 43°$, $a = 31$ ft, $b = 37$ ft
9. $A = 112.2°$, $a = 43.8$ cm, $b = 22.3$ cm
10. $A = 124.3°$, $a = 27.3$ cm, $b = 50.2$ cm
11. $C = 27° 50'$, $c = 347$ m, $b = 425$ m
12. $C = 51° 30'$, $c = 707$ m, $b = 821$ m
13. $B = 45° 10'$, $b = 1.79$ inches, $c = 1.12$ inches
14. $B = 62° 40'$, $b = 6.78$ inches, $c = 3.48$ inches
15. $B = 118°$, $b = 0.68$ cm, $a = 0.92$ cm
16. $B = 30°$, $b = 4.2$ cm, $a = 8.4$ cm
17. $A = 142°$, $b = 2.9$ yd, $a = 1.4$ yd
18. $A = 65°$, $b = 7.6$ yd, $a = 7.1$ yd
19. $C = 26.8°$, $c = 36.8$ km, $b = 36.8$ km
20. $C = 73.4°$, $c = 51.1$ km, $b = 92.4$ km

21. A 50-ft wire running from the top of a tent pole to the ground makes an angle of 58° with the ground. If the length of the tent pole is 44 ft, how far is it from the bottom of the tent pole to the point where the wire is fastened to the ground? (The tent pole is not necessarily perpendicular to the ground.)
22. A hot-air balloon is held at a constant altitude by two ropes that are anchored to the ground. One rope is 120 ft long and makes an angle of 65° with the ground. The other rope is 115 ft long. What is the distance between the points on the ground at which the two ropes are anchored?
23. A plane is headed due east with an airspeed of 340 mph. Its true course, however, is at 98° from due north. If the wind currents are a constant 55 mph, what are the possibilities for the ground speed of the plane?
24. A ship is headed due north at a constant 16 mph. Because of the ocean current, the true course of the ship is N 15° E. If the currents are a constant 14 mph, in what direction are the currents running?
25. After a wind storm, a farmer notices that his 32-ft windmill may be leaning, but he is not sure. From a point on the ground 30 ft from the base of the windmill, he finds that the angle of elevation to the top of the windmill is 48°. Is the windmill leaning? If so, what is the acute angle the windmill makes with the ground?

26. A boy is riding his motorcycle on a road that runs east and west. He leaves the road at a service station and rides 5.25 mi in the direction N 15.5° E. Then he turns to his right and rides 6.50 mi back to the road, where his motorcycle breaks down. How far will he have to walk to get back to the service station?

Review Problems

The problems that follow review material we covered in Section 6.2.

Find all solutions in the interval $0° \leq \theta < 360°$. If rounding is necessary, round to the nearest tenth of a degree.

27. $4 \sin \theta - \csc \theta = 0$ **28.** $2 \sin \theta - 1 = \csc \theta$
29. $2 \cos \theta - \sin 2\theta = 0$ **30.** $\cos 2\theta + 3 \cos \theta - 2 = 0$
31. $18 \sec^2 \theta - 17 \tan \theta \sec \theta - 12 = 0$ **32.** $7 \sin^2 \theta - 9 \cos 2\theta = 0$

Find all radian solutions using exact values only.

33. $2 \cos x - \sec x + \tan x = 0$ **34.** $2 \cos^2 x - \sin x = 1$
35. $\sin x + \cos x = 0$ **36.** $\sin x - \cos x = 1$

SECTION 7.3   # The Law of Cosines

In this section, we will derive another relationship that exists between the sides and angles in any triangle. It is called the *law of cosines* and is stated like this:

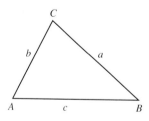

Law of Cosines
$$a^2 = b^2 + c^2 - 2bc \cos A$$
$$b^2 = a^2 + c^2 - 2ac \cos B$$
$$c^2 = a^2 + b^2 - 2ab \cos C$$

Figure 1

Derivation

To derive the formulas stated in the law of cosines, we apply the Pythagorean Theorem and some of our basic trigonometric identities. Applying the Pythagorean

Theorem to right triangle *BCD* in Figure 2, we have

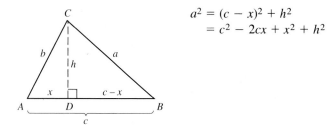

$$a^2 = (c - x)^2 + h^2$$
$$= c^2 - 2cx + x^2 + h^2$$

Figure 2

But from right triangle *ACD*, we have $x^2 + h^2 = b^2$, so

$$a^2 = c^2 - 2cx + b^2$$
$$= b^2 + c^2 - 2cx$$

Now, since $\cos A = x/b$, we have $x = b \cos A$, or

$$a^2 = b^2 + c^2 - 2bc \cos A$$

Applying the same sequence of substitutions and reasoning to the right triangles formed by the altitudes from vertices *A* and *B* will give us the other two formulas listed in the law of cosines.

We can use the law of cosines to solve triangles for which we are given two sides and the angle included between them (SAS) or triangles for which we are given all three sides (SSS).

Two Sides and the Included Angle

▶ **EXAMPLE 1**   Find the missing parts of triangle *ABC* if $A = 60°$, $b = 20$ inches, and $c = 30$ inches.

Solution   The solution process will include the use of both the law of cosines and the law of sines. We begin by using the law of cosines to find *a*.

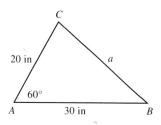

Figure 3

Side a

$$a^2 = b^2 + c^2 - 2bc \cos A \qquad \text{Law of cosines}$$
$$= 20^2 + 30^2 - 2(20)(30) \cos 60° \qquad \text{Substitute in given values}$$
$$= 400 + 900 - 1{,}200(0.5000) \qquad \text{Calculator}$$
$$a^2 = 700$$
$$a = 26 \text{ inches} \qquad \text{To the nearest integer}$$

Now that we have *a*, we can use the law of sines to solve for either *B* or *C*. When we have a choice of angles to solve for, and we are using the law of sines to do so, it is usually best to solve for the smaller angle. Since side *b* is smaller than side *c*, angle *B* will be smaller than angle *C*.

Angle B

$$\sin B = \frac{b \sin A}{a}$$
$$= \frac{20 \sin 60°}{26}$$
$$\sin B = 0.6662$$
$$\text{so} \quad B = 42° \qquad \text{To the nearest degree}$$

Note that we don't have to check $B' = 180° - 42° = 138°$ because we know *B* is an acute angle since it is smaller than angles *A* and *C*.

Angle C

$$C = 180° - (A + B)$$
$$= 180° - (60° + 42°)$$
$$= 78°$$

◀

▶ **EXAMPLE 2** The diagonals of a parallelogram are 24.2 centimeters and 35.4 centimeters and intersect at an angle of 65.5°. Find the length of the shorter side of the parallelogram.

Solution A diagram of the parallelogram is shown in Figure 4. We used the variable *x* to represent the length of the shorter side. Note also that we labeled half of each diagonal with its length to give us the sides of a triangle. (Recall that the diagonals of a parallelogram bisect each other.)

$$x^2 = (12.1)^2 + (17.7)^2 - 2(12.1)(17.7) \cos 65.5°$$
$$x^2 = 282.07$$
$$x = 16.8 \text{ centimeters} \qquad \text{To the nearest tenth}$$

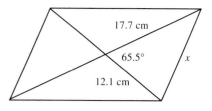

Figure 4

Next we will see how the law of cosines can be used to find the missing parts of a triangle for which all three sides are given.

Three Sides

To use the law of cosines to solve a triangle for which we are given all three sides, it is convenient to rewrite the equations with the cosines isolated on one side. Here is an equivalent form of the law of cosines.

$$\cos A = \frac{b^2 + c^2 - a^2}{2bc}$$

$$\cos B = \frac{a^2 + c^2 - b^2}{2ac}$$

$$\cos C = \frac{a^2 + b^2 - c^2}{2ab}$$

Here is how we arrived at the first of these formulas.

$$a^2 = b^2 + c^2 - 2bc \cos A$$

$$b^2 + c^2 - 2bc \cos A = a^2 \qquad \text{Exchange sides}$$

$$-2bc \cos A = -b^2 - c^2 + a^2 \qquad \text{Add } -b^2 \text{ and } -c^2 \text{ to both sides}$$

$$\cos A = \frac{b^2 + c^2 - a^2}{2bc} \qquad \text{Divide both sides by } -2bc$$

▶ **EXAMPLE 3**   Solve triangle ABC if $a = 34$ kilometers, $b = 20$ kilometers, and $c = 18$ kilometers.

Solution   We will use the law of cosines to solve for one of the angles and then use the law of sines to find one of the remaining angles. Since there is never any confusion as to whether an angle is acute or obtuse if we have its cosine (the cosine of an obtuse angle is negative), it is best to solve for the largest angle first.

Since the longest side is *a*, we solve for *A* first.

Angle A

$$\cos A = \frac{b^2 + c^2 - a^2}{2bc}$$

$$= \frac{20^2 + 18^2 - 34^2}{(2)(20)(18)}$$

$$\cos A = -0.6000$$

so $A = 127°$ To the nearest degree

There are two ways to find angle *C*. We can use either the law of sines or the law of cosines.

Angle C

Using the law of sines,

$$\sin C = \frac{c \sin A}{a}$$

$$= \frac{18 \sin 127°}{34}$$

$$\sin C = 0.4228$$

so $C = 25°$ To the nearest degree

Using the law of cosines,

$$\cos C = \frac{a^2 + b^2 - c^2}{2ab}$$

$$= \frac{34^2 + 20^2 - 18^2}{2(34)(20)}$$

$$\cos C = 0.9059$$

so $C = 25°$ To the nearest degree

Angle B

$$B = 180° - (A + C)$$
$$= 180° - (127° + 25°)$$
$$= 28°$$

◀

▶ **EXAMPLE 4**   A plane is flying with an airspeed of 185 miles per hour with heading 120°. The wind currents are running at a constant 32 miles per hour at 165° clockwise from due north. Find the true course and ground speed of the plane.

Solution   Figure 5 is a diagram of the situation with the vector **V** representing the airspeed and direction of the plane and **W** representing the speed and direction of the wind currents. From Figure 5, $\alpha = 180° - 120° = 60°$ and $\theta = 360° - (\alpha + 165°) = 135°$.

The magnitude of **V** + **W** can be found from the law of cosines.

$$|V + W|^2 = |V|^2 + |W|^2 - 2|V||W| \cos \theta$$
$$= 185^2 + 32^2 - 2(185)(32) \cos 135°$$
$$= 43,621$$

so $|V + W| = 210$ miles per hour   To two significant digits

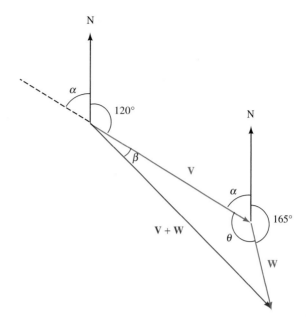

Figure 5

To find the direction of **V** + **W**, we first find β using the law of sines.

$$\frac{\sin \beta}{32} = \frac{\sin \theta}{210}$$

$$\sin \beta = \frac{32 \sin 135°}{210}$$

$$= 0.1077$$

so $\beta = 6°$ To the nearest degree

The true course is $120° + \beta = 120° + 6° = 126°$. The speed of the plane with respect to the ground is 210 miles per hour.   ◀

PROBLEM SET 7.3

Each problem below refers to triangle *ABC*.

1. If $a = 120$ inches, $b = 66$ inches, and $C = 60°$, find c.
2. If $a = 120$ inches, $b = 66$ inches, and $C = 120°$, find c.
3. If $a = 22$ yd, $b = 24$ yd, and $c = 26$ yd, find the largest angle.
4. If $a = 13$ yd, $b = 14$ yd, and $c = 15$ yd, find the largest angle.
5. If $b = 4.2$ m, $c = 6.8$ m, and $A = 116°$, find a.
6. If $a = 3.7$ m, $c = 6.4$ m, and $B = 23°$, find b.
7. If $a = 38$ cm, $b = 10$ cm, and $c = 31$ cm, find the largest angle.
8. If $a = 51$ cm, $b = 24$ cm, and $c = 31$ cm, find the largest angle.

Solve each triangle below.

9. $a = 410$ m, $c = 340$ m, $B = 151.5°$
10. $a = 76.3$ m, $c = 42.8$ m, $B = 16.3°$
11. $a = 0.48$ yd, $b = 0.63$ yd, $c = 0.75$ yd
12. $a = 48$ yd, $b = 75$ yd, $c = 63$ yd
13. $b = 0.923$ km, $c = 0.387$ km, $A = 43° 20'$
14. $b = 63.4$ km, $c = 75.2$ km, $A = 124° 40'$
15. $a = 4.38$ ft, $b = 3.79$ ft, $c = 5.22$ ft
16. $a = 832$ ft, $b = 623$ ft, $c = 345$ ft
17. Use the law of cosines to show that, if $A = 90°$, then $a^2 = b^2 + c^2$.
18. Use the law of cosines to show that, if $a^2 = b^2 + c^2$, then $A = 90°$.
19. The diagonals of a parallelogram are 56 inches and 34 inches and intersect at an angle of 120°. Find the length of the shorter side.
20. The diagonals of a parallelogram are 14 m and 16 m and intersect at an angle of 60°. Find the length of the longer side.
21. Two planes leave an airport at the same time. Their speeds are 130 mph and 150 mph, and the angle between their courses is 36°. How far apart are they after 1.5 hours?
22. Two ships leave a harbor entrance at the same time. The first ship is traveling at a constant 18 mph, while the second is traveling at a constant 22 mph. If the angle between their courses is 123°, how far apart are they after 2 hours?
23. Two planes take off at the same time from an airport. The first plane is flying at 246 mph on a course of 135.0°. The second plane is flying in the direction 175.0° at 357 mph. Assuming there are no wind currents blowing, how far apart are they after 2 hours?
24. Two ships leave the harbor at the same time. One ship is traveling at 14 mph on a course with a bearing of S 13° W, while the other is traveling at 12 mph on a course with a bearing of N 75° E. How far apart are they after 3 hours?
25. A plane is flying with an airspeed of 160 mph and heading of 150°. The wind currents are running at 35 mph at 165° clockwise from due north. Use vectors to find the true course and ground speed of the plane.
26. A plane is flying with an airspeed of 244 mph with heading 272.7°. The wind currents are running at a constant 45.7 mph in the direction 262.6°. Find the ground speed and true course of the plane.
27. A plane has an airspeed of 195 mph and a heading of 30.0°. The ground speed of the plane is 207 mph, and its true course is in the direction 34.0°. Find the speed and direction of the air currents, assuming they are constants.
28. The airspeed and heading of a plane are 140 mph and 130°, respectively. If the ground speed of the plane is 135 mph and its true course is 137°, find the speed and direction of the wind currents, assuming they are constants.

Review Problems

The problems that follow review material we covered in Section 6.3.

Find all solutions in radians using exact values only.

29. $\sin 3x = 1/2$
30. $\cos 4x = 0$
31. $\tan^2 3x = 1$
32. $\tan^2 4x = 1$

Find all degree solutions.

33. $2 \cos^2 3\theta - 9 \cos 3\theta + 4 = 0$
34. $3 \sin^2 2\theta - 2 \sin 2\theta - 5 = 0$
35. $\sin 4\theta \cos 2\theta + \cos 4\theta \sin 2\theta = -1$
36. $\cos 3\theta \cos 2\theta - \sin 3\theta \sin 2\theta = -1$

Solve each equation for θ if $0° \le \theta < 360°$.

37. $\sin \theta + \cos \theta = 1$ **38.** $\sin \theta - \cos \theta = 0$

SECTION 7.4 # The Area of a Triangle

In this section, we will derive three formulas for the area S of a triangle. We will start by deriving the formula used to find the area of a triangle for which two sides and the included angle are given.

Two Sides and the Included Angle

To derive our first formula, we begin with the general formula for the area of a triangle:

$$S = \frac{1}{2}(\text{base})(\text{height})$$

The base of triangle ABC in Figure 1 is c and the height is h. So the formula for S becomes, in this case,

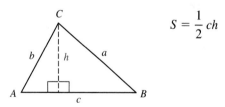

$$S = \frac{1}{2}ch$$

Figure 1

Suppose that, for triangle ABC, we are given the lengths of sides b and c and the measure of angle A. Then we can write $\sin A$ as

$$\sin A = \frac{h}{b}$$

or, by solving for h, $h = b \sin A$

Substituting this expression for h into the formula

$$S = \frac{1}{2}ch$$

we have

$$S = \frac{1}{2} bc \sin A$$

Applying the same kind of reasoning to the heights drawn from A and B, we also have

$$S = \frac{1}{2} ab \sin C$$

$$S = \frac{1}{2} ac \sin B$$

Each of these three formulas indicates that to find the area of a triangle for which we are given two sides and the angle included between them, we multiply half the product of the two sides by the sine of the angle included between them.

▶ **EXAMPLE 1**　Find the area of triangle ABC if $A = 35.1°$, $b = 2.43$ centimeters, and $c = 3.57$ centimeters.

Solution　Applying the first formula we derived, we have

$$S = \frac{1}{2} bc \sin A$$

$$= \frac{1}{2} (2.43)(3.57) \sin 35.1°$$

$$= \frac{1}{2} (2.43)(3.57)(0.5750)$$

$$= 2.49 \text{ centimeters}^2 \quad \text{To three significant digits} \qquad ◀$$

Two Angles and One Side

The next area formula we will derive is used to find the area of triangles for which we are given two angles and one side.

Suppose we were given angles A and B and side a in triangle ABC in Figure 1. We could easily solve for C by subtracting the sum of A and B from $180°$.

To find side b, we use the law of sines

$$\frac{b}{\sin B} = \frac{a}{\sin A}$$

Solving this equation for b would give us

$$b = \frac{a \sin B}{\sin A}$$

Substituting this expression for b into the formula

$$S = \frac{1}{2} ab \sin C$$

we have

$$S = \frac{1}{2} a \left(\frac{a \sin B}{\sin A} \right) \sin C$$

$$= \frac{a^2 \sin B \sin C}{2 \sin A}$$

A similar sequence of steps can be used to derive

$$S = \frac{b^2 \sin A \sin C}{2 \sin B}$$

and

$$S = \frac{c^2 \sin A \sin B}{2 \sin C}$$

The formula we use depends on the side we are given.

▶ **EXAMPLE 2**   Find the area of triangle ABC if $A = 24° \, 10'$, $B = 120° \, 40'$, and $a = 4.25$ feet.

Solution   We begin by finding C.

$$C = 180° - (24° \, 10' + 120° \, 40')$$
$$= 35° \, 10'$$

Now, applying the formula

$$S = \frac{a^2 \sin B \sin C}{2 \sin A}$$

with $a = 4.25$, $A = 24° \, 10'$, $B = 120° \, 40'$, and $C = 35° \, 10'$, we have

$$S = \frac{(4.25)^2 (\sin 120° \, 40')(\sin 35° \, 10')}{2 \sin 24° \, 10'}$$

$$= \frac{(4.25)^2 (0.8601)(0.5760)}{2(0.4094)}$$

$$= 10.9 \text{ feet}^2 \qquad \blacktriangleleft$$

Three Sides

The last area formula we will discuss is called Heron's formula. It is used to find the area of a triangle in which all three sides are known.

Heron's Formula

The area of a triangle with sides of length a, b, and c is given by

$$S = \sqrt{s(s - a)(s - b)(s - c)}$$

where s is half the perimeter of the triangle; that is,

$$s = \frac{1}{2}(a + b + c) \qquad \text{or} \qquad 2s = a + b + c$$

PROOF   We begin our proof by squaring both sides of the formula

$$S = \frac{1}{2} ab \sin C$$

to obtain

$$S^2 = \frac{1}{4} a^2 b^2 \sin^2 C$$

Next, we multiply both sides of the equation by $4/a^2 b^2$ to isolate $\sin^2 C$ on the right side.

$$\frac{4S^2}{a^2 b^2} = \sin^2 C$$

Replacing $\sin^2 C$ with $1 - \cos^2 C$ and then factoring as the difference of two squares, we have

$$= 1 - \cos^2 C$$
$$= (1 + \cos C)(1 - \cos C)$$

From the law of cosines we know that $\cos C = (a^2 + b^2 - c^2)/2ab$.

$$= \left[1 + \frac{a^2 + b^2 - c^2}{2ab} \right]\left[1 - \frac{a^2 + b^2 - c^2}{2ab} \right]$$

$$= \left[\frac{2ab + a^2 + b^2 - c^2}{2ab} \right]\left[\frac{2ab - a^2 - b^2 + c^2}{2ab} \right]$$

$$= \left[\frac{(a^2 + 2ab + b^2) - c^2}{2ab} \right]\left[\frac{c^2 - (a^2 - 2ab + b^2)}{2ab} \right]$$

$$= \left[\frac{(a + b)^2 - c^2}{2ab} \right]\left[\frac{c^2 - (a - b)^2}{2ab} \right]$$

Now we factor each numerator as the difference of two squares and multiply the denominators.

$$= \frac{[(a + b + c)(a + b - c)][(c + a - b)(c - a + b)]}{4a^2b^2}$$

Now, since $a + b + c = 2s$, it is also true that

$$a + b - c = a + b + c - 2c = 2s - 2c$$
$$c + a - b = a + b + c - 2b = 2s - 2b$$
$$c - a + b = a + b + c - 2a = 2s - 2a$$

Substituting these expressions into our last equation, we have

$$= \frac{2s(2s - 2c)(2s - 2b)(2s - 2a)}{4a^2b^2}$$

Factoring out a 2 from each term in the numerator and showing the left side of our equation along with the right side, we have

$$\frac{4S^2}{a^2b^2} = \frac{16s(s - a)(s - b)(s - c)}{4a^2b^2}$$

Multiplying both sides by $a^2b^2/4$ we have

$$S^2 = s(s - a)(s - b)(s - c)$$

Taking the square root of both sides of the equation, we have Heron's formula.

$$S = \sqrt{s(s - a)(s - b)(s - c)}$$

▶ **EXAMPLE 3**   Find the area of triangle ABC if $a = 12$ meters, $b = 14$ meters, and $c = 8.0$ meters.

Solution   We begin by calculating the formula for s, half the perimeter of ABC.

$$s = \frac{1}{2}(12 + 14 + 8)$$
$$= 17$$

Substituting this value of s into Heron's formula along with the given values of a, b, and c, we have

$$S = \sqrt{17(17 - 12)(17 - 14)(17 - 8)}$$
$$= \sqrt{17(5)(3)(9)}$$
$$= \sqrt{2,295}$$
$$= 48 \text{ meters}^2 \qquad \text{To two significant digits} \qquad ◀$$

PROBLEM SET 7.4

Each problem below refers to triangle *ABC*. In each case, find the area of the triangle.

1. $a = 50$ cm, $b = 70$ cm, $C = 60°$
2. $a = 10$ cm, $b = 12$ cm, $C = 120°$
3. $a = 41.5$ m, $c = 34.5$ m, $B = 151.5°$
4. $a = 76.3$ m, $c = 42.8$ m, $B = 16.3°$
5. $b = 0.923$ km, $c = 0.387$ km, $A = 43° 20'$
6. $b = 63.4$ km, $c = 75.2$ km, $A = 124° 40'$
7. $A = 46°$, $B = 95°$, $c = 6.8$ m
8. $B = 57°$, $C = 31°$, $a = 7.3$ m
9. $A = 42.5°$, $B = 71.4°$, $a = 210$ inches
10. $A = 110.4°$, $C = 21.8°$, $c = 240$ inches
11. $A = 43° 30'$, $C = 120° 30'$, $a = 3.48$ ft
12. $B = 14° 20'$, $C = 75° 40'$, $b = 2.72$ ft
13. $a = 44$ inches, $b = 66$ inches, $c = 88$ inches
14. $a = 23$ inches, $b = 34$ inches, $c = 45$ inches
15. $a = 4.8$ yd, $b = 6.3$ yd, $c = 7.5$ yd
16. $a = 48$ yd, $b = 75$ yd, $c = 63$ yd
17. $a = 4.38$ ft, $b = 3.79$ ft, $c = 5.22$ ft
18. $a = 8.32$ ft, $b = 6.23$ ft, $c = 3.45$ ft
19. Find the area of a parallelogram if the angle between two of the sides is $120°$ and the two sides are 15 inches and 12 inches.
20. Find the area of a parallelogram if the two sides measure 24.1 inches and 31.4 inches and the longer diagonal is 32.4 inches.
21. The area of a triangle is 40 cm². Find the length of the side included between the angles $A = 30°$ and $B = 50°$.
22. The area of a triangle is 80 in². Find the length of the side included between $A = 25°$ and $C = 110°$.

Review Problems

The problems that follow review material we covered in Section 6.4.

Eliminate the parameter *t* and graph the resulting equation.

23. $x = \cos t$, $y = \sin t$
24. $x = \cos t$, $y = -\sin t$
25. $x = 3 + 2 \sin t$, $y = 1 + 2 \cos t$
26. $x = \cos t - 3$, $y = \sin t + 2$

Eliminate the parameter *t*, but do not graph.

27. $x = 2 \tan t$, $y = 3 \sec t$
28. $x = 4 \cot t$, $y = 2 \csc t$
29. $x = \sin t$, $y = \cos 2t$
30. $x = \cos t$, $y = \cos 2t$

Vectors: An Algebraic Approach

In this section, we will take a second look at vectors, this time from an algebraic point of view. Much of the credit for this treatment of vectors is attributed to both Irish mathematician William Rowan Hamilton (1805–1865) and German mathematician Hermann Grassmann (1809–1877). Grassmann is famous for his book *Ausdehnungslehre (The Calculus of Extension),* which was first published in 1844. However, it wasn't until 50 years later that this work was fully accepted or the significance realized, when Albert Einstein used it in his theory of relativity.

As we mentioned in Section 2.5, any vector **V** can be written in terms of its horizontal and vertical component vectors, $\mathbf{V}_x$ and $\mathbf{V}_y$, respectively. In this section, we will expand our work with horizontal and vertical vectors in order to develop an algebraic form for vectors. To begin, we need to define two new vectors.

DEFINITION   The vector that extends from the origin to the point (1, 0) is called the *unit horizontal vector* and is denoted by **i**. The vector that extends from the origin to the point (0, 1) is called the *unit vertical vector* and is denoted by the vector **j**. Figure 1 shows the vectors **i** and **j**.

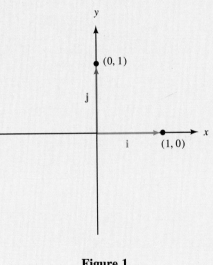

Figure 1

▶ **EXAMPLE 1**   Draw the vector that goes from the origin to the point (3, 4), and then write it in terms of its unit vectors **i** and **j**.

Solution   Figure 2 shows the vector **V** that extends from the origin to the point (3, 4).

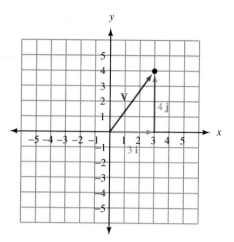

Figure 2
$$\mathbf{V} = 3\mathbf{i} + 4\mathbf{j}$$

◀

Notation For the vectors shown in Example 1, the magnitude of the horizontal vector $3\mathbf{i}$ is the coefficient 3, which we call the *horizontal component* of $\mathbf{V}$. Likewise, the coefficient 4 in the vertical vector $4\mathbf{j}$ is called the *vertical component* of $\mathbf{V}$.

Every vector $\mathbf{V}$ can be written in terms of horizontal and vertical components and the unit vectors $\mathbf{i}$ and $\mathbf{j}$, as Figure 3 illustrates.

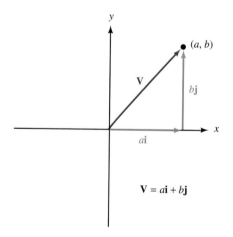

Figure 3

More Notation An alternate way to write the vector $\mathbf{V} = 3\mathbf{i} + 4\mathbf{j}$ is $\langle 3, 4 \rangle$, which is a more compact representation. When we use the notation $\langle 3, 4 \rangle$, it is understood that 3 is the horizontal component of the vector and 4 is its vertical component.

Magnitude

As you know from Section 2.5, the magnitude of a vector is its length. Referring to Figure 3, we can find the magnitude of the vector $\mathbf{V} = a\mathbf{i} + b\mathbf{j} = \langle a, b \rangle$ using the Pythagorean Theorem:

$$|\mathbf{V}| = \sqrt{a^2 + b^2}$$

▶ **EXAMPLE 2**   Find the magnitude of each of the following vectors.

   a. $\mathbf{V} = 6\mathbf{i} + 8\mathbf{j}$
   b. $|\mathbf{W}| = \langle 5, -12 \rangle$

Solution   To find the magnitude of each vector, we find the positive square root of the sum of the squares of the horizontal and vertical components.

   a. $\begin{aligned} |\mathbf{V}| &= \sqrt{6^2 + 8^2} \\ &= \sqrt{36 + 64} \\ &= \sqrt{100} \\ &= 10 \end{aligned}$

   b. $\begin{aligned} |\mathbf{W}| &= \sqrt{5^2 + (-12)^2} \\ &= \sqrt{25 + 144} \\ &= \sqrt{169} \\ &= 13 \end{aligned}$ ◀

Here is a summary of the information we have developed to this point.

Algebraic Vectors

If $\mathbf{i}$ is the unit vector from $(0, 0)$ to $(1, 0)$, and $\mathbf{j}$ is the unit vector from $(0, 0)$ to $(0, 1)$, then any vector $\mathbf{V}$ can be written as

$$\mathbf{V} = a\mathbf{i} + b\mathbf{j} = \langle a, b \rangle$$

where a and b are real numbers. The magnitude of $\mathbf{V}$ is

$$|\mathbf{V}| = \sqrt{a^2 + b^2}$$

▶ **EXAMPLE 3**   Vector $\mathbf{V}$ has its tail at the origin, and makes an angle of $35°$ with the positive x-axis. Its magnitude is 12. Write $\mathbf{V}$ in terms of the unit vectors $\mathbf{i}$ and $\mathbf{j}$.

Solution   Figure 4 is a diagram of $\mathbf{V}$. The horizontal and vertical components of $\mathbf{V}$ are a and b, respectively.

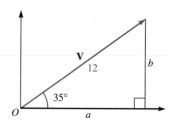

Figure 4

We find a and b using right triangle trigonometry.

$$a = 12 \cos 35° = 9.8$$
$$b = 12 \sin 35° = 6.9$$

Writing **V** in terms of the unit vectors **i** and **j** we have

$$\mathbf{V} = 9.8\mathbf{i} + 6.9\mathbf{j}$$ ◀

Addition and Subtraction with Algebraic Vectors

Adding and subtracting vectors written in terms of the unit vectors **i** and **j** is simply a matter of adding (or subtracting) the horizontal components and adding (or subtracting) the vertical components. Figure 5 shows the vector sum of vectors **U** and **V**.

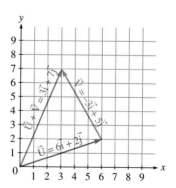

Figure 5

By simply counting squares on the grid you can convince yourself that the sum can be obtained by adding horizontal components and adding vertical components.

That is,

$$\mathbf{U} + \mathbf{V} = (6\mathbf{i} + 2\mathbf{j}) + (-3\mathbf{i} + 5\mathbf{j})$$
$$= (6 - 3)\mathbf{i} + (2 + 5)\mathbf{j}$$
$$= 3\mathbf{i} + 7\mathbf{j}$$

Figure 6 shows the difference of the vectors **U** and **V**.

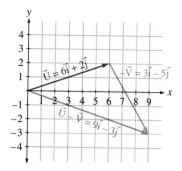

Figure 6

As the diagram in Figure 6 indicates, subtraction of algebraic vectors can be accomplished by subtracting corresponding components. That is,

$$\mathbf{U} - \mathbf{V} = (6\mathbf{i} + 2\mathbf{j}) - (-3\mathbf{i} + 5\mathbf{j})$$
$$= [6 - (-3)]\mathbf{i} + (2 - 5)\mathbf{j}$$
$$= 9\mathbf{i} - 3\mathbf{j}$$

Scalar Multiplication

To multiply a vector in component form by a scalar (real number) we multiply each component of the vector by the scalar. Figure 7 shows the vector $\mathbf{V} = 2\mathbf{i} + 3\mathbf{j}$ and the vector $3\mathbf{V}$. As you can see,

$$3\mathbf{V} = 3(2\mathbf{i} + 3\mathbf{j})$$
$$= 6\mathbf{i} + 9\mathbf{j}$$

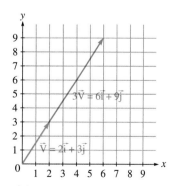

Figure 7

▶ **EXAMPLE 4** If **U** = 5**i** − 3**j** and **V** = −6**i** + 4**j**, find

a. 4**U** + 5**V**

b. 4**U** − 5**V**

c. |4**U** + 5**V**|

Solution

a. $4\mathbf{U} + 5\mathbf{V} = 4(5\mathbf{i} − 3\mathbf{j}) + 5(−6\mathbf{i} + 4\mathbf{j})$
$= (20\mathbf{i} − 12\mathbf{j}) + (−30\mathbf{i} + 20\mathbf{j})$
$= (20 − 30)\mathbf{i} + (−12 + 20)\mathbf{j}$
$= −10\mathbf{i} + 8\mathbf{j}$

b. $4\mathbf{U} − 5\mathbf{V} = 4(5\mathbf{i} − 3\mathbf{j}) − 5(−6\mathbf{i} + 4\mathbf{j})$
$= (20\mathbf{i} − 12\mathbf{j}) + (30\mathbf{i} − 20\mathbf{j})$
$= (20 + 30)\mathbf{i} + (−12 − 20)\mathbf{j}$
$= 50\mathbf{i} − 32\mathbf{j}$

c. $|4\mathbf{U} + 5\mathbf{V}| = \sqrt{(−10)^2 + 8^2}$
$= \sqrt{100 + 64}$
$= \sqrt{164}$
$= 2\sqrt{41}$ or 12.8 to the nearest tenth ◀

As an application of algebraic vectors, let's return to one of the static equilibrium problems we solved in Section 2.5.

▶ **EXAMPLE 5** Danny is 5 years old and weighs 42 pounds. He is sitting on a swing when his sister Stacey pulls him and the swing back horizontally through an angle of 30° and then stops. Find the tension in the ropes of the swing. (Figure 8 is a diagram of the situation.)

Solution   When we solved this problem in Section 2.5, we noted that there are three forces acting on Danny (and the swing), which we have labeled **W**, **H**, and **T**. Recall that the vector **W** is due to the force of gravity: its magnitude is |**W**| = 42 pounds, and its direction is straight down. The vector **H** represents the force with which Stacey is pulling Danny horizontally, and **T** is the force acting on Danny in the direction of the ropes.

If we place the origin of a coordinate system on the point at which the tails of the three vectors intersect, we can write each vector in terms of its magnitude and the unit vectors **i** and **j**. The coordinate system is shown in Figure 9. Since the direction along which **H** acts is in the positive x direction, we can write

$$\mathbf{H} = |\mathbf{H}|\mathbf{i}$$

Likewise, since the weight vector **W** is straight down, we can write

$$\mathbf{W} = −|\mathbf{W}|\mathbf{j} = −42\mathbf{j}$$

Using right triangle trigonometry, we write **T** in terms of its horizontal and vertical components:

$$\mathbf{T} = −|\mathbf{T}| \cos 60° \,\mathbf{i} + |\mathbf{T}| \sin 60° \,\mathbf{j}$$

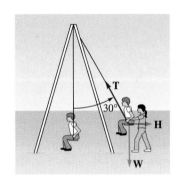

Figure 8

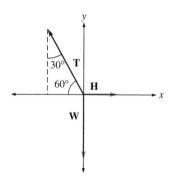

Figure 9

Since Danny and the swing are at rest, we have static equilibrium. Therefore, the sum of the vectors is 0.

$$\mathbf{T} + \mathbf{H} + \mathbf{W} = 0$$
$$-|\mathbf{T}| \cos 60° \, \mathbf{i} + |\mathbf{T}| \sin 60° \, \mathbf{j} + |\mathbf{H}|\mathbf{i} + (-42\mathbf{j}) = 0$$

Collecting all the **i** components and all the **j** components together, we have

$$(-|\mathbf{T}| \cos 60° + |\mathbf{H}|)\mathbf{i} + (|\mathbf{T}| \sin 60° - 42)\mathbf{j} = 0$$

The only way this can happen is if both components are 0. Setting the coefficient of **j** to 0, we have

$$|\mathbf{T}| \sin 60° - 42 = 0$$

$$|\mathbf{T}| = \frac{42}{\sin 60°} = 48 \text{ pounds to two significant digits}$$

Setting the coefficient of **i** to 0, and then substituting 48 for $|\mathbf{T}|$, we have

$$-|\mathbf{T}| \cos 60° + |\mathbf{H}| = 0$$
$$-48 \cos 60° + |\mathbf{H}| = 0$$
$$|\mathbf{H}| = 48 \cos 60° = 24 \text{ pounds}$$ ◀

The Dot Product

The *dot product* (or scalar product because the result is indeed a scalar and not a vector) is a form of multiplication with vectors. For our purposes, it will be useful when finding the angle between two vectors. Here is the definition of the dot product of two vectors.

> **DEFINITION**   The *dot product* of two vectors $\mathbf{U} = a\mathbf{i} + b\mathbf{j}$ and $\mathbf{V} = c\mathbf{i} + d\mathbf{j}$ is written $\mathbf{U} \cdot \mathbf{V}$ and is defined as follows:
>
> $$\mathbf{U} \cdot \mathbf{V} = (a\mathbf{i} + b\mathbf{j}) \cdot (c\mathbf{i} + d\mathbf{j})$$
> $$= (ac) + (bd)$$

As you can see, the dot product is a real number (scalar) not a vector.

▶ **EXAMPLE 6**   Find each of the following dot products.

a. $\mathbf{U} \cdot \mathbf{V}$ when $\mathbf{U} = \langle 3, 4 \rangle$ and $\mathbf{V} = \langle 2, 5 \rangle$
b. $\langle -1, 2 \rangle \cdot \langle 3, -5 \rangle$
c. $\mathbf{S} \cdot \mathbf{W}$ when $\mathbf{S} = 6\mathbf{i} + 3\mathbf{j}$ and $\mathbf{W} = 2\mathbf{i} - 7\mathbf{j}$

Solution For each problem, we simply multiply the coefficients a and c and add that result to the product of the coefficients b and d.

a. $\mathbf{U} \cdot \mathbf{V} = 3(2) + 4(5)$
$= 6 + 20$
$= 26$

b. $\langle -1, 2 \rangle \cdot \langle 3, -5 \rangle = -1(3) + 2(-5)$
$= -3 + (-10)$
$= -13$

c. $\mathbf{S} \cdot \mathbf{W} = 6(2) + 3(-7)$
$= 12 + (-21)$
$= -9$ ◀

Finding the Angle Between Two Vectors

One application of the dot product is finding the angle between two vectors. To do this, we will use an alternate form of the dot product, shown in the following theorem.

Theorem 7.1

The dot product of two vectors is equal to the product of their magnitudes multiplied by the cosine of the angle between them. That is, when θ is the angle between two nonzero vectors $\mathbf{U}$ and $\mathbf{V}$, then

$$\mathbf{U} \cdot \mathbf{V} = |\mathbf{U}||\mathbf{V}| \cos \theta$$

The proof of this theorem is derived from the law of cosines and is left as an exercise.

When we are given two vectors and asked to find the angle between them, we rewrite the formula in Theorem 7.1 by dividing each side by $|\mathbf{U}||\mathbf{V}|$. The result is

$$\cos \theta = \frac{\mathbf{U} \cdot \mathbf{V}}{|\mathbf{U}||\mathbf{V}|}$$

This formula is equivalent to our original formula, but is easier to work with when finding the angle between two vectors.

▶ **EXAMPLE 7** Find the angle between the vectors $\mathbf{U}$ and $\mathbf{V}$.

a. $\mathbf{U} = \langle 2, 3 \rangle$ and $\mathbf{V} = \langle -3, 2 \rangle$
b. $\mathbf{U} = 6\mathbf{i} - \mathbf{j}$ and $\mathbf{V} = \mathbf{i} + 4\mathbf{j}$

Solution

a. $\cos\theta = \dfrac{\mathbf{U}\cdot\mathbf{V}}{|\mathbf{U}||\mathbf{V}|}$

$\qquad = \dfrac{2(-3) + 3(2)}{\sqrt{2^2 + 3^2}\cdot\sqrt{(-3)^2 + 2^2}}$

$\qquad = \dfrac{-6 + 6}{\sqrt{13}\cdot\sqrt{13}}$

$\qquad = \dfrac{0}{13}$

$\cos\theta = 0$

$\qquad \theta = 90°$

b. $\cos\theta = \dfrac{\mathbf{U}\cdot\mathbf{V}}{|\mathbf{U}||\mathbf{V}|}$

$\qquad = \dfrac{6(1) + (-1)4}{\sqrt{6^2 + (-1)^2}\cdot\sqrt{1^2 + 4^2}}$

$\qquad = \dfrac{6 + (-4)}{\sqrt{37}\cdot\sqrt{17}}$

$\qquad = \dfrac{2}{25.08}$

$\cos\theta = 0.0797$

$\qquad \theta = 94.57°$ To the nearest hundredth of a degree   ◀

Perpendicular Vectors

If two nonzero vectors are perpendicular, then the angle between them is 90°. Since the cosine of 90° is always 0, the dot product of two perpendicular vectors must also be 0. This fact gives rise to the following theorem.

Theorem 7.2

If **U** and **V** are two nonzero vectors, then

$$\mathbf{U}\cdot\mathbf{V} = 0 \Leftrightarrow \mathbf{U}\perp\mathbf{V}$$

In Words: Two nonzero vectors are perpendicular if and only if their dot product is 0.

▶ **EXAMPLE 8** Which of the following vectors are perpendicular to each other?

$$\mathbf{U} = 8\mathbf{i} + 6\mathbf{j}$$
$$\mathbf{V} = 3\mathbf{i} - 4\mathbf{j}$$
$$\mathbf{W} = 4\mathbf{i} + 3\mathbf{j}$$

Solution   Find $\mathbf{U} \cdot \mathbf{V}$, $\mathbf{V} \cdot \mathbf{W}$, and $\mathbf{U} \cdot \mathbf{W}$. If the dot product is zero, then the two vectors are perpendicular.

$$
\begin{aligned}
\mathbf{U} \cdot \mathbf{V} &= 8(3) + 6(-4) \\
&= 24 - 24 \\
&= 0 \qquad \text{Therefore } \mathbf{U} \text{ and } \mathbf{V} \text{ are perpendicular}
\end{aligned}
$$

$$
\begin{aligned}
\mathbf{V} \cdot \mathbf{W} &= 3(4) + (-4)3 \\
&= 12 - 12 \\
&= 0 \qquad \text{Therefore } \mathbf{V} \text{ and } \mathbf{W} \text{ are perpendicular}
\end{aligned}
$$

$$
\begin{aligned}
\mathbf{U} \cdot \mathbf{W} &= 8(4) + 6(3) \\
&= 32 + 18 \\
&= 40 \qquad \text{Therefore } \mathbf{U} \text{ and } \mathbf{W} \text{ are not perpendicular} \qquad ◀
\end{aligned}
$$

PROBLEM SET 7.5

Draw the vector $\mathbf{V}$ that goes from the origin to the given point. Then write $\mathbf{V}$ in terms of the unit vectors $\mathbf{i}$ and $\mathbf{j}$.

1. $(4, 4)$ **2.** $(-4, -4)$
3. $(2, 5)$ **4.** $(5, 2)$
5. $(-3, 6)$ **6.** $(-6, 3)$
7. $(4, -5)$ **8.** $(5, -4)$
9. $(-1, -5)$ **10.** $(-5, -1)$

Find the magnitude of each of the following vectors.

11. $\mathbf{V} = 3\mathbf{i} + 4\mathbf{j}$ **12.** $\mathbf{V} = 6\mathbf{i} + 8\mathbf{j}$
13. $\mathbf{U} = 5\mathbf{i} - 12\mathbf{j}$ **14.** $\mathbf{U} = 20\mathbf{i} - 21\mathbf{j}$
15. $\mathbf{W} = \mathbf{i} + 2\mathbf{j}$ **16.** $\mathbf{W} = 3\mathbf{i} + \mathbf{j}$
17. $\langle -5, 6 \rangle$ **18.** $\langle -3, 7 \rangle$
19. $\langle 2, 0 \rangle$ **20.** $\langle 0, 5 \rangle$
21. $\langle -2, -5 \rangle$ **22.** $\langle -8, -3 \rangle$

For each pair of vectors, find $\mathbf{U} + \mathbf{V}$, $\mathbf{U} - \mathbf{V}$, $3\mathbf{U} + 2\mathbf{V}$, and $\mathbf{U} \cdot \mathbf{V}$.

23. $\mathbf{U} = \mathbf{i} + \mathbf{j}$ $\mathbf{V} = \mathbf{i} - \mathbf{j}$ **24.** $\mathbf{U} = -\mathbf{i} + \mathbf{j}$ $\mathbf{V} = \mathbf{i} + \mathbf{j}$
25. $\mathbf{U} = 6\mathbf{i}$ $\mathbf{V} = -8\mathbf{j}$ **26.** $\mathbf{U} = -3\mathbf{i}$ $\mathbf{V} = 5\mathbf{j}$
27. $\mathbf{U} = 2\mathbf{i} + 5\mathbf{j}$ $\mathbf{V} = 5\mathbf{i} + 2\mathbf{j}$ **28.** $\mathbf{U} = 5\mathbf{i} + 3\mathbf{j}$ $\mathbf{V} = 3\mathbf{i} + 5\mathbf{j}$

Find each of the following dot products.

29. $\langle 6, 6 \rangle \cdot \langle 3, 5 \rangle$ **30.** $\langle 3, 4 \rangle \cdot \langle 5, 5 \rangle$
31. $\langle -23, 4 \rangle \cdot \langle 15, -6 \rangle$ **32.** $\langle 11, -8 \rangle \cdot \langle 4, -7 \rangle$

Find the angle θ between the given vectors to the nearest tenth of a degree.

33. $U = 13i$ $\quad\quad V = -6j$
34. $U = -4i$ $\quad\quad V = 17j$
35. $U = 4i + 5j$ $\quad\quad V = 7i - 4j$
36. $U = -3i + 5j$ $\quad\quad V = 6i + 3j$
37. $U = 13i - 8j$ $\quad\quad V = 2i + 11j$
38. $U = 11i + 7j$ $\quad\quad V = -4i + 6j$

Show that each pair of vectors is perpendicular.

39. i and j
40. $i + j$ and $i - j$
41. $-i$ and j
42. $2i + j$ and $i - 2j$

43. In general, show that the vectors $V = ai + bj$ and $W = -bi + aj$ are always perpendicular.

44. Show that the slope of the line that contains $V = ai + bj$ is $m = \dfrac{-b}{a}$.

45. Use the diagram shown in Figure 10 below along with the law of cosines to prove Theorem 7.1. (Begin by writing $U = ai + bj$ and $V = ci + dj$.)

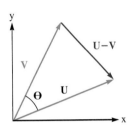

Figure 10

46. Use Theorem 7.1 to prove Theorem 7.2.

Review Problems

The problems that follow are problems you may have worked previously in Section 2.5 or Section 7.1. Solve each problem using the methods shown in Example 5 of this section.

47. A 10-pound weight is lying on a situp bench at the gym. If the bench is inclined at an angle of $15°$, there are three forces acting on the weight, as shown in Figure 11. Find the magnitude of N and the magnitude of F.

48. Repeat Problem 47 for a 25.0-pound weight and a bench inclined at $10.0°$.

49. Tyler and his cousin Kelly have attached a rope to the branch of a tree and tied a board to the other end to form a swing. Tyler stands on the board while his cousin pushes him through an angle of $25.5°$ and holds him there. If Tyler weighs 95.5 pounds, find the magnitude of the force Kelly must push with horizontally to keep Tyler in static equilibrium. See Figure 12.

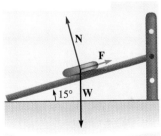

Figure 11

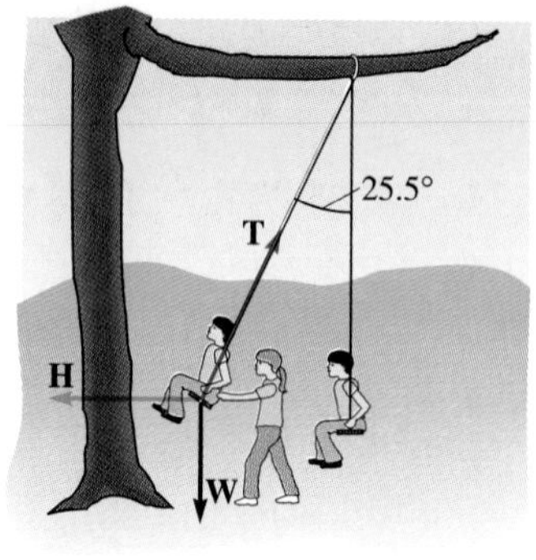

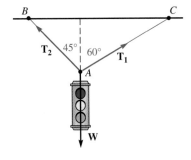

Figure 12

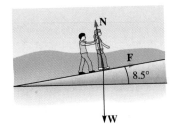

Figure 13

50. After they are done swinging, Tyler and Kelly decide to rollerskate. They come to a hill that is inclined at 8.5°. Tyler pushes Kelly halfway up the hill and then holds her there (Figure 13). If Kelly weighs 58.0 pounds, find the magnitude of the force Tyler must push with to keep Kelly from rolling down the hill. (We are assuming that the rollerskates make the hill into a frictionless surface so that the only force keeping Kelly from rolling backwards down the hill is the force Tyler is pushing with.)

51. A traffic light weighing 22 pounds is suspended by two wires as shown in Figure 14. Find the magnitude of the tension in wire AB, and the magnitude of the tension in wire AC.

Figure 15

Figure 14

52. A tightrope walker is standing still with one foot on the tightrope as shown in Figure 15. If the tightrope walker weighs 125 pounds, find the magnitudes of the tension in the rope toward each of the ends of the rope.

CHAPTER 7 SUMMARY

EXAMPLES

1. If $A = 30°$, $B = 70°$, and $a = 8.0$ centimeters in triangle ABC, then, by the law of sines,

$$b = \frac{a \sin B}{\sin A} = \frac{8 \sin 70°}{\sin 30°}$$

$$= 15 \text{ centimeters}$$

The Law of Sines [7.1]

For any triangle ABC, the following relationships are always true:

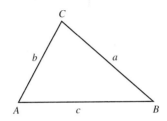

$$\frac{\sin A}{a} = \frac{\sin B}{b} = \frac{\sin C}{c}$$

or, equivalently,

$$\frac{a}{\sin A} = \frac{b}{\sin B} = \frac{c}{\sin C}$$

2. In triangle ABC, if $a = 54$ centimeters, $b = 62$ centimeters, and $A = 40°$, then

$$\sin B = \frac{b \sin A}{a} = \frac{62 \sin 40°}{54}$$

$$= 0.7380$$

Since $\sin B$ is positive for any angle in quadrant I or II, we have two possibilities for B:

$$B = 48° \quad \text{or} \quad B' = 180° - 48° = 132°$$

This indicates that two triangles exist, both of which fit the given information.

The Ambiguous Case [7.2]

When we are given two sides and an angle opposite one of them (SSA), we have several possibilities for the triangle or triangles that result. One of the possibilities is that no triangle will fit the given information. Another possibility is that two different triangles can be obtained from the given information, and a third possibility is that exactly one triangle will fit the given information. Because of these different possibilities, we call the situation where we are solving a triangle in which we are given two sides and the angle opposite one of them the *ambiguous case*.

3. In triangle ABC, if $a = 34$ kilometers, $b = 20$ kilometers, and $c = 18$ kilometers, then we can find A using the law of cosines.

$$\cos A = \frac{b^2 + c^2 - a^2}{2bc}$$

$$= \frac{20^2 + 18^2 - 34^2}{(2)(20)(18)}$$

$$\cos A = -0.6000$$

$$\text{so } A = 127°$$

The Law of Cosines [7.3]

In any triangle ABC, the following relationships are always true:

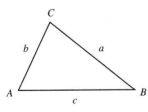

$$a^2 = b^2 + c^2 - 2bc \cos A$$
$$b^2 = a^2 + c^2 - 2ac \cos B$$
$$c^2 = a^2 + b^2 - 2ab \cos C$$

or, equivalently,

$$\cos A = \frac{b^2 + c^2 - a^2}{2bc}$$

$$\cos B = \frac{a^2 + c^2 - b^2}{2ac}$$

$$\cos C = \frac{a^2 + b^2 - c^2}{2ab}$$

4. For triangle ABC,

The Area of a Triangle [7.4]

The area of a triangle for which we are given two sides and the included angle is given by

a. If $a = 12$ centimeters, $b = 15$ centimeters, and $C = 20°$, then the area of ABC is

$$S = \frac{1}{2}(12)(15) \sin 20°$$

$$= 30.8 \text{ centimeters}^2 \text{ to the nearest tenth}$$

$$S = \frac{1}{2}ab \sin C$$

$$S = \frac{1}{2}ac \sin B$$

$$S = \frac{1}{2}bc \sin A$$

b. If $a = 24$ inches, $b = 14$ inches, and $c = 18$ inches, then the area of ABC is

The area of a triangle for which we are given all three sides is given by the formula

$$S = \sqrt{s(s - a)(s - b)(s - c)}$$

$$S = \sqrt{28(28 - 24)(28 - 14)(28 - 18)}$$
$$= \sqrt{28(4)(14)(10)}$$
$$= \sqrt{15,680}$$
$$= 125.2 \text{ inches}^2 \text{ to the nearest tenth}$$

where $s = \frac{1}{2}(a + b + c)$

c. If $A = 40°$, $B = 72°$, and $c = 45$ meters, then the area of ABC is

The area of a triangle for which we are given two angles and a side is given by

$$S = \frac{45^2 \sin 40° \sin 72°}{2 \sin 68°}$$

$$= \frac{2,025(0.6428)(0.9511)}{2(0.9272)}$$

$$= 667.6 \text{ meters}^2 \text{ to the nearest tenth}$$

$$S = \frac{a^2 \sin B \sin C}{2 \sin A}$$

$$S = \frac{b^2 \sin C \sin A}{2 \sin B}$$

$$S = \frac{c^2 \sin A \sin B}{2 \sin C}$$

5. The vector **V** that extends from the origin to the point $(-3, 4)$ is

Algebraic Vectors [7.5]

The vector that extends from the origin to the point $(1, 0)$ is called the *unit horizontal vector* and is denoted by **i**. The vector that extends from the origin to the point $(0, 1)$ is called the *unit vertical vector* and is denoted by **j**. Any nonzero vector **V** can be written

$$\mathbf{V} = -3\mathbf{i} + 4\mathbf{j} = \langle -3, 4 \rangle$$

1) In terms of unit vectors as

$$\mathbf{V} = a\mathbf{i} + b\mathbf{j}$$

2) In component form as

$$\mathbf{V} = \langle a, b \rangle$$

where a and b are real numbers.

6. The magnitude of
$\mathbf{V} = -3\mathbf{i} + 4\mathbf{j}$ is

$$|\mathbf{V}| = \sqrt{(-3)^2 + 4^2}$$
$$= \sqrt{25} = 5$$

Magnitude [7.5]

The magnitude of $\mathbf{V} = a\mathbf{i} + b\mathbf{j} = \langle a, b \rangle$ is

$$|\mathbf{V}| = \sqrt{a^2 + b^2}$$

7. If $\mathbf{U} = 6\mathbf{i} + 2\mathbf{j}$ and
$\mathbf{V} = -3\mathbf{i} + 5\mathbf{j}$, then

$$\mathbf{U} + \mathbf{V} = (6\mathbf{i} + 2\mathbf{j})$$
$$+ (-3\mathbf{i} + 5\mathbf{j})$$
$$= (6 - 3)\mathbf{i}$$
$$+ (2 + 5)\mathbf{j}$$
$$= 3\mathbf{i} + 7\mathbf{j}$$

$$\mathbf{U} - \mathbf{V} = (6\mathbf{i} + 2\mathbf{j})$$
$$- (-3\mathbf{i} + 5\mathbf{j})$$
$$= [6 - (-3)]\mathbf{i}$$
$$+ (2 - 5)\mathbf{j}$$
$$= 9\mathbf{i} - 3\mathbf{j}$$

Addition and Subtraction with Algebraic Vectors [7.5]

If $\mathbf{U} = a\mathbf{i} + b\mathbf{j}$ and $\mathbf{V} = c\mathbf{i} + d\mathbf{j}$, then vector addition and subtraction are defined as follows:

Addition: $\qquad \mathbf{U} + \mathbf{V} = (a + c)\mathbf{i} + (b + d)\mathbf{j}$

Subtraction: $\qquad \mathbf{U} - \mathbf{V} = (a - c)\mathbf{i} + (b - d)\mathbf{j}$

8. If $\mathbf{U} = -3\mathbf{i} + 4\mathbf{j}$ and $\mathbf{V} = 4\mathbf{i} + 3\mathbf{j}$, then

$$\mathbf{U} \cdot \mathbf{V} = -3(4) + 4(3) = 0$$

Dot Product [7.5]

The *dot product* of two vectors $\mathbf{U} = a\mathbf{i} + b\mathbf{j}$ and $\mathbf{V} = c\mathbf{i} + d\mathbf{j}$ is written $\mathbf{U} \cdot \mathbf{V}$ and is defined as follows:

$$\mathbf{U} \cdot \mathbf{V} = ac + bd$$

If θ is the angle between the two vectors, then it is also true that

$$\mathbf{U} \cdot \mathbf{V} = |\mathbf{U}||\mathbf{V}| \cos \theta$$

which we can solve for $\cos \theta$ to obtain the formula that allows us to find angle θ:

$$\cos \theta = \frac{\mathbf{U} \cdot \mathbf{V}}{|\mathbf{U}||\mathbf{V}|}$$

9. The vectors U and V in Example 8 above are perpendicular because their dot product is 0.

Perpendicular Vectors [7.5]

If $\mathbf{U}$ and $\mathbf{V}$ are two nonzero vectors, then

$$\mathbf{U} \cdot \mathbf{V} = 0 \Leftrightarrow \mathbf{U} \perp \mathbf{V}$$

In Words: Two nonzero vectors are perpendicular if and only if their dot product is 0.

CHAPTER 7 TEST

Problems 1 through 14 refer to triangle *ABC*, which is not necessarily a right triangle.

1. If $A = 32°$, $B = 70°$, and $a = 3.8$ inches, use the law of sines to find b.
2. If $B = 118°$, $C = 37°$, and $c = 2.9$ inches, use the law of sines to find b.
3. If $A = 38.2°$, $B = 63.4°$, and $c = 42.0$ cm, find all the missing parts.
4. If $A = 24.7°$, $C = 106.1°$, and $b = 34.0$ cm, find all the missing parts.
5. Use the law of sines to show that no triangle exists for which $A = 60°$, $a = 12$ inches, and $b = 42$ inches.
6. Use the law of sines to show that exactly one triangle exists for which $A = 42°$, $a = 29$ inches, and $b = 21$ inches.
7. Find two triangles for which $A = 51°$, $a = 6.5$ ft, and $b = 7.9$ ft.
8. Find two triangles for which $A = 26°$, $a = 4.8$ ft, and $b = 9.4$ ft.
9. If $C = 60°$, $a = 10$ cm, and $b = 12$ cm, use the law of cosines to find c.
10. If $C = 120°$, $a = 10$ cm, and $b = 12$ cm, use the law of cosines to find c.
11. If $a = 5$ km, $b = 7$ km, and $c = 9$ km, use the law of cosines to find C to the nearest tenth of a degree.
12. If $a = 10$ km, $b = 12$ km, and $c = 11$ km, use the law of cosines to find B to the nearest tenth of a degree.
13. Find all the missing parts if $a = 6.4$ m, $b = 2.8$ m, and $C = 119°$.
14. Find all the missing parts if $b = 3.7$ m, $c = 6.2$ m, and $A = 35°$.
15. The two equal sides of an isosceles triangle are each 38 cm. If the base measures 48 cm, find the measure of the two equal angles.
16. A lamp pole casts a shadow 53 ft long when the angle of elevation of the sun is 48°. Find the height of the lamp pole to the nearest foot.
17. A man standing near a building notices that the angle of elevation to the top of the building is 64°. He then walks 240 ft farther away from the building and finds the angle of elevation to the top to be 43°. How tall is the building?
18. The diagonals of a parallelogram are 26.8 m and 39.4 m. If they meet at an angle of 134.5°, find the length of the shorter side of the parallelogram.
19. Suppose Figure 1 is an exaggerated diagram of a plane flying above the earth. When the plane is 4.55 mi above point *B*, the pilot finds angle *A* to be 90.8°. Assuming that the radius of the earth is 3,960 mi, what is the distance from point *A* to point *B* along the circumference of the earth? (*Hint:* Find angle *C* first; then use the formula for arc length.)
20. A man wandering in the desert walks 3.3 mi in the direction S 44° W. He then turns and walks 2.2 mi in the direction N 55° W. At that time, how far is he from his starting point, and what is his bearing from his starting point?
21. Two guy wires from the top of a tent pole are anchored to the ground on each side of the pole by two stakes so that the two stakes and the tent pole lie along the same line. One of the wires is 56 ft long and makes an angle of 47° with the ground. The other wire is 65 ft long and makes an angle of 37° with the ground. How far apart are the stakes that hold the wires to the ground?

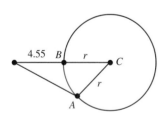

4.55

Figure 1

22. A plane is headed due east with an airspeed of 345 mph. Its true course, however, is at 95.5° from due north. If the wind currents are a constant 55.0 mph, what are the possibilities for the ground speed of the plane?

23. To estimate the height of a tree, two people position themselves 25 ft apart. From the first person, the bearing of the tree is N 48° E and the angle of elevation to the top of the tree is 73°. If the bearing of the tree from the second person is N 38° W, estimate the height of the tree to the nearest foot.

24. A plane flying with an airspeed of 325 mph is headed in the direction N 87.6° W. The wind currents are running at a constant 65.4 mph at 262.6°. Find the ground speed and true course of the plane.

Let $\mathbf{U} = 5\mathbf{i} + 12\mathbf{j}$, $\mathbf{V} = -4\mathbf{i} + \mathbf{j}$, and $\mathbf{W} = \mathbf{i} - 4\mathbf{j}$, and find

25. $|\mathbf{U}|$
26. $3\mathbf{U} + 5\mathbf{V}$
27. $3\mathbf{U} - 5\mathbf{V}$
28. $|2\mathbf{V} - \mathbf{W}|$
29. $\mathbf{V} \cdot \mathbf{W}$
30. The angle between $\mathbf{U}$ and $\mathbf{V}$ to the nearest tenth of a degree.
31. Find the area of the triangle in Problem 3.
32. Find the area of the triangle in Problem 4.
33. Find the area of the triangle in Problem 9.
34. Find the area of the triangle in Problem 10.
35. Find the area of the triangle in Problem 11.
36. Find the area of the triangle in Problem 12.

8

Complex Numbers and Polar Coordinates

I I I I I I I I I I

▶ ### To the Student

The first four sections of this chapter are a study of complex numbers. You may have already studied some of this material in algebra. If so, the first section of this chapter will be review for you. The material on complex numbers that may not be review for you is based on a new definition that makes it possible to look at complex numbers from a trigonometric point of view. In Section 8.2, we use this new definition to write complex numbers in what is called trigonometric form. Once we can write complex numbers in trigonometric form, many of the operations with complex numbers, such as multiplication, become much easier.

In the last two sections of this chapter, we will study polar coordinates. Polar coordinates are used to name points in the plane and are an alternative to rectangular coordinates. The definition for the polar coordinates of a point in the plane is based on our original definition for sine and cosine.

Complex Numbers

The study of complex numbers is a relatively new field in mathematics. The first attempt to consider them came in 1545 when Jerome Cardan published his book *Ars Magna* (the actual title is longer than this and translates as *The Great Art of Solving*

Algebraic Equations). Here is a problem Cardan proposes in his book:

> If someone says to you, divide 10 into two parts, one of
> which multiplied into the other shall produce . . . 40, it is
> evident that this case or question is impossible.
> Nevertheless, we shall solve it in this fashion.

In our present-day notation, the problem can be solved with a system of equations:

$$x + y = 10$$
$$xy = 40$$

Solving the first equation for y, we have $y = 10 - x$. Substituting this value of y into the second equation, we have

$$x(10 - x) = 40$$

The equation is quadratic. We write it in standard form and apply the quadratic formula.

$$0 = x^2 - 10x + 40$$
$$x = \frac{10 \pm \sqrt{100 - 4(1)(40)}}{2}$$
$$= \frac{10 \pm \sqrt{100 - 160}}{2}$$
$$= \frac{10 \pm \sqrt{-60}}{2}$$
$$= \frac{10 \pm 2\sqrt{-15}}{2}$$
$$= 5 \pm \sqrt{-15}$$

This is as far as Cardan could take the problem because he did not know what to do with the square root of a negative number. We handle this situation by using *complex numbers*. Our work with complex numbers is based on the following definition.

DEFINITION   The number i is such that $i^2 = -1$. (That is, i is the number whose square is -1.)

The number i is not a real number. We can use it to write square roots of negative numbers without a negative sign. To do so, we reason that if $a > 0$, then $\sqrt{-a} = \sqrt{ai^2} = i\sqrt{a}$.

▶ **EXAMPLE 1**   Write each expression in terms of i.

$$\textbf{a. } \sqrt{-9} \qquad \textbf{b. } \sqrt{-12} \qquad \textbf{c. } \sqrt{-17}$$

Solution

a. $\sqrt{-9} = i\sqrt{9} = 3i$
b. $\sqrt{-12} = i\sqrt{12} = 2i\sqrt{3}$
c. $\sqrt{-17} = i\sqrt{17}$ ◀

Note   In order to simplify expressions that contain square roots of negative numbers by using the properties of radicals developed in algebra, it is necessary to write each square root in terms of i before applying the properties of radicals. For example,

this is correct: $\sqrt{-4}\sqrt{-9} = (i\sqrt{4})(i\sqrt{9}) = (2i)(3i) = 6i^2 = -6$
this is incorrect: $\sqrt{-4}\sqrt{-9} = \sqrt{-4(-9)} = \sqrt{36} = 6$

Remember, the properties of radicals you developed in algebra hold only for expressions in which the numbers under the radical sign are positive. When the radicals contain negative numbers, you must first write each radical in terms of i and then simplify.

Next, we use i to write a definition for complex numbers.

> **DEFINITION**   A *complex number* is any number that can be written in the form
>
> $$a + bi$$
>
> where a and b are real numbers and $i^2 = -1$. The form $a + bi$ is called *standard form* for complex numbers. The number a is called the *real part* of the complex number. The number b is called the *imaginary part* of the complex number. If $b = 0$, then $a + bi = a$, which is a real number. If $a = 0$ and $b \neq 0$, then $a + bi = bi$, which is called an *imaginary number*.

▶ **EXAMPLE 2**

a. The number $3 + 2i$ is a complex number in standard form. The number 3 is the real part, and the number 2 (not $2i$) is the imaginary part.
b. The number $-7i$ is a complex number because it can be written as $0 + (-7)i$. The real part is 0. The imaginary part is -7. The number $-7i$ is also an imaginary number since $a = 0$ and $b \neq 0$.
c. The number 4 is a complex number because it can be written as $4 + 0i$. The real part is 4 and the imaginary part is 0.   ◀

From part c in Example 2, it is apparent that real numbers are also complex numbers. The real numbers are a subset of the complex numbers.

Equality for Complex Numbers

> **DEFINITION** Two complex numbers are equal if and only if their real parts are equal and their imaginary parts are equal. That is, for real numbers a, b, c, and d,
>
> $$a + bi = c + di \text{ if and only if } a = c \text{ and } b = d$$

▶ **EXAMPLE 3** Find x and y if $(-3x - 9) + 4i = 6 + (3y - 2)i$.

Solution The real parts are $-3x - 9$ and 6. The imaginary parts are 4 and $3y - 2$.

$$
\begin{array}{ccc}
-3x - 9 = 6 & \text{and} & 4 = 3y - 2 \\
-3x = 15 & & 6 = 3y \\
x = -5 & & y = 2
\end{array}
$$ ◀

Addition and Subtraction of Complex Numbers

> **DEFINITION** If $z_1 = a_1 + b_1 i$ and $z_2 = a_2 + b_2 i$ are complex numbers, then the sum and difference of z_1 and z_2 are defined as follows:
>
> $$z_1 + z_2 = (a_1 + b_1 i) + (a_2 + b_2 i) = (a_1 + a_2) + (b_1 + b_2)i$$
> $$z_1 - z_2 = (a_1 + b_1 i) - (a_2 + b_2 i) = (a_1 - a_2) + (b_1 - b_2)i$$

As you can see, we add and subtract complex numbers in the same way we would add and subtract polynomials: by combining similar terms.

▶ **EXAMPLE 4** If $z_1 = 3 - 5i$ and $z_2 = -6 - 2i$, find $z_1 + z_2$ and $z_1 - z_2$.

Solution

$$z_1 + z_2 = (3 - 5i) + (-6 - 2i) = -3 - 7i$$
$$z_1 - z_2 = (3 - 5i) - (-6 - 2i) = 9 - 3i$$ ◀

Powers of i

If we assume the properties of exponents hold when the base is i, we can write any integer power of i as i, -1, $-i$, or 1. Using the fact that $i^2 = -1$, we have

$$i^1 = i$$
$$i^2 = -1$$
$$i^3 = i^2 \cdot i = -1(i) = -i$$
$$i^4 = i^2 \cdot i^2 = -1(-1) = 1$$

Since $i^4 = 1$, i^5 will simplify to i and we will begin repeating the sequence $i, -1,$ $-i, 1$ as we increase our exponent by one each time.

$$i^5 = i^4 \cdot i = 1(i) = i$$
$$i^6 = i^4 \cdot i^2 = 1(-1) = -1$$
$$i^7 = i^4 \cdot i^3 = 1(-i) = -i$$
$$i^8 = i^4 \cdot i^4 = 1(1) = 1$$

We can simplify higher powers of i by writing them in terms of i^4 since i^4 is always 1.

▶ **EXAMPLE 5**   Simplify each power of i.

a. $i^{20} = (i^4)^5 = 1^5 = 1$
b. $i^{23} = (i^4)^5 \cdot i^3 = 1(-i) = -i$
c. $i^{30} = (i^4)^7 \cdot i^2 = 1(-1) = -1$ ◀

Multiplication and Division with Complex Numbers

> **DEFINITION**   If $z_1 = a_1 + b_1 i$ and $z_2 = a_2 + b_2 i$ are complex numbers, then their product is defined as follows:
>
> $$z_1 z_2 = (a_1 + b_1 i)(a_2 + b_2 i)$$
> $$= (a_1 a_2 - b_1 b_2) + (a_1 b_2 + a_2 b_1)i$$

This formula is simply a result of binomial multiplication and is much less complicated than it looks. Since complex numbers have the form of binomials with i as the variable, we can multiply two complex numbers using the same methods we use to multiply binomials and not have another formula to memorize.

▶ **EXAMPLE 6**   Multiply $(3 - 4i)(2 - 5i)$.

Solution   Multiplying as if these were two binomials, we have

$$(3 - 4i)(2 - 5i) = 3 \cdot 2 - 3 \cdot 5i - 2 \cdot 4i + 4i \cdot 5i$$
$$= 6 - 15i - 8i + 20i^2$$
$$= 6 - 23i + 20i^2$$

Now, since $i^2 = -1$, we can simplify further.

$$= 6 - 23i + 20(-1)$$
$$= 6 - 23i - 20$$
$$= -14 - 23i$$ ◀

▶ **EXAMPLE 7**   Multiply $(4 - 5i)(4 + 5i)$.

Solution   This product has the form $(a - b)(a + b)$, which we know results in the difference of two squares, $a^2 - b^2$.

$$(4 - 5i)(4 + 5i) = 4^2 - (5i)^2$$
$$= 16 - 25i^2$$
$$= 16 - 25(-1)$$
$$= 16 + 25$$
$$= 41 \quad \blacktriangleleft$$

The product of the two complex numbers $4 - 5i$ and $4 + 5i$ is the real number 41. This fact is very useful and leads to the following definition.

DEFINITION   The complex numbers $a + bi$ and $a - bi$ are called *complex conjugates*. Their product is the real number $a^2 + b^2$. Here's why:

$$(a + bi)(a - bi) = a^2 - (bi)^2$$
$$= a^2 - b^2i^2$$
$$= a^2 - b^2(-1)$$
$$= a^2 + b^2$$

The fact that the product of two complex conjugates is a real number is the key to division with complex numbers.

▶ **EXAMPLE 8**   Divide $\dfrac{5i}{2 - 3i}$.

Solution   We want to find a complex number in standard form that is equivalent to the quotient $5i/(2 - 3i)$. To do so, we need to replace the denominator with a real number. We can accomplish this by multiplying both the numerator and the denominator by $2 + 3i$, which is the conjugate of $2 - 3i$.

$$\frac{5i}{2 - 3i} = \frac{5i}{2 - 3i} \cdot \frac{(2 + 3i)}{(2 + 3i)}$$
$$= \frac{5i(2 + 3i)}{(2 - 3i)(2 + 3i)}$$
$$= \frac{10i + 15i^2}{4 - 9i^2}$$
$$= \frac{10i + 15(-1)}{4 - 9(-1)}$$
$$= \frac{-15 + 10i}{13}$$
$$= -\frac{15}{13} + \frac{10}{13}i$$

Notice that we have written our answer in standard form. The real part is $-15/13$ and the imaginary part is $10/13$. ◀

▶ **EXAMPLE 9** In the introduction to this section, we found that $x = 5 + \sqrt{-15}$ is one part of a solution to the system of equations

$$x + y = 10$$
$$xy = 40$$

Find the value of y that accompanies this value of x. Then show that together they form a solution to the system of equations that describes Cardan's problem.

Solution First we write the number $x = 5 + \sqrt{-15}$ with our complex number notation as $x = 5 + i\sqrt{15}$. Now, to find the value of y that accompanies this value of x, we substitute $5 + i\sqrt{15}$ for x in the equation $x + y = 10$ and solve for y:

$$\text{When} \qquad x = 5 + i\sqrt{15}$$
$$\text{then} \qquad x + y = 10$$
$$\text{becomes} \qquad 5 + i\sqrt{15} + y = 10$$
$$y = 5 - i\sqrt{15}$$

All we have left to do is show that these values of x and y satisfy the second equation, $xy = 40$. We do so by substitution:

$$\text{If} \qquad x = 5 + i\sqrt{15} \qquad \text{and} \qquad y = 5 - i\sqrt{15}$$
$$\text{then the equation} \qquad xy = 40$$
$$\text{becomes} \qquad (5 + i\sqrt{15})(5 - i\sqrt{15}) = 40$$
$$5^2 - (i\sqrt{15})^2 = 40$$
$$25 - i^2(\sqrt{15})^2 = 40$$
$$25 - (-1)(15) = 40$$
$$25 + 15 = 40$$
$$40 = 40$$

Since our last statement is a true statement, our two expressions for x and y together form a solution to the equation. ◀

PROBLEM SET 8.1

Write each expression in terms of i.

1. $\sqrt{-16}$ **2.** $\sqrt{-49}$ **3.** $\sqrt{-121}$ **4.** $\sqrt{-400}$
5. $\sqrt{-18}$ **6.** $\sqrt{-45}$ **7.** $\sqrt{-8}$ **8.** $\sqrt{-20}$

Write in terms of i and then simplify.

9. $\sqrt{-4} \cdot \sqrt{-9}$ **10.** $\sqrt{-25} \cdot \sqrt{-1}$ **11.** $\sqrt{-1} \cdot \sqrt{-9}$ **12.** $\sqrt{-16} \cdot \sqrt{-4}$

Find x and y so that each of the following equations is true:

13. $4 + 7i = 6x - 14yi$ **14.** $2 - 5i = -x + 10yi$
15. $(5x + 2) - 7i = 4 + (2y + 1)i$ **16.** $(7x - 1) + 4i = 2 + (5y + 2)i$
17. $(x^2 - 6) + 9i = x + y^2i$ **18.** $(x^2 - 2x) + y^2i = 8 + (2y - 1)i$

Find all x and y between 0 and 2π so that each of the following equations is true:

19. $\cos x + i \sin y = \sin x + i$

20. $\sin x + i \cos y = -\cos x - i$

21. $(\sin^2 x + 1) + i \tan y = 2 \sin x + i$

22. $(\cos^2 x + 1) + i \tan y = 2 \cos x - i$

Combine the following complex numbers:

23. $(7 + 2i) + (3 - 4i)$

24. $(3 - 5i) + (2 + 4i)$

25. $(6 + 7i) - (4 + i)$

26. $(5 + 2i) - (3 + 6i)$

27. $(7 - 3i) - (4 + 10i)$

28. $(11 - 6i) - (2 - 4i)$

29. $(3 \cos x + 4i \sin y) + (2 \cos x - 7i \sin y)$

30. $(2 \cos x - 3i \sin y) + (3 \cos x - 2i \sin y)$

31. $[(3 + 2i) - (6 + i)] + (5 + i)$

32. $[(4 - 5i) - (2 + i)] + (2 + 5i)$

33. $(7 - 4i) - [(-2 + i) - (3 + 7i)]$

34. $(10 - 2i) - [(2 + i) - (3 - i)]$

Simplify each power of i.

35. i^{12} **36.** i^{13} **37.** i^{14} **38.** i^{15}

39. i^{32} **40.** i^{34} **41.** i^{33} **42.** i^{35}

Find the following products:

43. $-6i(3 - 8i)$

44. $6i(3 + 8i)$

45. $(2 - 4i)(3 + i)$

46. $(2 + 4i)(3 - i)$

47. $(3 + 2i)^2$

48. $(3 - 2i)^2$

49. $(5 + 4i)(5 - 4i)$

50. $(4 + 5i)(4 - 5i)$

51. $(7 + 2i)(7 - 2i)$

52. $(2 + 7i)(2 - 7i)$

53. $2i(3 + i)(2 + 4i)$

54. $3i(1 + 2i)(3 + i)$

55. $3i(1 + i)^2$

56. $4i(1 - i)^2$

Find the following quotients. Write all answers in standard form for complex numbers.

57. $\dfrac{2i}{3 + i}$ **58.** $\dfrac{3i}{2 + i}$ **59.** $\dfrac{2 + 3i}{2 - 3i}$ **60.** $\dfrac{3 + 2i}{3 - 2i}$

61. $\dfrac{5 - 2i}{i}$ **62.** $\dfrac{5 - 2i}{-i}$ **63.** $\dfrac{2 + i}{5 - 6i}$ **64.** $\dfrac{5 + 4i}{3 + 6i}$

Let $z_1 = 2 + 3i$, $z_2 = 2 - 3i$, and $z_3 = 4 + 5i$ and find

65. $z_1 z_2$ **66.** $z_2 z_1$ **67.** $z_1 z_3$ **68.** $z_3 z_1$

69. $2z_1 + 3z_2$ **70.** $3z_1 + 2z_2$ **71.** $z_3(z_1 + z_2)$ **72.** $z_3(z_1 - z_2)$

73. Assume x represents a real number and multiply $(x + 3i)(x - 3i)$.

74. Assume x represents a real number and multiply $(x - 4i)(x + 4i)$.

75. Show that $x = 2 + 3i$ is a solution to the equation $x^2 - 4x + 13 = 0$.

76. Show that $x = 3 + 2i$ is a solution to the equation $x^2 - 6x + 13 = 0$.

77. Show that $x = a + bi$ is a solution to the equation $x^2 - 2ax + (a^2 + b^2) = 0$.

78. Show that $x = a - bi$ is a solution to the equation $x^2 - 2ax + (a^2 + b^2) = 0$.

Use the method shown in the introduction to this section to solve each system of equations.

79. $x + y = 8$
$\quad\quad xy = 20$
81. $2x + y = 4$
$\quad\quad xy = 8$

80. $x - y = 10$
$\quad\quad xy = -40$
82. $3x + y = 6$
$\quad\quad xy = 9$

83. If z is a complex number, show that the product of z and its conjugate is a real number.
84. If z is a complex number, show that the sum of z and its conjugate is an imaginary number.
85. Is addition of complex numbers a commutative operation? That is, if z_1 and z_2 are two complex numbers, is it always true that $z_1 + z_2 = z_2 + z_1$?
86. Is subtraction with complex numbers a commutative operation?

Review Problems

The problems that follow review material we covered in Sections 1.3, 3.1, and Chapter 7. Reviewing these problems will help you with some of the material in the next section.

Find $\sin \theta$ and $\cos \theta$ if the given point lies on the terminal side of θ.

87. $(3, -4)$ **88.** $(-5, 12)$ **89.** (a, b) **90.** $(1, -1)$

Find θ between $0°$ and $360°$ if

91. $\sin \theta = \dfrac{1}{\sqrt{2}}$ and $\cos \theta = -\dfrac{1}{\sqrt{2}}$

92. $\sin \theta = \dfrac{1}{2}$ and θ terminates in QII

Solve triangle ABC if

93. $A = 73.1°$, $b = 243$ cm, and $c = 157$ cm
94. $B = 24.2°$, $C = 63.8°$, and $b = 5.92$ inches
95. $a = 42.1$ m, $b = 56.8$ m, and $c = 63.4$ m
96. $B = 32.8°$, $a = 625$ ft, and $b = 521$ ft

SECTION 8.2 ## Trigonometric Form for Complex Numbers

As you know, the quadratic formula, $x = \dfrac{-b \pm \sqrt{b^2 - 4ac}}{2a}$, can be used to solve any quadratic equation in the form $ax^2 + bx + c = 0$. In his book *Ars Magna*, Jerome Cardan gives a similar formula that can be used to solve certain cubic equations. Here it is in our notation:

If $\quad x^3 = ax + b$

then $\quad x = \sqrt[3]{\dfrac{b}{2} + \sqrt{\left(\dfrac{b}{2}\right)^2 - \left(\dfrac{a}{3}\right)^3}} + \sqrt[3]{\dfrac{b}{2} - \sqrt{\left(\dfrac{b}{2}\right)^2 - \left(\dfrac{a}{3}\right)^3}}$

This formula is known as Cardan's formula. In his book, Cardan attempts to use his formula to solve the equation

$$x^3 = 15x + 4$$

This equation has the form $x^3 = ax + b$, where $a = 15$ and $b = 4$. Substituting these values for a and b in Cardan's formula, we have

$$x = \sqrt[3]{\frac{4}{2} + \sqrt{\left(\frac{4}{2}\right)^2 - \left(\frac{15}{3}\right)^3}} + \sqrt[3]{\frac{4}{2} - \sqrt{\left(\frac{4}{2}\right)^2 - \left(\frac{15}{3}\right)^3}}$$

$$= \sqrt[3]{2 + \sqrt{4 - 125}} + \sqrt[3]{2 - \sqrt{4 - 125}}$$

$$= \sqrt[3]{2 + \sqrt{-121}} + \sqrt[3]{2 - \sqrt{-121}}$$

Cardan couldn't go any further than this because he didn't know what to do with $\sqrt{-121}$, and therefore he couldn't take the cube root of an expression that contained this number. In this section, we will take the first step in finding cube roots of complex numbers by learning how to write complex numbers in what is called *trigonometric form*. Before we do so, let's look at a definition that will give us a visual representation for complex numbers.

> **DEFINITION**   The graph of the complex number $x + yi$ is a vector (arrow) that extends from the origin out to the point (x, y).

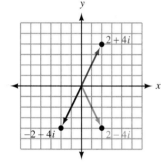

Figure 1

▶ **EXAMPLE 1**   Graph each complex number.

$$2 + 4i, \qquad -2 - 4i, \qquad 2 - 4i$$

Solution   The graphs are shown in Figure 1. Notice how the graphs of $2 + 4i$ and $2 - 4i$, which are conjugates, have symmetry about the x-axis. Note also that the graphs of $2 + 4i$ and $-2 - 4i$, which are opposites, have symmetry about the origin.   ◀

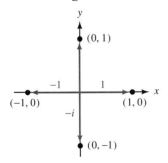

Figure 2

▶ **EXAMPLE 2**   Graph the complex numbers 1, i, -1, and $-i$.

Solution   Here are the four complex numbers written in standard form.

$$1 = 1 + 0i \qquad\qquad i = 0 + i$$
$$-1 = -1 + 0i \qquad\quad -i = 0 - i$$

The graph of each is shown in Figure 2.   ◀

If we write a real number as a complex number in standard form, its graph will fall on the x-axis. Therefore, we call the x-axis the *real axis* when we are graphing complex numbers. Likewise, because the imaginary numbers i and $-i$ fall on the y-axis, we call the y-axis the *imaginary axis* when we are graphing complex numbers.

DEFINITION The *absolute value* or *modulus* of the complex number $z = x + yi$ is the distance from the origin to the point (x, y). If this distance is denoted by r, then

$$r = |z| = |x + yi| = \sqrt{x^2 + y^2}$$

▶ **EXAMPLE 3** Find the modulus of each of the complex numbers $5i$, 7, and $3 + 4i$.

Solution Writing each number in standard form and then applying the definition of modulus, we have

$$\text{For} \quad z = 5i = 0 + 5i, \quad r = |z| = |0 + 5i| = \sqrt{0^2 + 5^2} = 5$$
$$\text{For} \quad z = 7 = 7 + 0i, \quad r = |z| = |7 + 0i| = \sqrt{7^2 + 0^2} = 7$$
$$\text{For} \quad z = 3 + 4i, \quad r = |z| = |3 + 4i| = \sqrt{3^2 + 4^2} = 5 \quad ◀$$

DEFINITION The *argument* of the complex number $z = x + yi$ is the smallest positive angle θ from the positive real axis to the graph of z.

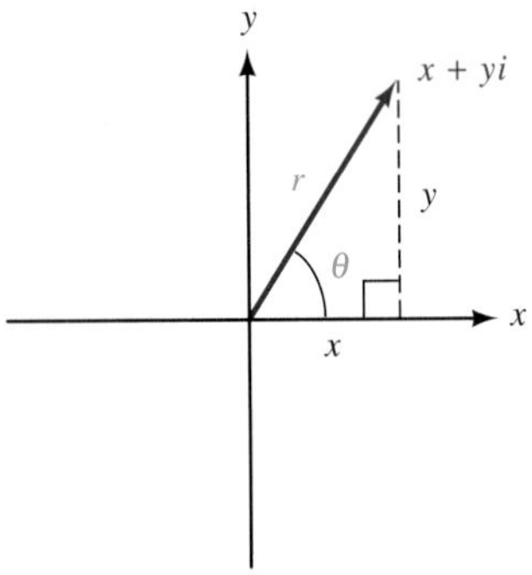

Figure 3

Figure 3 illustrates the relationships between the complex number $z = x + yi$, its graph, and the modulus r and argument θ of z. From Figure 3 we see that

$$\cos \theta = \frac{x}{r} \quad \text{or} \quad x = r \cos \theta$$

and

$$\sin \theta = \frac{y}{r} \quad \text{or} \quad y = r \sin \theta$$

We can use this information to write z in terms of r and θ.

$$\begin{aligned} z &= x + yi \\ &= r \cos \theta + (r \sin \theta)i \\ &= r \cos \theta + ri \sin \theta \\ &= r(\cos \theta + i \sin \theta) \end{aligned}$$

This last expression is called the *trigonometric form* for z. The formal definition follows.

DEFINITION   If $z = x + yi$ is a complex number in standard form, then the *trigonometric form* for z is given by

$$z = r(\cos \theta + i \sin \theta)$$

where r is the modulus of z and θ is the argument of z.

We can convert back and forth between standard form and trigonometric form by using the relationships that follow:

For $z = x + yi = r(\cos \theta + i \sin \theta)$

$r = \sqrt{x^2 + y^2}$ and θ is such that

$$\cos \theta = \frac{x}{r}, \quad \sin \theta = \frac{y}{r}, \quad \text{and} \quad \tan \theta = \frac{y}{x}$$

$-1 + i = \sqrt{2}(\cos 135° + i \sin 135°)$

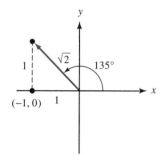

Figure 4

▶ **EXAMPLE 4**   Write $z = -1 + i$ in trigonometric form.

Solution   We have $x = -1$ and $y = 1$; therefore,

$$r = \sqrt{(-1)^2 + 1^2} = \sqrt{2}$$

Angle θ is the smallest positive angle for which $\cos \theta = x/r = -1/\sqrt{2}$ and $\sin \theta = y/r = 1/\sqrt{2}$ ($\tan \theta = y/x = 1/-1 = -1$). Therefore, θ must be $135°$.

Using these values of r and θ in the formula for trigonometric form, we have

$$z = r(\cos \theta + i \sin \theta)$$
$$= \sqrt{2}(\cos 135° + i \sin 135°)$$

The graph of z is shown in Figure 4.   ◀

$2(\cos 60° + i \sin 60°)$
$= 1 + i\sqrt{3}$

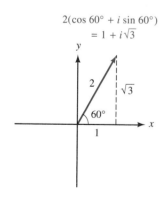

Figure 5

▶ **EXAMPLE 5**   Write $z = 2(\cos 60° + i \sin 60°)$ in standard form.

Solution   Using exact values for $\cos 60°$ and $\sin 60°$, we have

$$z = 2(\cos 60° + i \sin 60°)$$
$$= 2\left(\frac{1}{2} + i\frac{\sqrt{3}}{2}\right)$$
$$= 1 + i\sqrt{3}$$

The graph of z is shown in Figure 5.   ◀

As you can see, converting from trigonometric form to standard form is usually more direct than converting from standard form to trigonometric form.

▶ **EXAMPLE 6**   In the introduction to this section, we mentioned two complex numbers that Jerome Cardan had difficulty with: $2 + \sqrt{-121}$ and $2 - \sqrt{-121}$. Write each of these numbers in trigonometric form.

Solution First we write them as complex numbers

$$2 + \sqrt{-121} = 2 + 11i$$
$$2 - \sqrt{-121} = 2 - 11i$$

The modulus of each is

$$r = \sqrt{2^2 + 11^2} = \sqrt{125} = 5\sqrt{5}$$

For $2 + 11i$, we have $\tan \theta = \dfrac{11}{2}$. Using a calculator and rounding to the nearest hundredth of a degree, we find that $\theta = 79.70°$. Therefore,

$$2 + 11i = 5\sqrt{5}(\cos 79.70° + i \sin 79.70°)$$

For $2 - 11i$, we have $\tan \theta = -\dfrac{11}{2}$, giving us $\theta = 280.30°$ to the nearest hundredth of a degree. Therefore,

$$2 - 11i = 5\sqrt{5}(\cos 280.30° + i \sin 280.30°)$$

Note that if we were to take the trigonometric form of this last number and simplify it using decimals and a calculator, the result would not be exactly the same as the original number $2 - 11i$. That is,

$$5\sqrt{5}(\cos 280.30° + i \sin 280.30°) = 11.18(0.18 - i\,0.98)$$
$$= 2.01 - 10.96i$$

The difference is due to the rounding we did to obtain $280.30°$ and the decimal approximation to $5\sqrt{5}$. ◄

PROBLEM SET 8.2

Graph each complex number. In each case, give the absolute value of the number.

1. $3 + 4i$ **2.** $3 - 4i$ **3.** $1 + i$ **4.** $1 - i$

5. $-5i$ **6.** $4i$ **7.** 2 **8.** -4

9. $-4 - 3i$ **10.** $-3 - 4i$

Graph each complex number along with its opposite and conjugate.

11. $2 - i$ **12.** $2 + i$ **13.** $4i$ **14.** $-3i$

15. -3 **16.** 5 **17.** $-5 - 2i$ **18.** $-2 - 5i$

Write each complex number in standard form.

19. $2(\cos 30° + i \sin 30°)$ **20.** $4(\cos 30° + i \sin 30°)$

21. $4(\cos 120° + i \sin 120°)$ **22.** $8(\cos 120° + i \sin 120°)$

23. $\cos 210° + i \sin 210°$ **24.** $\cos 240° + i \sin 240°$

25. $\cos 315° + i \sin 315°$ **26.** $\sqrt{2}(\cos 315° + i \sin 315°)$

Use a calculator to help write each complex number in standard form. Round the numbers in your answers to the nearest hundredth.

27. $10(\cos 12° + i \sin 12°)$
28. $100(\cos 70° + i \sin 70°)$
29. $100(\cos 143° + i \sin 143°)$
30. $100(\cos 171° + i \sin 171°)$
31. $\cos 205° + i \sin 205°$
32. $\cos 261° + i \sin 261°$
33. $10(\cos 342° + i \sin 342°)$
34. $10(\cos 318° + i \sin 318°)$

Write each complex number in trigonometric form. In each case, begin by sketching the graph to help find the argument θ.

35. $-1 + i$ **36.** $1 + i$ **37.** $1 - i$ **38.** $-1 - i$
39. $3 + 3i$ **40.** $5 + 5i$ **41.** $8i$ **42.** $-8i$
43. -9 **44.** 2 **45.** $-2 + 2i\sqrt{3}$ **46.** $-2\sqrt{3} + 2i$

Write each complex number in trigonometric form. Round all angles to the nearest hundredth of a degree.

47. $3 + 4i$ **48.** $3 - 4i$ **49.** $20 + 21i$ **50.** $21 - 20i$
51. $7 - 24i$ **52.** $8 - 15i$ **53.** $11 + 2i$ **54.** $11 - 2i$

55. We know that $2i \cdot 3i = 6i^2 = -6$. Change $2i$ and $3i$ to trigonometric form, and then show that their product in trigonometric form is still -6.
56. Change $4i$ and 2 to trigonometric form and then multiply. Show that this product is $8i$.
57. Show that $2(\cos 30° + i \sin 30°)$ and $2[\cos(-30°) + i \sin(-30°)]$ are conjugates.
58. Show that $2(\cos 60° + i \sin 60°)$ and $2[\cos(-60°) + i \sin(-60°)]$ are conjugates.
59. Show that if $z = \cos \theta + i \sin \theta$, then $|z| = 1$.
60. Show that if $z = \cos \theta - i \sin \theta$, then $|z| = 1$.

Review Problems

The problems that follow review material we covered in Sections 5.2 and 7.2. Reviewing the problems from 5.2 will help you understand the next section.

61. Use the formula for $\cos(A + B)$ to find the exact value of $\cos 75°$.
62. Use the formula for $\sin(A + B)$ to find the exact value of $\sin 75°$.

Let $\sin A = 3/5$ with A in QI and $\sin B = 5/13$ with B in QI and find

63. $\sin(A + B)$ **64.** $\cos(A + B)$

Simplify each expression to a single trigonometric function.

65. $\sin 30° \cos 90° + \cos 30° \sin 90°$ **66.** $\cos 30° \cos 90° - \sin 30° \sin 90°$
67. $\cos 18° \cos 32° - \sin 18° \sin 32°$ **68.** $\sin 18° \cos 32° + \cos 18° \sin 32°$

In triangle ABC, $A = 45.6°$ and $b = 567$ inches. Find B for each value of a below. (You may get one or two values for B, or you may find that no triangle fits the given description.)

69. $a = 234$ inches

70. $a = 678$ inches

71. $a = 456$ inches

72. $a = 789$ inches

Research Project

73. With the publication of *Ars Magna*, a dispute arose between Cardan and another mathematician named Tartaglia. What was that dispute? In your opinion who was at fault?

SECTION 8.3

Products and Quotients in Trigonometric Form

Multiplication and division with complex numbers becomes a very simple process when the numbers are written in trigonometric form. Let's state the rule for finding the product of two complex numbers written in trigonometric form as a theorem and then prove the theorem.

Theorem (Multiplication)

If

$$z_1 = r_1(\cos \theta_1 + i \sin \theta_1)$$

and

$$z_2 = r_2(\cos \theta_2 + i \sin \theta_2)$$

are two complex numbers in trigonometric form, then their product, $z_1 z_2$, is

$$z_1 z_2 = [r_1(\cos \theta_1 + i \sin \theta_1)][r_2(\cos \theta_2 + i \sin \theta_2)]$$
$$= r_1 r_2[\cos (\theta_1 + \theta_2) + i \sin (\theta_1 + \theta_2)]$$

In Words: To multiply two complex numbers in trigonometric form, multiply absolute values and add angles.

PROOF We begin by multiplying algebraically. Then we simplify our product by using the sum formulas we introduced in Section 5.2.

$$z_1 z_2 = [r_1(\cos \theta_1 + i \sin \theta_1)][r_2(\cos \theta_2 + i \sin \theta_2)]$$
$$= r_1 r_2(\cos \theta_1 + i \sin \theta_1)(\cos \theta_2 + i \sin \theta_2)$$
$$= r_1 r_2(\cos \theta_1 \cos \theta_2 + i \cos \theta_1 \sin \theta_2 + i \sin \theta_1 \cos \theta_2 + i^2 \sin \theta_1 \sin \theta_2)$$
$$= r_1 r_2[\cos \theta_1 \cos \theta_2 + i(\cos \theta_1 \sin \theta_2 + \sin \theta_1 \cos \theta_2) - \sin \theta_1 \sin \theta_2]$$
$$= r_1 r_2[(\cos \theta_1 \cos \theta_2 - \sin \theta_1 \sin \theta_2) + i(\sin \theta_1 \cos \theta_2 + \cos \theta_1 \sin \theta_2)]$$
$$= r_1 r_2[\cos (\theta_1 + \theta_2) + i \sin (\theta_1 + \theta_2)]$$

This completes our proof. As you can see, to multiply two complex numbers in trigonometric form, we multiply absolute values, $r_1 r_2$, and add angles, $\theta_1 + \theta_2$.

▶ **EXAMPLE 1**   Find the product of $3(\cos 40° + i \sin 40°)$ and $5(\cos 10° + i \sin 10°)$.

Solution   Applying the formula from our theorem on products, we have

$$[3(\cos 40° + i \sin 40°)][5(\cos 10° + i \sin 10°)]$$
$$= 3 \cdot 5[\cos (40° + 10°) + i \sin (40° + 10°)]$$
$$= 15(\cos 50° + i \sin 50°)$$ ◀

▶ **EXAMPLE 2**   Find the product of $z_1 = 1 + i\sqrt{3}$ and $z_2 = -\sqrt{3} + i$ in standard form, and then write z_1 and z_2 in trigonometric form and find their product again.

Solution   Leaving each complex number in standard form and multiplying, we have

$$z_1 z_2 = (1 + i\sqrt{3})(-\sqrt{3} + i)$$
$$= -\sqrt{3} + i - 3i + i^2\sqrt{3}$$
$$= -2\sqrt{3} - 2i$$

Changing z_1 and z_2 to trigonometric form and multiplying looks like this:

$$z_1 = 1 + i\sqrt{3} = 2(\cos 60° + i \sin 60°)$$
$$z_2 = -\sqrt{3} + i = 2(\cos 150° + i \sin 150°)$$
$$z_1 z_2 = [2(\cos 60° + i \sin 60°)][2(\cos 150° + i \sin 150°)]$$
$$= 4(\cos 210° + i \sin 210°)$$

To compare our two products, we convert our product in trigonometric form to standard form.

$$4(\cos 210° + i \sin 210°) = 4\left(-\frac{\sqrt{3}}{2} - \frac{1}{2}i\right)$$
$$= -2\sqrt{3} - 2i$$

As you can see, both methods of multiplying complex numbers produce the same result.   ◀

The next theorem is an extension of the work we have done so far with multiplication. We will not give a formal proof of the theorem.

DeMoivre's Theorem

If $z = r(\cos \theta + i \sin \theta)$ is a complex number in trigonometric form and n is an integer, then

$$z^n = [r(\cos \theta + i \sin \theta)]^n = r^n(\cos n\theta + i \sin n\theta)$$

The theorem seems reasonable after the work we have done with multiplication. For example, if n is 2,

$$[r(\cos \theta + i \sin \theta)]^2 = r(\cos \theta + i \sin \theta)r(\cos \theta + i \sin \theta)$$
$$= r^2(\cos 2\theta + i \sin 2\theta)$$

▶ **EXAMPLE 3** Find $(1 + i)^{10}$.

Solution First we write $1 + i$ in trigonometric form:

$$1 + i = \sqrt{2}(\cos 45° + i \sin 45°)$$

Then we use DeMoivre's Theorem to raise this expression to the 10th power.

$$(1 + i)^{10} = [\sqrt{2}(\cos 45° + i \sin 45°)]^{10}$$
$$= (\sqrt{2})^{10}(\cos 10 \cdot 45° + i \sin 10 \cdot 45°)$$
$$= 32(\cos 450° + i \sin 450°)$$

which we can simplify to

$$= 32(\cos 90° + i \sin 90°)$$

since 90° and 450° are coterminal. In standard form, our result is

$$= 32(0 + i)$$
$$= 32i$$

That is,

$$(1 + i)^{10} = 32i \qquad \blacktriangleleft$$

Since multiplication with complex numbers in trigonometric form is accomplished by multiplying absolute values and adding angles, we should expect that division is accomplished by dividing absolute values and subtracting angles.

Theorem (Division)

If

$$z_1 = r_1(\cos \theta_1 + i \sin \theta_1)$$

and

$$z_2 = r_2(\cos \theta_2 + i \sin \theta_2)$$

are two complex numbers in trigonometric form, then their quotient, z_1/z_2, is

$$\frac{z_1}{z_2} = \frac{r_1(\cos \theta_1 + i \sin \theta_1)}{r_2(\cos \theta_2 + i \sin \theta_2)}$$

$$= \frac{r_1}{r_2}[\cos(\theta_1 - \theta_2) + i \sin(\theta_1 - \theta_2)]$$

PROOF   As was the case with division of complex numbers in standard form, the major step in this proof is multiplying the numerator and denominator of our quotient by the conjugate of the denominator.

$$\frac{r_1(\cos \theta_1 + i \sin \theta_1)}{r_2(\cos \theta_2 + i \sin \theta_2)}$$

$$= \frac{r_1(\cos \theta_1 + i \sin \theta_1)}{r_2(\cos \theta_2 + i \sin \theta_2)} \cdot \frac{(\cos \theta_2 - i \sin \theta_2)}{(\cos \theta_2 - i \sin \theta_2)}$$

$$= \frac{r_1(\cos \theta_1 + i \sin \theta_1)(\cos \theta_2 - i \sin \theta_2)}{r_2(\cos^2 \theta_2 + \sin^2 \theta_2)}$$

$$= \frac{r_1}{r_2} (\cos \theta_1 \cos \theta_2 - i \cos \theta_1 \sin \theta_2 + i \sin \theta_1 \cos \theta_2 - i^2 \sin \theta_1 \sin \theta_2)$$

$$= \frac{r_1}{r_2} [(\cos \theta_1 \cos \theta_2 + \sin \theta_1 \sin \theta_2) + i(\sin \theta_1 \cos \theta_2 - \cos \theta_1 \sin \theta_2)]$$

$$= \frac{r_1}{r_2} [\cos (\theta_1 - \theta_2) + i \sin (\theta_1 - \theta_2)]$$

▶ **EXAMPLE 4**   Find the quotient when $20(\cos 75° + i \sin 75°)$ is divided by $4(\cos 40° + i \sin 40°)$.

Solution   We divide according to the formula given in our theorem on division.

$$\frac{20(\cos 75° + i \sin 75°)}{4(\cos 40° + i \sin 40°)} = \frac{20}{4} [\cos (75° - 40°) + i \sin (75° - 40°)]$$

$$= 5(\cos 35° + i \sin 35°) \qquad ◀$$

▶ **EXAMPLE 5**   Divide $z_1 = 1 + i\sqrt{3}$ by $z_2 = \sqrt{3} + i$ and leave the answer in standard form. Then change each to trigonometric form and divide again.

Solution   Dividing in standard form, we have

$$\frac{z_1}{z_2} = \frac{1 + i\sqrt{3}}{\sqrt{3} + i}$$

$$= \frac{1 + i\sqrt{3}}{\sqrt{3} + i} \cdot \frac{\sqrt{3} - i}{\sqrt{3} - i}$$

$$= \frac{\sqrt{3} - i + 3i - i^2\sqrt{3}}{3 + 1}$$

$$= \frac{2\sqrt{3} + 2i}{4}$$

$$= \frac{\sqrt{3}}{2} + \frac{1}{2} i$$

Changing z_1 and z_2 to trigonometric form and dividing again, we have

$$z_1 = 1 + i\sqrt{3} = 2(\cos 60° + i \sin 60°)$$
$$z_2 = \sqrt{3} + i = 2(\cos 30° + i \sin 30°)$$
$$\frac{z_1}{z_2} = \frac{2(\cos 60° + i \sin 60°)}{2(\cos 30° + i \sin 30°)}$$
$$= \frac{2}{2}[\cos(60° - 30°) + i \sin(60° - 30°)]$$
$$= \cos 30° + i \sin 30°$$

which, in standard form, is

$$\frac{\sqrt{3}}{2} + \frac{1}{2}i$$

◀

PROBLEM SET 8.3

Multiply. Leave all answers in trigonometric form.

1. $3(\cos 20° + i \sin 20°) \cdot 4(\cos 30° + i \sin 30°)$
2. $5(\cos 15° + i \sin 15°) \cdot 2(\cos 25° + i \sin 25°)$
3. $7(\cos 110° + i \sin 110°) \cdot 8(\cos 47° + i \sin 47°)$
4. $9(\cos 115° + i \sin 115°) \cdot 4(\cos 51° + i \sin 51°)$
5. $2(\cos 135° + i \sin 135°) \cdot 2(\cos 45° + i \sin 45°)$
6. $2(\cos 120° + i \sin 120°) \cdot 4(\cos 30° + i \sin 30°)$

Find the product $z_1 z_2$ in standard form. Then write z_1 and z_2 in trigonometric form and find their product again. Finally, convert the answer that is in trigonometric form to standard form to show that the two products are equal.

7. $z_1 = 1 + i, z_2 = -1 + i$
8. $z_1 = 1 + i, z_2 = 2 + 2i$
9. $z_1 = 1 + i\sqrt{3}, z_2 = -\sqrt{3} + i$
10. $z_1 = -1 + i\sqrt{3}, z_2 = \sqrt{3} + i$
11. $z_1 = 3i, z_2 = -4i$
12. $z_1 = 2i, z_2 = -5i$
13. $z_1 = 1 + i, z_2 = 4i$
14. $z_1 = 1 + i, z_2 = 3i$
15. $z_1 = -5, z_2 = 1 + i\sqrt{3}$
16. $z_1 = -3, z_2 = \sqrt{3} + i$

Use DeMoivre's Theorem to find each of the following. Write your answer in standard form.

17. $[2(\cos 10° + i \sin 10°)]^6$
18. $[4(\cos 15° + i \sin 15°)]^3$
19. $(\cos 12° + i \sin 12°)^{10}$
20. $(\cos 18° + i \sin 18°)^{10}$
21. $[3(\cos 60° + i \sin 60°)]^4$
22. $[3(\cos 30° + i \sin 30°)]^4$
23. $[\sqrt{2}(\cos 45° + i \sin 45°)]^{10}$
24. $[\sqrt{2}(\cos 70° + i \sin 70°)]^6$
25. $(1 + i)^4$
26. $(1 + i)^5$
27. $(-\sqrt{3} + i)^4$
28. $(\sqrt{3} + i)^4$
29. $(1 + i)^6$
30. $(-1 + i)^8$
31. $(-2 + 2i)^3$
32. $(-2 - 2i)^3$

Divide. Leave your answers in trigonometric form.

33. $\dfrac{20(\cos 75° + i \sin 75°)}{5(\cos 40° + i \sin 40°)}$

34. $\dfrac{30(\cos 80° + i \sin 80°)}{10(\cos 30° + i \sin 30°)}$

35. $\dfrac{18(\cos 51° + i \sin 51°)}{12(\cos 32° + i \sin 32°)}$

36. $\dfrac{21(\cos 63° + i \sin 63°)}{14(\cos 44° + i \sin 44°)}$

37. $\dfrac{4(\cos 90° + i \sin 90°)}{8(\cos 30° + i \sin 30°)}$

38. $\dfrac{6(\cos 120° + i \sin 120°)}{8(\cos 90° + i \sin 90°)}$

Find the quotient z_1/z_2 in standard form. Then write z_1 and z_2 in trigonometric form and find their quotient again. Finally, convert the answer that is in trigonometric form to standard form to show that the two quotients are equal.

39. $z_1 = 2 + 2i, z_2 = 1 + i$
40. $z_1 = 2 - 2i, z_2 = 1 - i$
41. $z_1 = \sqrt{3} + i, z_2 = 2i$
42. $z_1 = 1 + i\sqrt{3}, z_2 = 2i$
43. $z_1 = 4 + 4i, z_2 = 2 - 2i$
44. $z_1 = 6 + 6i, z_2 = -3 - 3i$
45. $z_1 = 8, z_2 = -4$
46. $z_1 = -6, z_2 = 3$

Convert all complex numbers to trigonometric form and then simplify each expression. Write all answers in standard form.

47. $\dfrac{(1 + i)^4(2i)^2}{-2 + 2i}$

48. $\dfrac{(\sqrt{3} + i)^4(2i)^5}{(1 + i)^{10}}$

49. $\dfrac{(1 + i\sqrt{3})^4(\sqrt{3} - i)^2}{(1 - i\sqrt{3})^3}$

50. $\dfrac{(2 + 2i)^5(-3 + 3i)^3}{(\sqrt{3} + i)^{10}}$

51. Show that $x = 2(\cos 60° + i \sin 60°)$ is a solution to the quadratic equation $x^2 - 2x + 4 = 0$ by replacing x with $2(\cos 60° + i \sin 60°)$ and simplifying.

52. Show that $x = 2(\cos 300° + i \sin 300°)$ is a solution to the equation $x^2 - 2x + 4 = 0$.

53. Show that $w = 2(\cos 15° + i \sin 15°)$ is a fourth root of $z = 8 + 8i\sqrt{3}$ by raising w to the fourth power and simplifying to get z. (The number w is a fourth root of z, $w = z^{1/4}$, if the fourth power of w is z, $w^4 = z$.)

54. Show that $x = 1/2 + (\sqrt{3}/2)i$ is a cube root of -1.

DeMoivre's Theorem can be used to find reciprocals of complex numbers. Recall from algebra that the reciprocal of x is $1/x$, which can be expressed as x^{-1}. Use this fact, along with DeMoivre's Theorem, to find the reciprocal of each number below.

55. $1 + i$ **56.** $1 - i$ **57.** $\sqrt{3} - i$ **58.** $\sqrt{3} + i$

Review Problems

The problems that follow review material we covered in Sections 5.3, 5.4, 7.1, and 7.3.

If $\cos A = -1/3$ and A is between $90°$ and $180°$, find

59. $\cos 2A$ **60.** $\sin 2A$ **61.** $\sin \dfrac{A}{2}$ **62.** $\cos \dfrac{A}{2}$

63. $\csc \dfrac{A}{2}$ **64.** $\sec \dfrac{A}{2}$ **65.** $\tan 2A$ **66.** $\tan \dfrac{A}{2}$

67. A crew member on a fishing boat traveling due north off the coast of California observes that the bearing of Morro Rock from the boat is N 35° E. After sailing another 9.2 mi, the crew member looks back to find that the bearing of Morro Rock from the ship is S 27° E. At that time, how far is the boat from Morro Rock?

68. A tent pole is held in place by two guy wires on opposite sides of the pole. One of the guy wires makes an angle of 43.2° with the ground and is 10.1 ft long. How long is the other guy wire if it makes an angle of 34.5° with the ground?

69. A plane is flying with an airspeed of 170 mph with a heading of 112°. The wind currents are a constant 28 mph in the direction of due north. Find the true direction and ground speed of the plane.

70. If a parallelogram has sides of 33 cm and 22 cm that meet at an angle of 111°, how long is the longer diagonal?

SECTION 8.4 # Roots of a Complex Number

What is it about mathematics that draws some people toward it? In many cases, it is the fact that we can describe the world around us with mathematics; for some people, mathematics gives a clearer picture of the world in which we live. In other cases, the attraction is within mathematics itself. That is, for some people, mathematics itself is attractive, regardless of its connection to the real world. What you will find in this section is something in this second category. It is a property that real and complex numbers contain that is, by itself, surprising and attractive for people who enjoy mathematics. It has to do with roots of real and complex numbers. Here it is:

If we solve the equation $x^2 = 25$, our solutions will be square roots of 25. Likewise, if we solve $x^3 = 8$, our solutions will be cube roots of 8. Further, the solutions to $x^4 = 81$ will be fourth roots of 81.

Without showing the work involved, here are the solutions to these three equations:

Equations	Solutions
$x^2 = 25$	-5 and 5
$x^3 = 8$	2, $-1 + i\sqrt{3}$, and $-1 - i\sqrt{3}$
$x^4 = 81$	-3, 3, $-3i$, and $3i$

The number 25 has two square roots, 8 has three cube roots, and 81 has four fourth roots. As you will see as we progress through this section, the numbers we used in the equations given are unimportant; every real (or complex) number has exactly two square roots, three cube roots, four fourth roots. In fact, every real (or complex) number has exactly n distinct nth roots, a surprising and attractive fact about numbers. The key to finding these roots is trigonometric form for complex numbers.

Suppose that z and w are complex numbers such that w is the nth root of z. That is,

$$w = \sqrt[n]{z}$$

If we raise both sides of this last equation to the nth power, we have

$$w^n = z$$

Now suppose $z = r(\cos \theta + i \sin \theta)$ and $w = s(\cos \alpha + i \sin \alpha)$. Substituting these expressions into the equation $w^n = z$, we have

$$[s(\cos \alpha + i \sin \alpha)]^n = r(\cos \theta + i \sin \theta)$$

We can rewrite the left side of this last equation using DeMoivre's Theorem.

$$s^n(\cos n\,\alpha + i \sin n\,\alpha) = r(\cos \theta + i \sin \theta)$$

The only way these two expressions can be equal is if their absolute values are equal and their angles are coterminal.

Absolute Values Equal	*Angles Coterminal*
$s^n = r$	$n\,\alpha = \theta + 360°k$ $k = $ an integer

Solving for s and α, we have

$$s = r^{1/n} \qquad\qquad \alpha = \frac{\theta + 360°k}{n}$$

To summarize, we find the nth roots of a complex number by first finding the real nth root of the absolute value and then adding multiples of 360° to θ and dividing the result by n. The multiples of 360° that we add on will range from 360°(0) up to 360°$(n - 1)$. After that we start repeating angles.

Theorem (Roots)

The nth roots of the complex number

$$z = r(\cos \theta + i \sin \theta)$$

are given by

$$w_k = r^{1/n}\left[\cos \frac{\theta + 360°k}{n} + i \sin \frac{\theta + 360°k}{n}\right]$$

where $k = 0, 1, 2, \ldots, n - 1$.

▶ **EXAMPLE 1**   Find the 4 fourth roots of $z = 16(\cos 60° + i \sin 60°)$.

Solution   According to the formula given in our theorem on roots, the 4 fourth roots will be

$$w_k = 16^{1/4}\left[\cos \frac{60° + 360°k}{4} + i \sin \frac{60° + 360°k}{4}\right] \qquad k = 0, 1, 2, 3$$

$$= 2[\cos (15° + 90°k) + i \sin (15° + 90°k)]$$

Replacing k with 0, 1, 2, and 3, we have

$$w_0 = 2(\cos 15° + i \sin 15°) \qquad \text{when } k = 0$$
$$w_1 = 2(\cos 105° + i \sin 105°) \qquad \text{when } k = 1$$
$$w_2 = 2(\cos 195° + i \sin 195°) \qquad \text{when } k = 2$$
$$w_3 = 2(\cos 285° + i \sin 285°) \qquad \text{when } k = 3$$

It is interesting to note the relationships among the graphs of these four roots. As Figure 1 indicates, the graphs of the four roots are evenly distributed around the coordinate plane.

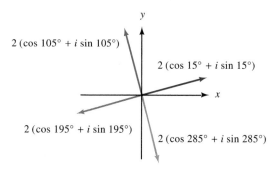

Figure 1   ◀

▶ **EXAMPLE 2**   Solve $x^3 + 1 = 0$.

Solution   Adding -1 to each side of the equation, we have

$$x^3 = -1$$

The solutions to this equation are the cube roots of -1. We already know that one of the cube roots of -1 is -1. There are two other complex cube roots as well. To find them, we write -1 in trigonometric form and then apply the formula from our theorem on roots. Writing -1 in trigonometric form, we have

$$-1 = 1(\cos 180° + i \sin 180°)$$

The 3 cube roots are given by

$$w_k = 1^{1/3} \left[\cos \frac{180° + 360°k}{3} + i \sin \frac{180° + 360°k}{3} \right]$$
$$= \cos (60° + 120°k) + i \sin (60° + 120°k)$$

where $k = 0$, 1, and 2. Replacing k with 0, 1, and 2 and then simplifying each result, we have

$$w_0 = \cos 60° + i \sin 60° = \frac{1}{2} + \frac{\sqrt{3}}{2} i \qquad \text{when } k = 0$$
$$w_1 = \cos 180° + i \sin 180° = -1 \qquad \text{when } k = 1$$

$$w_2 = \cos 300° + i \sin 300° = \frac{1}{2} - \frac{\sqrt{3}}{2}i \qquad \text{when } k = 2$$

The graphs of these three roots are shown in Figure 2.

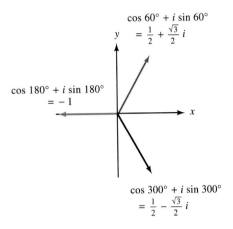

Figure 2

Note that the two complex roots are conjugates. Let's check root w_0 by cubing it.

$$
\begin{aligned}
w_0{}^3 &= (\cos 60° + i \sin 60°)^3 \\
&= \cos 3 \cdot 60° + i \sin 3 \cdot 60° \\
&= \cos 180° + i \sin 180° \\
&= -1
\end{aligned}
$$

◀

You may have noticed that the equation in Example 2 can be solved by algebraic methods. Let's solve $x^3 + 1 = 0$ by factoring and compare our results to those in Example 2. [Recall the formula for factoring the sum of two cubes as follows: $a^3 + b^3 = (a + b)(a^2 - ab + b^2)$.]

$$x^3 + 1 = 0$$
$$(x + 1)(x^2 - x + 1) = 0 \qquad \text{Factor the left side}$$
$$x + 1 = 0 \qquad x^2 - x + 1 = 0 \qquad \text{Set factors equal to 0}$$

The first equation gives us $x = -1$ for a solution. To solve the second equation, we use the quadratic formula.

$$
\begin{aligned}
x &= \frac{-(-1) \pm \sqrt{(-1)^2 - 4(1)(1)}}{2(1)} \\
&= \frac{1 \pm \sqrt{-3}}{2} \\
&= \frac{1 \pm i\sqrt{3}}{2}
\end{aligned}
$$

This last expression gives us two solutions; they are

$$\frac{1}{2} + \frac{\sqrt{3}}{2}\, i \qquad \text{and} \qquad \frac{1}{2} - \frac{\sqrt{3}}{2}\, i$$

As you can see, the three solutions we found in Example 2 using trigonometry match the three solutions found using algebra. Notice that the algebraic method of solving the equation $x^3 + 1 = 0$ depends on our ability to factor the left side of the equation. The advantage to the trigonometric method is that it is independent of our ability to factor.

▶ **EXAMPLE 3**   Solve the equation $x^4 - 2\sqrt{3}x^2 + 4 = 0$.

Solution   The equation is quadratic in x^2. We can solve for x^2 by applying the quadratic formula.

$$x^2 = \frac{2\sqrt{3} \pm \sqrt{12 - 4(1)(4)}}{2}$$

$$= \frac{2\sqrt{3} \pm \sqrt{-4}}{2}$$

$$= \frac{2\sqrt{3} \pm 2i}{2}$$

$$= \sqrt{3} \pm i$$

The two solutions for x^2 are $\sqrt{3} + i$ and $\sqrt{3} - i$, which we write in trigonometric form as follows:

$$x^2 = \sqrt{3} + i \qquad\qquad \text{or} \qquad x^2 = \sqrt{3} - i$$
$$= 2(\cos 30° + i \sin 30°) \qquad\qquad = 2(\cos 330° + i \sin 330°)$$

Now each of these expressions has two square roots, each of which is a solution to our original equation.

When $x^2 = 2(\cos 30° + i \sin 30°)$

$$x = 2^{1/2} \left[\cos \frac{30° + 360°k}{2} + i \sin \frac{30° + 360°k}{2} \right] \quad \text{for} \quad k = 0 \text{ and } 1$$

When $k = 0$,   we have $x = 2^{1/2}(\cos 15° + i \sin 15°)$
When $k = 1$,   we have $x = 2^{1/2}(\cos 195° + i \sin 195°)$

When $x^2 = 2(\cos 330° + i \sin 330°)$

$$x = 2^{1/2} \left[\cos \frac{330° + 360°k}{2} + i \sin \frac{330° + 360°k}{2} \right] \quad \text{for} \quad k = 0 \text{ and } 1$$

When $k = 0$,   we have $x = 2^{1/2}(\cos 165° + i \sin 165°)$
When $k = 1$,   we have $x = 2^{1/2}(\cos 345° + i \sin 345°)$

Using a calculator and rounding to the nearest hundredth, we can write decimal approximations to each of these four solutions.

Solutions

Trigonometric form		Decimal approximation
$2^{1/2}(\cos 15° + i \sin 15°)$	$=$	$1.37 + 0.37i$
$2^{1/2}(\cos 165° + i \sin 165°)$	$=$	$-1.37 + 0.37i$
$2^{1/2}(\cos 195° + i \sin 195°)$	$=$	$-1.37 - 0.37i$
$2^{1/2}(\cos 345° + i \sin 345°)$	$=$	$1.37 - 0.37i$

◀

PROBLEM SET 8.4

Find two square roots for each of the following complex numbers. Leave your answers in trigonometric form. In each case, graph the two roots.

1. $4(\cos 30° + i \sin 30°)$ **2.** $16(\cos 30° + i \sin 30°)$
3. $25(\cos 210° + i \sin 210°)$ **4.** $9(\cos 310° + i \sin 310°)$
5. $49(\cos 180° + i \sin 180°)$ **6.** $81(\cos 150° + i \sin 150°)$

Find two square roots for each of the following complex numbers. Write your answers in standard form.

7. $2 + 2i\sqrt{3}$ **8.** $-2 + 2i\sqrt{3}$ **9.** $4i$ **10.** $-4i$
11. -25 **12.** 25 **13.** $1 + i\sqrt{3}$ **14.** $1 - i\sqrt{3}$

Find three cube roots for each of the following complex numbers. Leave your answers in trigonometric form.

15. $8(\cos 210° + i \sin 210°)$ **16.** $27(\cos 303° + i \sin 303°)$
17. $4\sqrt{3} + 4i$ **18.** $-4\sqrt{3} + 4i$
19. -27 **20.** 8
21. $64i$ **22.** $-64i$

Solve each equation.

23. $x^3 - 27 = 0$ **24.** $x^3 + 8 = 0$ **25.** $x^4 - 16 = 0$ **26.** $x^4 + 81 = 0$

27. Find 4 fourth roots of $z = 16(\cos 120° + i \sin 120°)$. Write each root in standard form.
28. Find 4 fourth roots of $z = \cos 240° + i \sin 240°$. Leave your answers in trigonometric form.
29. Find 5 fifth roots of $z = 10^5(\cos 15° + i \sin 15°)$. Write each root in trigonometric form and then give a decimal approximation, accurate to the nearest hundredth, for each one.

30. Find 5 fifth roots of $z = 10^{10}(\cos 75° + i \sin 75°)$. Write each root in trigonometric form and then give a decimal approximation, accurate to the nearest hundredth, for each one.

31. Find 6 sixth roots of $z = -1$. Leave your answers in trigonometric form. Graph all six roots on the same coordinate system.

32. Find 6 sixth roots of $z = 1$. Leave your answers in trigonometric form. Graph all six roots on the same coordinate system.

Solve each of the following equations. Leave your solutions in trigonometric form.

33. $x^4 - 2x^2 + 4 = 0$

34. $x^4 + 2x^2 + 4 = 0$

35. $x^4 + 2x^2 + 2 = 0$

36. $x^4 - 2x^2 + 2 = 0$

Review Problems

The problems that follow review material we covered in Sections 4.2, 4.3, and 7.4.

Graph each equation on the given interval.

37. $y = -2 \sin(-3x),\ 0 \le x \le 2\pi$

38. $y = -2 \cos(-3x),\ 0 \le x \le 2\pi$

Graph one complete cycle of each of the following:

39. $y = -\cos\left(2x + \dfrac{\pi}{2}\right)$

40. $y = -\cos\left(2x - \dfrac{\pi}{2}\right)$

41. $y = 3 \sin\left(\dfrac{\pi}{3}x - \dfrac{\pi}{3}\right)$

42. $y = 3 \cos\left(\dfrac{\pi}{3}x - \dfrac{\pi}{3}\right)$

Find the area of triangle ABC if

43. $A = 56.2°,\ b = 2.65$ cm, and $c = 3.84$ cm

44. $B = 21.8°,\ a = 44.4$ cm, and $c = 22.2$ cm

45. $a = 2.3$ ft, $b = 3.4$ ft, and $c = 4.5$ ft

46. $a = 5.4$ ft, $b = 4.3$ ft, and $c = 3.2$ ft

Research Project

47. Recall from the introduction to Section 8.2 that Jerome Cardan's solutions to the equation $x^3 = 15x + 4$ were given by

$$x = \sqrt[3]{2 + 11i} + \sqrt[3]{2 - 11i}$$

Let's assume that the cube roots shown above are complex conjugates. If they are, then we can simplify our work by noticing that

$$x = \sqrt[3]{2 + 11i} + \sqrt[3]{2 - 11i} = a + bi + a - bi = 2a$$

which means that we simply double the real part of each cube root of $2 + 11i$ to find the solutions to $x^3 = 15x + 4$. Now, to end our work with Cardan, find the 3 cube roots of $2 + 11i$. Then, noting the discussion above, use the 3 cube roots to solve the equation $x^3 = 15x + 4$. Write your answers accurate to the nearest thousandth.

Polar Coordinates

Up to this point in our study of trigonometry, whenever we have given the position of a point in the plane, we have used rectangular coordinates. That is, we give the position of points in the plane relative to a set of perpendicular axes. In a way, the rectangular coordinate system is a map that tells us how to get anywhere in the plane by traveling so far horizontally and so far vertically relative to the origin of our coordinate system. For example, to reach the point whose address is (2, 1), we start at the origin and travel 2 units forward and then 1 unit up. In this section, we will see that there are other ways to get to the point (2, 1). In particular, we can travel $\sqrt{5}$ units on the terminal side of an angle of 30° in standard position. This type of map is the basis of what we call *polar coordinates:* The address of each point in the plane is given by an ordered pair (r, θ), where r is a directed distance on the terminal side of standard position angle θ.

▶ **EXAMPLE 1** A point lies at (4, 4) on a rectangular coordinate system. Give its address in polar coordinates (r, θ).

Solution We want to reach the same point by traveling r units on the terminal side of a standard position angle θ. Figure 1 shows the point (4, 4), along with the distance r and angle θ.

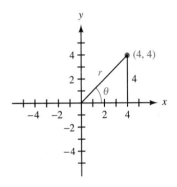

Figure 1

The triangle formed is a 45°–45°–90° right triangle. Therefore, r is $4\sqrt{2}$, and θ is 45°. In rectangular coordinates, the address of our point is (4, 4). In polar coordinates, the address is $(4\sqrt{2}, 45°)$. ◀

Now let's formalize our ideas about polar coordinates.

Figure 2

The foundation of the polar coordinate system is a ray called the *polar axis,* whose initial point is called the *pole.* (See Figure 2.) In the polar coordinate system, points are named by ordered pairs (r, θ) in which r is the directed distance from the pole on the terminal side of an angle θ, the initial side of which is the polar axis. If the terminal side of θ has been rotated in a counterclockwise direction from the polar axis, then θ is a positive angle. Otherwise, θ is a negative angle. For example, the point (5, 30°) is 5 units from the pole along a ray that has been rotated 30° from the polar axis, as shown in Figure 3.

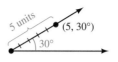

Figure 3

To simplify matters, in this book we place the pole at the origin of our rectangular coordinate system and take the polar axis to be the positive *x*-axis.

The circular grids used in Example 2 are helpful when graphing points given in polar coordinates. The lines on the grid are multiples of 15°, and the circles have radii of 1, 2, 3, 4, 5, and 6.

▶ **EXAMPLE 2** Graph the points (3, 45°), (2, 120°), (−4, 60°), and (−5, 150°) on a polar coordinate system.

Solution To graph (3, 45°), we locate the point that is 3 units from the origin along the terminal side of 45°, as shown in Figure 4. The point (2, 120°) is 2 units out on the terminal side of 120°, as Figure 5 indicates.

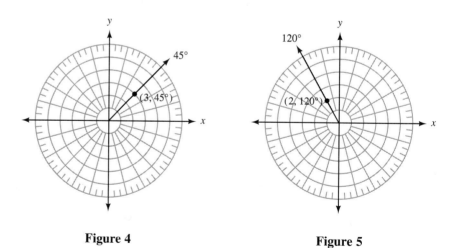

Figure 4   **Figure 5**

As you can see from Figures 4 and 5, if *r* is positive, we locate the point (*r*, *θ*) along the terminal side of *θ*. The next two points we will graph have negative values of *r*. To graph a point (*r*, *θ*) in which *r* is negative, we look for the point on the *projection* of the terminal side of *θ* through the origin.

To graph $(-4, 60°)$, we locate the point that is 4 units from the origin on the projection of $60°$ through the origin. Figure 6 shows the graph of $(-4, 60°)$.

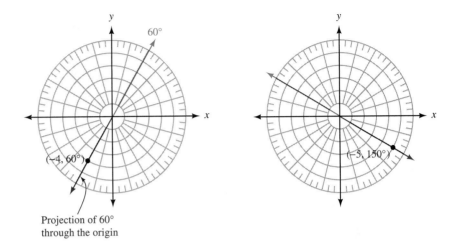

Figure 6   **Figure 7**

To graph $(-5, 150°)$, we look for the point that is 5 units from the origin along the projection of $150°$ through the origin. (See Figure 7.)   ◀

In rectangular coordinates, each point in the plane is named by a unique ordered pair (x, y). That is, no point can be named by two different ordered pairs. The same is not true of points named by polar coordinates. As Example 3 illustrates, the polar coordinates of a point are not unique.

▶ **EXAMPLE 3**   Give three other ordered pairs that name the same point as $(3, 60°)$.

Solution   As Figure 8 illustrates, the points $(-3, 240°)$, $(-3, -120°)$, and $(3, -300°)$ all name the point $(3, 60°)$. There are actually an infinite number of ordered pairs that name the point $(3, 60°)$. Any angle that is coterminal with $60°$ will have its terminal side pass through $(3, 60°)$. Therefore, all points of the form

$$(3, 60° + 360°k) \qquad k = \text{an integer}$$

will name the point $(3, 60°)$.

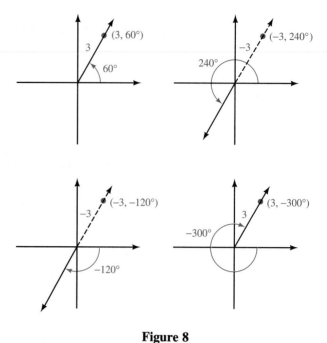

Figure 8

Polar Coordinates and Rectangular Coordinates

To derive the relationship between polar coordinates and rectangular coordinates, we consider a point P with rectangular coordinates (x, y) and polar coordinates (r, θ).

To convert back and forth between polar and rectangular coordinates, we simply use the relationships that exist among x, y, r, and θ in Figure 9.

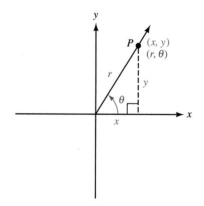

Figure 9

To Convert from Rectangular Coordinates to Polar Coordinates

Let

$$r = \pm\sqrt{x^2 + y^2} \qquad \text{and} \qquad \tan\theta = \frac{y}{x}$$

where the sign of r and the choice of θ place the point (r, θ) in the same quadrant as (x, y).

To Convert from Polar Coordinates to Rectangular Coordinates

Let

$$x = r\cos\theta \qquad \text{and} \qquad y = r\sin\theta$$

The process of converting to rectangular coordinates is simply a matter of substituting r and θ into the equations given above. To convert to polar coordinates, we have to choose θ and the sign of r so the point (r, θ) is in the same quadrant as the point (x, y).

▶ **EXAMPLE 4**   Convert to rectangular coordinates.

a. $(4, 30°)$ **b.** $(-\sqrt{2}, 135°)$ **c.** $(3, 270°)$

Solution   To convert from polar coordinates to rectangular coordinates, we substitute the given values of r and θ into the equations

$$x = r\cos\theta \qquad \text{and} \qquad y = r\sin\theta$$

Here are the conversions for each point along with the graphs in both rectangular and polar coordinates.

a. $x = 4\cos 30° = 4\left(\dfrac{\sqrt{3}}{2}\right) = 2\sqrt{3}$

$y = 4\sin 30° = 4\left(\dfrac{1}{2}\right) = 2$

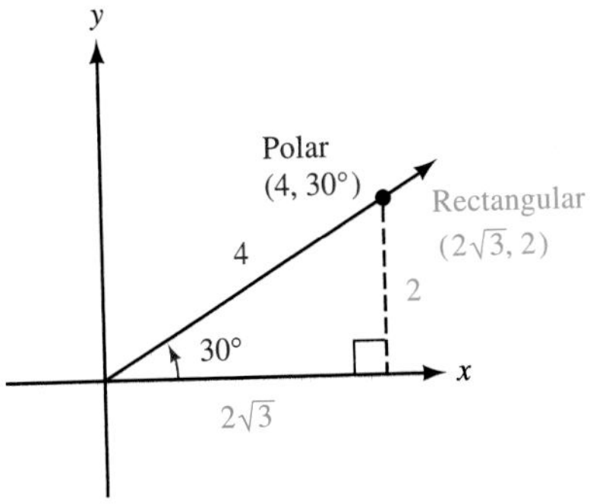

Figure 10

The point $(2\sqrt{3}, 2)$ in rectangular coordinates is equivalent to $(4, 30°)$ in polar coordinates.

b. $x = -\sqrt{2} \cos 135° = -\sqrt{2} \left(-\dfrac{1}{\sqrt{2}} \right) = 1$

$y = -\sqrt{2} \sin 135° = -\sqrt{2} \left(\dfrac{1}{\sqrt{2}} \right) = -1$

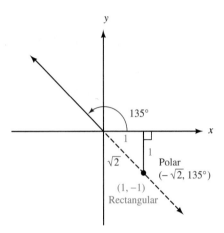

Figure 11

The point $(1, -1)$ in rectangular coordinates is equivalent to $(-\sqrt{2}, 135°)$ in polar coordinates.

c. $x = 3 \cos 270° = 3(0) = 0$

$y = 3 \sin 270° = 3(-1) = -3$

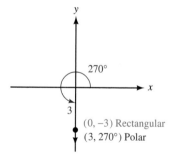

Figure 12

The point $(0, -3)$ in rectangular coordinates is equivalent to $(3, 270°)$ in polar coordinates. ◀

▶ **EXAMPLE 5**   Convert to polar coordinates.

a. (3, 3)   **b.** (−2, 0)   **c.** $(-1, \sqrt{3})$

Solution

a. Since x is 3 and y is 3, we have

$$r = \pm\sqrt{9 + 9} = \pm 3\sqrt{2} \qquad \text{and} \qquad \tan\theta = \frac{3}{3} = 1$$

Since (3, 3) is in quadrant 1, we can choose $r = 3\sqrt{2}$ and $\theta = 45°$. Remember, there are an infinite number of ordered pairs in polar coordinates that name the point (3, 3). The point $(3\sqrt{2}, 45°)$ is just one of them. Generally, we choose r and θ so that r is positive and θ is between 0° and 360°.

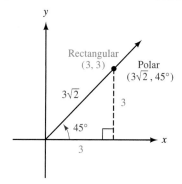

Figure 13

b. We have $x = -2$ and $y = 0$, so

$$r = \pm\sqrt{4 + 0} = \pm 2 \qquad \text{and} \qquad \tan\theta = \frac{0}{-2} = 0$$

Since (−2, 0) is on the negative x-axis, we can choose $r = 2$ and $\theta = 180°$ to get the point (2, 180°).

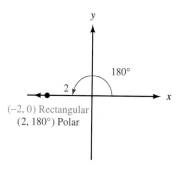

Figure 14

c. Since $x = -1$ and $y = \sqrt{3}$, we have

$$r = \pm\sqrt{1 + 3} = \pm 2 \quad \text{and} \quad \tan\theta = \frac{\sqrt{3}}{-1}$$

Since $(-1, \sqrt{3})$ is in quadrant II, we can let $r = 2$ and $\theta = 120°$. In polar coordinates, the point is $(2, 120°)$.

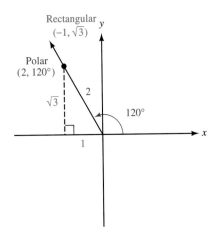

Figure 15

Equations in Polar Coordinates

Equations in polar coordinates have variables r and θ instead of x and y. The conversions we used to change ordered pairs from polar coordinates to rectangular coordinates and from rectangular coordinates to polar coordinates are the same ones we use to convert back and forth between equations given in polar coordinates and those in rectangular coordinates.

▶ **EXAMPLE 6** Change $r^2 = 9 \sin 2\theta$ to rectangular coordinates.

Solution Before we substitute to clear the equation of r and θ, we must use a double-angle identity to write $\sin 2\theta$ in terms of $\sin\theta$ and $\cos\theta$.

$$r^2 = 9 \sin 2\theta$$

$$r^2 = 9 \cdot 2 \sin\theta \cos\theta \qquad \text{Double-angle identity}$$

$$r^2 = 18 \cdot \frac{y}{r} \cdot \frac{x}{r} \qquad \text{Substitute } y/r \text{ for } \sin\theta \text{ and } x/r \text{ for } \cos\theta$$

$$r^2 = \frac{18xy}{r^2} \qquad \text{Multiply}$$

$$r^4 = 18xy \qquad \text{Multiply both sides by } r^2$$

$$(x^2 + y^2)^2 = 18xy \qquad \text{Substitute } x^2 + y^2 \text{ for } r^2$$

▶ **EXAMPLE 7**   Change $x + y = 4$ to polar coordinates.

Solution   Since $x = r \cos \theta$ and $y = r \sin \theta$, we have

$$r \cos \theta + r \sin \theta = 4$$

$$r(\cos \theta + \sin \theta) = 4 \qquad\qquad \text{Factor out } r$$

$$r = \frac{4}{\cos \theta + \sin \theta} \qquad \text{Divide both sides by } \cos \theta + \sin \theta$$

The last equation gives us r in terms of θ.   ◀

PROBLEM SET 8.5

Graph each ordered pair on a polar coordinate system.

1. $(2, 45°)$ **2.** $(3, 60°)$ **3.** $(3, 150°)$ **4.** $(4, 135°)$
5. $(1, -225°)$ **6.** $(2, -240°)$ **7.** $(-3, 45°)$ **8.** $(-4, 60°)$
9. $(-4, -210°)$ **10.** $(-5, -225°)$ **11.** $(-2, 0°)$ **12.** $(-2, 270°)$

For each ordered pair, give three other ordered pairs with θ between $-360°$ and $360°$ that name the same point.

13. $(2, 60°)$ **14.** $(1, 30°)$ **15.** $(5, 135°)$ **16.** $(3, 120°)$
17. $(-3, 30°)$ **18.** $(-2, 45°)$

Convert to rectangular coordinates. Use exact values.

19. $(2, 60°)$ **20.** $(-2, 60°)$ **21.** $(3, 270°)$ **22.** $(1, 180°)$
23. $(\sqrt{2}, -135°)$ **24.** $(\sqrt{2}, -225°)$ **25.** $(-4\sqrt{3}, 30°)$ **26.** $(4\sqrt{3}, -30°)$

Convert to polar coordinates with $r \geq 0$ and θ between $0°$ and $360°$.

27. $(-3, 3)$ **28.** $(-3, -3)$ **29.** $(-2\sqrt{3}, 2)$ **30.** $(2, -2\sqrt{3})$
31. $(2, 0)$ **32.** $(-2, 0)$ **33.** $(-\sqrt{3}, -1)$ **34.** $(-1, -\sqrt{3})$

Convert to polar coordinates. Use a calculator to find θ to the nearest tenth of a degree. Keep r positive and θ between $0°$ and $360°$.

35. $(3, 4)$ **36.** $(4, 3)$ **37.** $(-1, 2)$ **38.** $(1, -2)$
39. $(-2, -3)$ **40.** $(-3, -2)$

Write each equation with rectangular coordinates.

41. $r^2 = 9$ **42.** $r^2 = 4$ **43.** $r = 6 \sin \theta$ **44.** $r = 6 \cos \theta$
45. $r^2 = 4 \sin 2\theta$ **46.** $r^2 = 4 \cos 2\theta$
47. $r(\cos \theta + \sin \theta) = 3$ **48.** $r(\cos \theta - \sin \theta) = 2$

Write each equation in polar coordinates.

49. $x - y = 5$ **50.** $x + y = 5$ **51.** $x^2 + y^2 = 4$ **52.** $x^2 + y^2 = 9$
53. $x^2 + y^2 = 6x$ **54.** $x^2 + y^2 = 4x$
55. $y = x$ **56.** $y = -x$

Review Problems

The problems that follow review material we covered in Sections 4.2 and 4.4. Reviewing these problems will help you with the next section.

Graph one complete cycle of each equation.

57. $y = 6 \sin x$ **58.** $y = 6 \cos x$ **59.** $y = 4 \sin 2x$ **60.** $y = 2 \sin 4x$
61. $y = 4 + 2 \sin x$ **62.** $y = 4 + 2 \cos x$

SECTION 8.6 ## Equations in Polar Coordinates and Their Graphs

Over 2,000 years ago, Archimedes described a curve as starting at a point and moving out along a half-line at a constant rate. The endpoint of the half-line was anchored at the initial point and was rotating about it at a constant rate also. The curve, called the spiral of Archimedes, is shown in Figure 1, with a rectangular coordinate system superimposed on it so that the origin of the coordinate system coincides with the initial point on the curve.

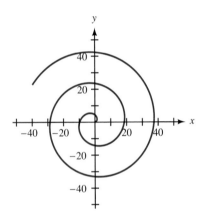

Figure 1

In rectangular coordinates, the equation that describes this curve is

$$\sqrt{x^2 + y^2} = 3 \tan^{-1} \frac{y}{x}$$

If we were to superimpose a polar coordinate system on the curve in Figure 1, instead of the rectangular coordinate system, then the equation in polar coordinates that describes the curve would be simply

$$r = 3\theta$$

As you can see, the equations for some curves are best given in terms of polar coordinates.

In this section, we will consider the graphs of polar equations. The solutions to these equations are ordered pairs (r, θ), where r and θ are the polar coordinates we defined in Section 8.5.

▶ **EXAMPLE 1**   Sketch the graph of $r = 6 \sin \theta$.

Solution   We can find ordered pairs (r, θ) that satisfy the equation by making a table. The table is a little different from the ones we made for rectangular coordinates. With polar coordinates, we substitute in convenient values for θ and then use the equation to find corresponding values of r. Let's use multiples of $30°$ and $45°$ for θ.

Table 1

θ	$r = 6 \sin \theta$	r	(r, θ)
$0°$	$r = 6 \sin 0° = 0$	0	$(0, 0°)$
$30°$	$r = 6 \sin 30° = 3$	3	$(3, 30°)$
$45°$	$r = 6 \sin 45° = 4.2$	4.2	$(4.2, 45°)$
$60°$	$r = 6 \sin 60° = 5.2$	5.2	$(5.2, 60°)$
$90°$	$r = 6 \sin 90° = 6$	6	$(6, 90°)$
$120°$	$r = 6 \sin 120° = 5.2$	5.2	$(5.2, 120°)$
$135°$	$r = 6 \sin 135° = 4.2$	4.2	$(4.2, 135°)$
$150°$	$r = 6 \sin 150° = 3$	3	$(3, 150°)$
$180°$	$r = 6 \sin 180° = 0$	0	$(0, 180°)$
$210°$	$r = 6 \sin 210° = -3$	-3	$(-3, 210°)$
$225°$	$r = 6 \sin 225° = -4.2$	-4.2	$(-4.2, 225°)$
$240°$	$r = 6 \sin 240° = -5.2$	-5.2	$(-5.2, 240°)$
$270°$	$r = 6 \sin 270° = -6$	-6	$(-6, 270°)$
$300°$	$r = 6 \sin 300° = -5.2$	-5.2	$(-5.2, 300°)$
$315°$	$r = 6 \sin 315° = -4.2$	-4.2	$(-4.2, 315°)$
$330°$	$r = 6 \sin 330° = -3$	-3	$(-3, 330°)$
$360°$	$r = 6 \sin 360° = 0$	0	$(0, 360°)$

Note   If we were to continue past $360°$ with values of θ, we would simply start to repeat the values of r we have already obtained, since $\sin \theta$ is periodic with period $360°$.

Plotting each point on a polar coordinate system and then drawing a smooth curve through them, we have the graph in Figure 2.

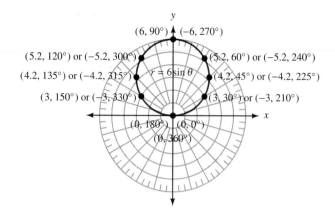

Figure 2   ◄

Could we have found the graph of $r = 6 \sin \theta$ in Example 1 without making a table? The answer is yes, there are a couple of other ways to do so. One way is to convert to rectangular coordinates and see if we recognize the graph from the rectangular equation. We begin by replacing $\sin \theta$ with y/r.

$$r = 6\,\frac{y}{r} \qquad \sin \theta = \frac{y}{r}$$

$$r^2 = 6y \qquad \text{Multiply both sides by } r$$

$$x^2 + y^2 = 6y \qquad r^2 = x^2 + y^2$$

The equation is now written in terms of rectangular coordinates. If we add $-6y$ to both sides and then complete the square on y, we will obtain the rectangular equation of a circle with center at $(0, 3)$ and a radius of 3.

$$x^2 + y^2 - 6y = 0 \qquad \text{Add } -6y \text{ to both sides}$$

$$x^2 + y^2 - 6y + 9 = 9 \qquad \begin{array}{l}\text{Complete the square on } y \text{ by adding 9}\\ \text{to both sides}\end{array}$$

$$x^2 + (y - 3)^2 = 3^2 \qquad \text{Standard form for the equation of a circle}$$

This method of graphing, by changing to rectangular coordinates, works well only in some cases. Many of the equations we will encounter in polar coordinates do not have graphs that are recognizable in rectangular form.

In Example 2, we will look at another method of graphing polar equations that does not depend on the use of a table.

▶ **EXAMPLE 2**   Sketch the graph of $r = 4 \sin 2\theta$.

Solution   One way to visualize the relationship between r and θ as given by the equation $r = 4 \sin 2\theta$ is to sketch the graph of $y = 4 \sin 2x$ on a rectangular coordinate system. (Since we have been using degree measure for our angles in polar coordinates, we will label the *x*-axis for the graph of $y = 4 \sin 2x$ in degrees rather than radians as we usually do.) The graph of $y = 4 \sin 2x$ will have an amplitude of 4 and a period of $360°/2 = 180°$. Figure 3 shows the graph of $y = 4 \sin 2x$ between $0°$ and $360°$.

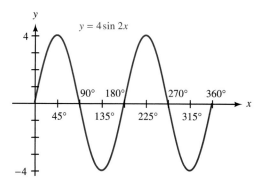

Figure 3

As you can see in Figure 3, as *x* goes from $0°$ to $45°$, *y* goes from 0 to 4. This means that, for the equation $r = 4 \sin 2\theta$, as θ goes from $0°$ to $45°$, *r* will go from 0 up to 4. A diagram of this is shown in Figure 4.

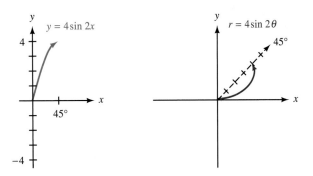

Figure 4

As *x* continues from $45°$ to $90°$, *y* decreases from 4 down to 0. Likewise, as θ rotates through $45°$ to $90°$, *r* will decrease from 4 down to 0. A diagram of this is shown in Figure 5. The numbers 1 and 2 in Figure 5 indicate the order in which those sections of the graph are drawn.

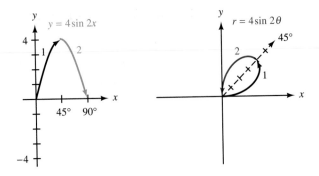

Figure 5

If we continue to reason in this manner, we will obtain a good sketch of the graph of $r = 4 \sin 2\theta$ by watching how y is affected by changes in x on the graph of $y = 4 \sin 2x$. Table 2 summarizes this information, and Figure 6 contains the graphs of both $y = 4 \sin 2x$ and $r = 4 \sin 2\theta$.

Table 2

Reference number on graphs	Variations in x (or θ)	Corresponding variations in y (or r)
1	0° to 45°	0 to 4
2	45° to 90°	4 to 0
3	90° to 135°	0 to −4
4	135° to 180°	−4 to 0
5	180° to 225°	0 to 4
6	225° to 270°	4 to 0
7	270° to 315°	0 to −4
8	315° to 360°	−4 to 0

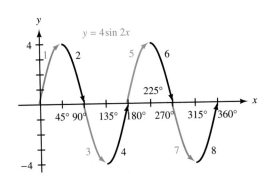
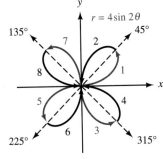

Figure 6

▶ **EXAMPLE 3**   Sketch the graph of $r = 4 + 2 \sin \theta$.

Solution   The graph of $r = 4 + 2 \sin \theta$ (Figure 8) is obtained by first graphing $y = 4 + 2 \sin x$ (Figure 7) and then noticing the relationship between variations in x and the corresponding variations in y. These variations are equivalent to those that exist between θ and r.

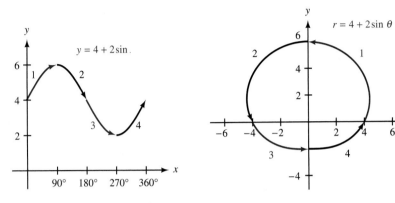

Figure 7   **Figure 8**

Table 3

Reference number on graphs	Variations in x (or θ)	Corresponding variations in y (or r)
1	0° to 90°	4 to 6
2	90° to 180°	6 to 4
3	180° to 270°	4 to 2
4	270° to 360°	2 to 4

◀

Although the method of graphing presented in Examples 2 and 3 is sometimes difficult to comprehend at first, with a little practice it becomes much easier. In any case, the usual alternative is to make a table and plot points until the shape of the curve can be recognized. Probably the best way to graph these equations is to use a combination of both methods.

Here are some other common graphs in polar coordinates along with the equations that produce them. When you start graphing some of the equations in Problem Set 8.6, you may want to keep these graphs handy for reference. It is sometimes easier to get started when you can anticipate the general shape of the curve.

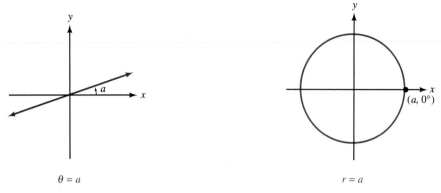

$\theta = a$

Figure 9

$r = a$

Figure 10

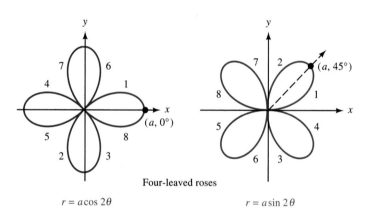

Four-leaved roses

$r = a \cos 2\theta$

$r = a \sin 2\theta$

Figure 11

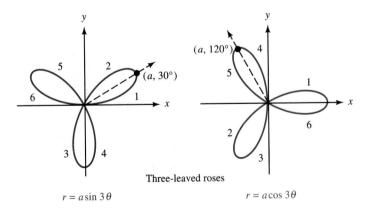

Three-leaved roses

$r = a \sin 3\theta$

$r = a \cos 3\theta$

Figure 12

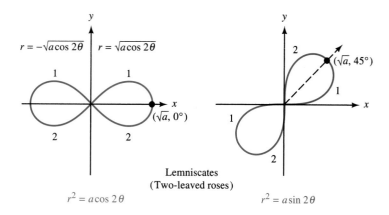

Lemniscates
(Two-leaved roses)

$r^2 = a\cos 2\theta$ $\qquad\qquad\qquad\qquad r^2 = a\sin 2\theta$

Figure 13

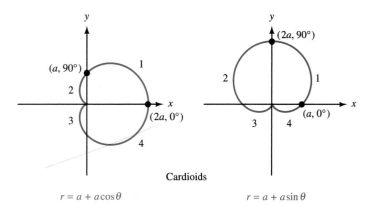

Cardioids

$r = a + a\cos\theta$ $\qquad\qquad\qquad\qquad r = a + a\sin\theta$

Figure 14

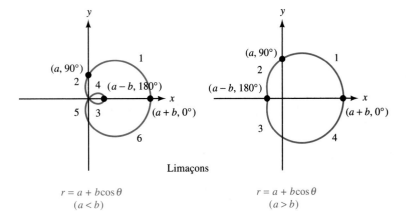

Limaçons

$r = a + b\cos\theta$ $\qquad\qquad\qquad\qquad r = a + b\cos\theta$
$(a < b)$ $\qquad\qquad\qquad\qquad\qquad\quad (a > b)$

Figure 15

PROBLEM SET 8.6

Sketch the graph of each equation by making a table using values of θ that are multiples of $45°$.

1. $r = 6 \cos \theta$ **2.** $r = 4 \sin \theta$

3. $r = \sin 3\theta$ **4.** $r = \cos 3\theta$

Graph each equation.

5. $r = 3$ **6.** $r = 2$

7. $\theta = 45°$ **8.** $\theta = 135°$

9. $r = 3 \sin \theta$ **10.** $r = 3 \cos \theta$

11. $r = 4 + 2 \sin \theta$ **12.** $r = 4 + 2 \cos \theta$

13. $r = 2 + 4 \cos \theta$ **14.** $r = 2 + 4 \sin \theta$

15. $r = 2 + 2 \sin \theta$ **16.** $r = 2 + 2 \cos \theta$

17. $r^2 = 9 \sin 2\theta$ **18.** $r^2 = 4 \cos 2\theta$

19. $r = 2 \sin 2\theta$ **20.** $r = 2 \cos 2\theta$

21. $r = 4 \cos 3\theta$ **22.** $r = 4 \sin 3\theta$

Convert each equation to polar coordinates and then sketch the graph.

23. $x^2 + y^2 = 16$ **24.** $x^2 + y^2 = 25$

25. $x^2 + y^2 = 6x$ **26.** $x^2 + y^2 = 6y$

27. $(x^2 + y^2)^2 = 2xy$ **28.** $(x^2 + y^2)^2 = x^2 - y^2$

Change each equation to rectangular coordinates and then graph.

29. $r(2 \cos \theta + 3 \sin \theta) = 6$ **30.** $r(3 \cos \theta - 2 \sin \theta) = 6$

31. $r(1 - \cos \theta) = 1$ **32.** $r(1 - \sin \theta) = 1$

33. $r = 4 \sin \theta$ **34.** $r = 6 \cos \theta$

35. Graph $r = 2 \sin \theta$ and $r = 2 \cos \theta$ and then name two points they have in common.

36. Graph $r = 2 + 2 \cos \theta$ and $r = 2 - 2 \cos \theta$ and name three points they have in common.

Review Problems

The problems that follow review material we covered in Section 4.5.

Graph each equation.

37. $y = \sin x - \cos x, 0 \le x \le 4\pi$ **38.** $y = \cos x - \sin x, 0 \le x \le 4\pi$

39. $y = x + \sin \pi x, 0 \le x \le 8$ **40.** $y = x + \cos \pi x, 0 \le x \le 8$

41. $y = 3 \sin x + \cos 2x, 0 \le x \le 4\pi$ **42.** $y = \sin x + \dfrac{1}{2} \cos 2x, 0 \le x \le 4\pi$

CHAPTER 8 SUMMARY

EXAMPLES

1. Each of the following is a complex number.

$$5 + 4i$$
$$-\sqrt{3} + i$$
$$7i$$
$$-8$$

The number $7i$ is complex because

$$7i = 0 + 7i$$

The number -8 is complex because

$$-8 = -8 + 0i$$

2. If $3x + 2i = 12 - 4yi$, then

$$3x = 12 \quad \text{and} \quad 2 = -4y$$
$$x = 4 \qquad\qquad y = -1/2$$

3. If $z_1 = 2 - i$ and $z_2 = 4 + 3i$, then

$$z_1 + z_2 = 6 + 2i$$

$$z_1 - z_2 = -2 - 4i$$

$$z_1 z_2 = (2 - i)(4 + 3i)$$
$$= 8 + 6i - 4i - 3i^2$$
$$= 11 + 2i$$

The conjugate of $4 + 3i$ is $4 - 3i$ and

$$(4 + 3i)(4 - 3i) = 16 + 9$$
$$= 25$$

Definitions [8.1]

The number i is such that $i^2 = -1$. If $a > 0$, the expression $\sqrt{-a}$ can be written as $\sqrt{ai^2} = i\sqrt{a}$.

A *complex number* is any number that can be written in the form

$$a + bi$$

where a and b are real numbers and $i^2 = -1$. The number a is called the *real part* of the complex number, and b is called the *imaginary part*. The form $a + bi$ is called *standard form*.

All real numbers are also complex numbers since they can be put in the form $a + bi$, where $b = 0$. If $a = 0$ and $b \neq 0$, then $a + bi$ is called an *imaginary number*.

Equality for Complex Numbers [8.1]

Two complex numbers are equal if and only if their real parts are equal and their imaginary parts are equal. That is,

$$a + bi = c + di \quad \text{if and only if} \quad a = c \quad \text{and} \quad b = d$$

Operations on Complex Numbers in Standard Form [8.1]

If $z_1 = a_1 + b_1 i$ and $z_2 = a_2 + b_2 i$ are two complex numbers in standard form, then the following definitions and operations apply.

Addition
$$z_1 + z_2 = (a_1 + a_2) + (b_1 + b_2)i$$

Add real parts; add imaginary parts.

Subtraction
$$z_1 - z_2 = (a_1 - a_2) + (b_1 - b_2)i$$

Subtract real parts; subtract imaginary parts.

Multiplication
$$z_1 z_2 = (a_1 a_2 - b_1 b_2) + (a_1 b_2 + a_2 b_1)i$$

In actual practice, simply multiply as you would multiply two binomials.

Conjugates
The conjugate of $a + bi$ is $a - bi$. Their product is the real number $a^2 + b^2$.

$$\frac{z_1}{z_2} = \frac{2 - i}{4 + 3i} \cdot \frac{4 - 3i}{4 - 3i}$$

$$= \frac{5 - 10i}{25}$$

$$= \frac{1}{5} - \frac{2}{5} i$$

Division

Multiply the numerator and denominator of the quotient by the conjugate of the denominator.

4. $i^{20} = (i^4)^5 = 1$

$i^{21} = (i^4)^5 \cdot i = i$

$i^{22} = (i^4)^5 \cdot i^2 = -1$

$i^{23} = (i^4)^5 \cdot i^3 = -i$

Powers of i [8.1]

If n is an integer, then i^n can always be simplified to i, -1, $-i$, or 1.

5.   The graph of $4 + 3i$ is

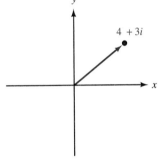

Graphing Complex Numbers [8.2]

The graph of the complex number $z = x + yi$ is the arrow (vector) that extends from the origin to the point (x, y).

6.   If $z = \sqrt{3} + i$, then

$$|z| = |\sqrt{3} + i| = \sqrt{3 + 1} = 2$$

Absolute Value of a Complex Number [8.2]

The *absolute value* (or *modulus*) of the complex number $z = x + yi$ is the distance from the origin to the point (x, y). If this distance is denoted by r, then

$$r = |z| = |x + yi| = \sqrt{x^2 + y^2}$$

7.   For $z = \sqrt{3} + i$, θ is the smallest positive angle for which

$$\sin \theta = \frac{1}{2} \text{ and } \cos \theta = \frac{\sqrt{3}}{2}$$

which means $\theta = 30°$

Argument of a Complex Number [8.2]

The *argument* of the complex number $z = x + yi$ is the smallest positive angle from the positive x-axis to the graph of z. If the argument of z is denoted by θ, then

$$\sin \theta = \frac{y}{r}, \quad \cos \theta = \frac{x}{r}, \quad \text{and} \quad \tan \theta = \frac{y}{x}$$

8.   If $z = \sqrt{3} + i$, then in trigonometric form

$$z = 2(\cos 30° + i \sin 30°)$$

Trigonometric Form of a Complex Number [8.2]

The complex number $z = x + yi$ is written in trigonometric form when it is written as

$$z = r(\cos \theta + i \sin \theta)$$

where r is the absolute value of z and θ is the argument of z.

9.   If
$z_1 = 8(\cos 40° + i \sin 40°)$ and
$z_2 = 4(\cos 10° + i \sin 10°)$,

then

$z_1 z_2 = 32(\cos 50° + i \sin 50°)$

$\dfrac{z_1}{z_2} = 2(\cos 30° + i \sin 30°)$

Products and Quotients in Trigonometric Form [8.3]

If $z_1 = r_1(\cos \theta_1 + i \sin \theta_1)$ and $z_2 = r_2(\cos \theta_2 + i \sin \theta_2)$ are two complex numbers in trigonometric form, then their product is

$$z_1 z_2 = r_1 r_2 [\cos (\theta_1 + \theta_2) + i \sin (\theta_1 + \theta_2)]$$

and their quotient is

$$\frac{z_1}{z_2} = \frac{r_1}{r_2} [\cos (\theta_1 - \theta_2) + i \sin (\theta_1 - \theta_2)]$$

10.   If
$z = \sqrt{2}(\cos 30° + i \sin 30°)$,

then

$z^{10} = \sqrt{2}^{10}(\cos 10 \cdot 30° + i \sin 10 \cdot 30°)$
$= 32(\cos 300° + i \sin 300°)$

DeMoivre's Theorem [8.3]

If $z = r(\cos \theta + i \sin \theta)$ is a complex number in trigonometric form and n is an integer, then

$$z^n = r^n(\cos n\theta + i \sin n\theta)$$

11.   The 3 cube roots of
$z = 8(\cos 60° + i \sin 60°)$ are
given by

$$w_1 = 8^{1/3} \left(\cos \frac{60° + 360°k}{3} \right.$$
$$\left. + i \sin \frac{60° + 360°k}{3} \right)$$
$$= 2[\cos (20° + 120°k)$$
$$+ i \sin (20° + 120°k)]$$

where $k = 0, 1, 2$. That is,

$w_0 = 2(\cos 20° + i \sin 20°)$
$w_1 = 2(\cos 140° + i \sin 140°)$
$w_2 = 2(\cos 260° + i \sin 260°)$

Roots of a Complex Number [8.4]

The n nth roots of the complex number $z = r(\cos \theta + i \sin \theta)$ are given by the formula

$$w_k = r^{1/n} \left(\cos \frac{\theta + 360°k}{n} + i \sin \frac{\theta + 360°k}{n} \right)$$

where $k = 0, 1, 2, \ldots, n - 1$. That is, the n nth roots are

$$w_0 = r^{1/n} \left(\cos \frac{\theta}{n} + i \sin \frac{\theta}{n} \right)$$

$$w_1 = r^{1/n} \left(\cos \frac{\theta + 360°}{n} + i \sin \frac{\theta + 360°}{n} \right)$$

$$w_2 = r^{1/n} \left(\cos \frac{\theta + 720°}{n} + i \sin \frac{\theta + 720°}{n} \right)$$

$$\vdots$$

$$w_{n-1} = r^{1/n} \left(\cos \frac{\theta + 360°(n - 1)}{n} + i \sin \frac{\theta + 360°(n - 1)}{n} \right)$$

12.

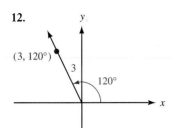

$(3, 120°)$

Polar Coordinates [8.5]

The ordered pair (r, θ) names the point that is r units from the origin along the terminal side of angle θ in standard position. The coordinates r and θ are said to be the *polar coordinates* of the point they name.

13.   Convert $(-\sqrt{2}, 135°)$ to rectangular coordinates.

$x = r \cos \theta = -\sqrt{2} \cos 135° = 1$

$y = r \sin \theta = -\sqrt{2} \sin 135° = -1$

Polar Coordinates and Rectangular Coordinates [8.5]

To derive the relationship between polar coordinates and rectangular coordinates, we consider a point P with rectangular coordinates (x, y) and polar coordinates (r, θ).

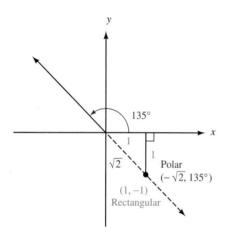

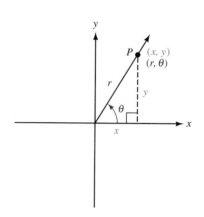

14.   Change $x + y = 4$ to polar coordinates.

Since $x = r \cos \theta$ and $y = r \sin \theta$, we have

$r \cos \theta + r \sin \theta = 4$

$r(\cos \theta + \sin \theta) = 4$ Factor out r

Divide both sides by $\cos \theta + \sin \theta$

$r = \dfrac{4}{\cos \theta + \sin \theta}$

The last equation gives us r in terms of θ.

Equations in Polar Coordinates [8.6]

Equations in polar coordinates have variables r and θ instead of x and y. The conversions we use to change ordered pairs from polar coordinates to rectangular coordinates and from rectangular coordinates to polar coordinates are the same ones we use to convert back and forth between equations given in polar coordinates and those in rectangular coordinates.

CHAPTER 8 TEST

Write in terms of i.

1. $\sqrt{-25}$

2. $\sqrt{-12}$

Find x and y so that each of the following equations is true.

3. $7x - 6i = 14 - 3yi$

4. $(x^2 - 3x) + 16i = 10 + 8yi$

Combine the following complex numbers:

5. $(6 - 3i) + [(4 - 2i) - (3 + i)]$

6. $(7 + 3i) - [(2 + i) - (3 - 4i)]$

Simplify each power of i.

7. i^{16}

8. i^{17}

Multiply. Leave your answer in standard form.

9. $(8 + 5i)(8 - 5i)$

10. $(3 + 5i)^2$

Divide. Write all answers in standard form.

11. $\dfrac{5 - 4i}{2i}$

12. $\dfrac{6 + 5i}{6 - 5i}$

For each of the following complex numbers give: (a) the absolute value; (b) the opposite; and (c) the conjugate.

13. $3 + 4i$ **14.** $3 - 4i$ **15.** $8i$ **16.** $-4($

Write each complex number in standard form.

17. $8(\cos 330° + i \sin 330°)$

18. $2(\cos 135° + i \sin 135°)$

Write each complex number in trigonometric form.

19. $2 + 2i$ **20.** $-\sqrt{3} + i$ **21.** $5i$ **22.** -3

Multiply or divide as indicated. Leave your answers in trigonometric form.

23. $5(\cos 25° + i \sin 25°) \cdot 3(\cos 40° + i \sin 40°)$

24. $\dfrac{10(\cos 50° + i \sin 50°)}{2(\cos 20° + i \sin 20°)}$

25. $[2(\cos 10° + i \sin 10°)]^5$

26. $[3(\cos 20° + i \sin 20°)]^4$

27. Find two square roots of $z = 49(\cos 50° + i \sin 50°)$. Leave your answer in trigonometric form.

28. Find 4 fourth roots for $z = 2 + 2i\sqrt{3}$. Leave your answer in trigonometric form.

Solve each equation. Write your solutions in trigonometric form.

29. $x^4 - 2\sqrt{3}x^2 + 4 = 0$ **30.** $x^3 = -1$

For each of the following points, name two other ordered pairs in polar coordinates that name the same point and then convert each of the original ordered pairs to rectangular coordinates.

31. $(4, 225°)$ **32.** $(-6, 60°)$

Convert to polar coordinates with r positive and θ between $0°$ and $360°$.

33. $(-3, 3)$ **34.** $(0, 5)$

Write each equation with rectangular coordinates.

35. $r = 6 \sin \theta$ **36.** $r = \sin 2\theta$

Write each equation in polar coordinates.

37. $x + y = 2$ **38.** $x^2 + y^2 = 8y$

Graph each equation.

39. $r = 4$ **40.** $\theta = 45°$
41. $r = 4 + 2 \cos \theta$ **42.** $r = \sin 2\theta$

Logarithms

I   I   I   I   I   I   I   I   I   I

▶ **To the Student**

Logarithms are exponents. The properties of logarithms are actually the properties of exponents. There are many applications of logarithms to both sciences and higher mathematics. For example, the pH of a liquid is defined in terms of logarithms. (That's the same pH that is given on the label of many hair conditioners.) The Richter scale for measuring earthquake intensity is a logarithmic scale, as is the decibel scale used for measuring the intensity of sound.

SECTION A.1 **Logarithms are Exponents**

As you know from your work in algebra, equations of the form

$$y = b^x \qquad (b > 0, b \neq 1)$$

are called exponential functions. The inverse of an exponential function is a logarithmic function. Since the equation of the inverse of a function can be obtained by exchanging x and y in the equation of the original function, the inverse of an exponential function must have the form

$$x = b^y \qquad (b > 0, b \neq 1)$$

Now this last equation is actually the equation of a logarithmic function, as the following definition indicates:

> DEFINITION   The expression $y = \log_b x$ is read "y is the logarithm to the base b of x" and is equivalent to the expression
>
> $$x = b^y \qquad (b > 0, b \neq 1)$$
>
> *In Words:* We say "y is the number we raise b to in order to get x."

Notation   When an expression is in the form $x = b^y$, it is said to be in exponential form. If an expression is in the form $y = \log_b x$, it is said to be in logarithmic form.

Here are some equivalent statements written in both forms.

Exponential form		Logarithmic form
$8 = 2^3$	$\Leftrightarrow$	$\log_2 8 = 3$
$25 = 5^2$	$\Leftrightarrow$	$\log_5 25 = 2$
$0.1 = 10^{-1}$	$\Leftrightarrow$	$\log_{10} 0.1 = -1$
$\dfrac{1}{8} = 2^{-3}$	$\Leftrightarrow$	$\log_2 \dfrac{1}{8} = -3$
$r = z^s$	$\Leftrightarrow$	$\log_z r = s$

▶ **EXAMPLE 1**   Solve for x: $\log_3 x = -2$.

Solution   In exponential form, the equation looks like this:

$$x = 3^{-2}$$

$$\text{or} \quad x = \frac{1}{9}$$

The solution is $\frac{1}{9}$.   ◀

▶ **EXAMPLE 2**   Solve: $\log_x 4 = 3$.

Solution   Again we use the definition of logarithms to write the expression in exponential form:

$$4 = x^3$$

Taking the cube root of both sides, we have

$$\sqrt[3]{4} = \sqrt[3]{x^3}$$
$$x = \sqrt[3]{4}$$

The solution set is $\{\sqrt[3]{4}\}$.   ◀

▶ **EXAMPLE 3**   Solve: $\log_8 4 = x$.

Solution   We write the expression again in exponential form:

$$4 = 8^x$$

Since both 4 and 8 can be written as powers of 2, we write them in terms of powers of 2:

$$2^2 = (2^3)^x$$
$$2^2 = 2^{3x}$$

The only way the left and right sides of this last line can be equal is if the exponents are equal—that is, if

$$2 = 3x$$

or $$x = \frac{2}{3}$$

The solution is $\frac{2}{3}$. We check as follows:

$$\log_8 4 = \frac{2}{3} \Leftrightarrow 4 = 8^{2/3}$$
$$4 = (\sqrt[3]{8})^2$$
$$4 = 2^2$$
$$4 = 4$$

The solution checks when used in the original equation.   ◀

Graphing Logarithmic Functions

Graphing logarithmic functions can be done using the graphs of exponential functions and the fact that the graphs of inverse functions have symmetry about the line $y = x$. Here's an example to illustrate.

▶ **EXAMPLE 4**   Graph the equation $y = \log_2 x$.

Solution   The equation $y = \log_2 x$ is, by definition, equivalent to the exponential equation

$$x = 2^y$$

which is the equation of the inverse of the function

$$y = 2^x$$

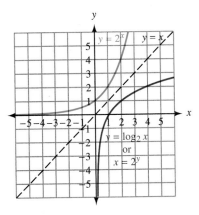

Figure 1

We simply reflect the graph of $y = 2^x$ about the line $y = x$ to get the graph of $x = 2^y$, which is also the graph of $y = \log_2 x$. (See Figure 1.)

It is apparent from the graph that $y = \log_2 x$ is a function, since no vertical line will cross its graph in more than one place. The same is true for all logarithmic equations of the form $y = \log_b x$ where b is a positive number other than 1. Note also that the graph of $y = \log_b x$ will always appear to the right of the y-axis, meaning that x will always be positive in the expression $y = \log_b x$.   ◄

Two Special Identities

If b is a positive real number other than 1, then each of the following is a consequence of the definition of a logarithm:

$$(1)\quad b^{\log_b x} = x \qquad \text{and} \qquad (2)\quad \log_b b^x = x$$

The justifications for these identities are similar. Let's consider only the first one. Consider the expression

$$y = \log_b x$$

By definition, it is equivalent to

$$x = b^y$$

Substituting $\log_b x$ for y in the last line gives us

$$x = b^{\log_b x}$$

The next examples in this section show how these two special properties can be used to simplify expressions involving logarithms.

▶ **EXAMPLE 5**   Simplify $\log_2 8$.

Solution   Substitute 2^3 for 8:

$$\log_2 8 = \log_2 2^3$$
$$= 3 \qquad\qquad ◄$$

▶ **EXAMPLE 6**   Simplify $\log_{10} 10{,}000$.

Solution   10,000 can be written as 10^4:

$$\log_{10} 10{,}000 = \log_{10} 10^4$$
$$= 4 \qquad\qquad ◄$$

▶ **EXAMPLE 7**   Simplify $\log_b b$ $(b > 0, b \neq 1)$.

Solution   Since $b^1 = b$, we have

$$\log_b b = \log_b b^1$$
$$= 1 \qquad\qquad ◄$$

▶ **EXAMPLE 8**   Simplify $\log_b 1$ $(b > 0, b \neq 1)$.

Solution   Since $1 = b^0$, we have

$$\begin{aligned} \log_b 1 &= \log_b b^0 \\ &= 0 \end{aligned}$$ ◀

▶ **EXAMPLE 9**   Simplify $\log_4(\log_5 5)$.

Solution   Since $\log_5 5 = 1$,

$$\begin{aligned} \log_4(\log_5 5) &= \log_4 1 \\ &= 0 \end{aligned}$$ ◀

One application of logarithms is in measuring the magnitude of an earthquake. If an earthquake has a shockwave T times greater than the smallest shockwave that can be measured on a seismograph, then the magnitude M of the earthquake, as measured on the Richter scale, is given by the formula

$$M = \log_{10} T$$

(When we talk about the size of a shockwave, we are talking about its amplitude. The amplitude of a wave is half the difference between its highest point and its lowest point.)

To illustrate the discussion, an earthquake that produces a shockwave that is 10,000 times greater than the smallest shockwave measurable on a seismograph will have a magnitude M on the Richter scale of

$$M = \log_{10} 10{,}000 = 4$$

▶ **EXAMPLE 10**   If an earthquake has a magnitude of $M = 5$ on the Richter scale, what can you say about the size of its shockwave?

Solution   To answer this question, we put $M = 5$ into the formula $M = \log_{10} T$ to obtain

$$5 = \log_{10} T$$

Writing this expression in exponential form, we have

$$T = 10^5 = 100{,}000$$

We can say that an earthquake that measures 5 on the Richter scale has a shockwave 100,000 times greater than the smallest shockwave measurable on a seismograph. ◀

From Example 10 and the discussion that preceded it, we find that an earthquake of magnitude 5 has a shockwave that is 10 times greater than an earthquake of magnitude 4 because 100,000 is 10 times 10,000. If we were to find the relative size of the shockwave of an earthquake of magnitude $M = 6$, we would find it to be 10 times as large as that of an earthquake of magnitude 5 and 100 times as large as that of an earthquake of magnitude 4.

PROBLEM SET A.1

Write each of the following expressions in logarithmic form.

1. $2^4 = 16$
2. $3^2 = 9$
3. $0.01 = 10^{-2}$
4. $0.001 = 10^{-3}$
5. $2^{-5} = \frac{1}{32}$
6. $4^{-2} = \frac{1}{16}$
7. $\left(\frac{1}{2}\right)^{-3} = 8$
8. $\left(\frac{1}{3}\right)^{-2} = 9$

Write each of the following expressions in exponential form.

9. $\log_{10} 100 = 2$
10. $\log_2 32 = 5$
11. $\log_8 1 = 0$
12. $\log_9 9 = 1$
13. $\log_{10} 0.001 = -3$
14. $\log_{10} 0.0001 = -4$
15. $\log_5 \frac{1}{25} = -2$
16. $\log_3 \frac{1}{81} = -4$

Solve each of the following equations for x.

17. $\log_3 x = 2$
18. $\log_2 x = -4$
19. $\log_8 2 = x$
20. $\log_{25} 5 = x$
21. $\log_x 4 = 2$
22. $\log_x 16 = 4$

Sketch the graph of each of the following logarithmic equations.

23. $y = \log_3 x$
24. $y = \log_{1/2} x$
25. $y = \log_{1/3} x$
26. $y = \log_4 x$

Simplify each of the following.

27. $\log_2 16$
28. $\log_3 9$
29. $\log_{25} 125$
30. $\log_9 27$
31. $\log_{10} 1{,}000$
32. $\log_{10} 10{,}000$
33. $\log_3 3$
34. $\log_4 4$
35. $\log_5 1$
36. $\log_{10} 1$
37. $\log_3 (\log_6 6)$
38. $\log_5 (\log_3 3)$
39. $\log_4 [\log_2 (\log_2 16)]$
40. $\log_4 [\log_3(\log_2 8)]$

In chemistry, the pH of a solution is defined in terms of logarithms as pH $= -\log_{10}$ [H$^+$] where [H$^+$] is the concentration of the hydrogen ion in solution. An acid solution has a pH lower than 7, and a basic solution has a pH higher than 7.

41. In distilled water, the concentration of the hydrogen ion is [H$^+$] $= 10^{-7}$. What is the pH?
42. Find the pH of a bottle of vinegar, if the concentration of the hydrogen ion is [H$^+$] $= 10^{-3}$.
43. A hair conditioner has a pH of 6. Find the concentration of the hydrogen ion, [H$^+$], in the conditioner.
44. If a glass of orange juice has a pH of 4, what is the concentration of the hydrogen ion, [H$^+$], in the orange juice?
45. Find the magnitude M of an earthquake with a shockwave that measures $T = 100$ on a seismograph.

46. Find the magnitude M of an earthquake with a shockwave that measures $T = 100{,}000$ on a seismograph.

47. If an earthquake has a magnitude of 8 on the Richter scale, how many times greater is its shockwave than the smallest shockwave measurable on a seismograph?

48. If an earthquake has a magnitude of 6 on the Richter scale, how many times greater is its shockwave than the smallest shockwave measurable on a seismograph?

SECTION A.2 **Properties of Logarithms**

For the following three properties, x, y, and b are all positive real numbers, $b \neq 1$, and r is any real number.

Property 1

$\log_b (xy) = \log_b x + \log_b y$
 In Words: The logarithm of a *product* is the *sum* of the logarithms.

Property 2

$\log_b \left(\dfrac{x}{y} \right) = \log_b x - \log_b y$

 In Words: The logarithm of a *quotient* is the *difference* of the logarithms.

Property 3

$\log_b x^r = r \log_b x$
 In Words: The logarithm of a number raised to a *power* is the *product* of the power and the logarithm of the number.

Proof of Property 1   To prove Property 1, we simply apply the first identity for logarithms given in the preceding section:

$$b^{\log_b xy} = xy = (b^{\log_b x})(b^{\log_b y}) = b^{\log_b x + \log_b y}$$

Since the first and last expressions are equal and the bases are the same, the exponents $\log_b xy$ and $\log_b x + \log_b y$ must be equal. Therefore,

$$\log_b xy = \log_b x + \log_b y$$

The proofs of Properties 2 and 3 proceed in much the same manner, so we will omit them here. The examples that follow show how the three properties can be used.

▶ **EXAMPLE 1**   Expand, using the properties of logarithms: $\log_5 \dfrac{3xy}{z}$.

Solution   Applying Property 2, we can write the quotient of $3xy$ and z in terms of a difference:

$$\log_5 \frac{3xy}{z} = \log_5 3xy - \log_5 z$$

Applying Property 1 to the product $3xy$, we write it in terms of addition:

$$\log_5 \frac{3xy}{z} = \log_5 3 + \log_5 x + \log_5 y - \log_5 z$$ ◀

▶ **EXAMPLE 2**   Expand, using the properties of logarithms:

$$\log_2 \frac{x^4}{\sqrt{y} \cdot z^3}$$

Solution   We write $\sqrt{y}$ as $y^{1/2}$ and apply the properties:

$$\log_2 \frac{x^4}{\sqrt{y} \cdot z^3} = \log_2 \frac{x^4}{y^{1/2} z^3} \qquad\qquad \sqrt{y} = y^{1/2}$$

$$= \log_2 x^4 - \log_2 (y^{1/2} \cdot z^3) \qquad \text{Property 2}$$
$$= \log_2 x^4 - (\log_2 y^{1/2} + \log_2 z^3) \qquad \text{Property 1}$$
$$= \log_2 x^4 - \log_2 y^{1/2} - \log_2 z^3 \qquad \text{Remove parentheses}$$

$$= 4 \log_2 x - \frac{1}{2} \log_2 y - 3 \log_2 z \qquad \text{Property 3}$$ ◀

We can also use the three properties to write an expression in expanded form as just one logarithm.

▶ **EXAMPLE 3**   Write as a single logarithm:

$$2 \log_{10} a + 3 \log_{10} b - \frac{1}{3} \log_{10} c$$

Solution   We begin by applying Property 3:

$$2 \log_{10} a + 3 \log_{10} b - \frac{1}{3} \log_{10} c$$

$$= \log_{10} a^2 + \log_{10} b^3 - \log_{10} c^{1/3} \qquad \text{Property 3}$$
$$= \log_{10} (a^2 \cdot b^3) - \log_{10} c^{1/3} \qquad \text{Property 1}$$

$$= \log_{10} \frac{a^2 b^3}{c^{1/3}} \qquad\qquad\qquad \text{Property 2}$$

$$= \log_{10} \frac{a^2 b^3}{\sqrt[3]{c}} \qquad\qquad\qquad c^{1/3} = \sqrt[3]{c}$$ ◀

The properties of logarithms along with the definition of logarithms are useful in solving equations that involve logarithms.

▶ **EXAMPLE 4**   Solve for x: $\log_2 (x + 2) + \log_2 x = 3$.

Solution   Applying Property 1 to the left side of the equation allows us to write it as a single logarithm:

$$\log_2 (x + 2) + \log_2 x = 3$$
$$\log_2 [(x + 2)(x)] = 3$$

The last line can be written in exponential form using the definition of logarithms:

$$(x + 2)(x) = 2^3$$

Solve as usual:

$$x^2 + 2x = 8$$
$$x^2 + 2x - 8 = 0$$
$$(x + 4)(x - 2) = 0$$
$$x + 4 = 0 \quad \text{or} \quad x - 2 = 0$$
$$x = -4 \quad \text{or} \quad x = 2$$

In the previous section, we noted the fact that x in the expression $y = \log_b x$ cannot be a negative number. Since substitution of $x = -4$ into the original equation gives

$$\log_2 (-2) + \log_2 (-4) = 3$$

which contains logarithms of negative numbers, we cannot use -4 as a solution. The solution set is $\{2\}$.   ◀

PROBLEM SET A.2

Use the three properties of logarithms given in this section to expand each expression as much as possible.

1. $\log_3 4x$
2. $\log_2 5x$
3. $\log_6 \dfrac{5}{x}$
4. $\log_3 \dfrac{x}{5}$

5. $\log_2 y^5$
6. $\log_7 y^3$
7. $\log_9 \sqrt[3]{z}$
8. $\log_8 \sqrt{z}$

9. $\log_6 x^2 y^3$
10. $\log_{10} x^2 y^4$
11. $\log_5 \sqrt{x} \cdot y^4$
12. $\log_8 \sqrt[3]{xy^6}$

13. $\log_b \dfrac{xy}{z}$
14. $\log_b \dfrac{3x}{y}$
15. $\log_{10} \dfrac{4}{xy}$
16. $\log_{10} \dfrac{5}{4y}$

17. $\log_{10} \dfrac{x^2 y}{\sqrt{z}}$
18. $\log_{10} \dfrac{\sqrt{x} \cdot y}{z^3}$
19. $\log_{10} \dfrac{x^3 \sqrt{y}}{z^4}$
20. $\log_{10} \dfrac{x^4 \sqrt[3]{y}}{\sqrt{z}}$

21. $\log_b \sqrt[3]{\dfrac{x^2 y}{z^4}}$
22. $\log_b \sqrt[4]{\dfrac{x^4 y^3}{z^5}}$

Write each expression as a single logarithm.

23. $\log_b x + \log_b z$

24. $\log_b x - \log_b z$

25. $2 \log_3 x - 3 \log_3 y$

26. $4 \log_2 x + 5 \log_2 y$

27. $\frac{1}{2} \log_{10} x + \frac{1}{3} \log_{10} y$

28. $\frac{1}{3} \log_{10} x - \frac{1}{4} \log_{10} y$

29. $3 \log_2 x + \frac{1}{2} \log_2 y - \log_2 z$

30. $2 \log_3 x + 3 \log_3 y - \log_3 z$

31. $\frac{1}{2} \log_2 x - 3 \log_2 y - 4 \log_2 z$

32. $3 \log_{10} x - \log_{10} y - \log_{10} z$

33. $\frac{3}{2} \log_{10} x - \frac{3}{4} \log_{10} y - \frac{4}{5} \log_{10} z$

34. $3 \log_{10} x - \frac{4}{3} \log_{10} y - 5 \log_{10} z$

Solve each of the following equations.

35. $\log_2 x + \log_2 3 = 1$

36. $\log_2 x - \log_2 3 = 1$

37. $\log_3 x - \log_3 2 = 2$

38. $\log_3 x + \log_3 2 = 2$

39. $\log_3 x + \log_3 (x - 2) = 1$

40. $\log_6 x + \log_6 (x - 1) = 1$

41. $\log_3 (x + 3) - \log_3 (x - 1) = 1$

42. $\log_4 (x - 2) - \log_4 (x + 1) = 1$

43. $\log_2 x + \log_2 (x - 2) = 3$

44. $\log_4 x + \log_4 (x + 6) = 2$

45. $\log_8 x + \log_8 (x - 3) = \frac{2}{3}$

46. $\log_{27} x + \log_{27} (x + 8) = \frac{2}{3}$

47. $\log_5 \sqrt{x} + \log_5 \sqrt{6x + 5} = 1$

48. $\log_2 \sqrt{x} + \log_2 \sqrt{6x + 5} = 1$

49. The formula $M = 0.21(\log_{10} a - \log_{10} b)$ is used in the food processing industry to find the number of minutes M of heat processing a certain food should undergo at 250°F to reduce the probability of survival of *C. botulinum* spores. The letter a represents the total number of spores per can before heating, and b represents the total number of spores per can after heating. Find M if $a = 1$ and $b = 10^{-12}$. Then find M using the same values for a and b in the formula

$$M = 0.21 \log_{10} \frac{a}{b}.$$

50. The formula $N = \log_{10} \dfrac{P_1}{P_2}$ is used in radio electronics to find the ratio of the acoustic powers of two electric circuits in terms of their electric powers. Find N if P_1 is 100 and P_2 is 1. Then use the same two values of P_1 and P_2 to find N in the formula $N = \log_{10} P_1 - \log_{10} P_2$.

51. Use the properties of logarithms to show that $\log_{10} (8.43 \times 10^2)$ can be written as $2 + \log_{10} 8.43$.

52. Use the properties of logarithms to show that $\log_{10} (2.76 \times 10^3)$ can be written as $3 + \log_{10} 2.76$.

53. Use the properties of logarithms to show that the formula $\log_{10} A = \log_{10} 100(1.06)^t$ can be written as $\log_{10} A = 2 + t \log_{10} 1.06$.

54. Show that the formula $\log_{10} A = \log_{10} 3(2)^{t/5,600}$ can be written as $\log_{10} A = \log_{10} 3 + \dfrac{t}{5,600} \log_{10} 2$.

SECTION A.3   ## Common Logarithms and Natural Logarithms

There are two kinds of logarithms that occur more frequently than other logarithms. Logarithms with a base of 10 are common because our number system is a base 10 number system. For this reason, we call base 10 logarithms *common logarithms*.

Common Logarithms

> **DEFINITION**   A *common logarithm* is a logarithm with a base of 10. Since common logarithms are used so frequently, it is customary, in order to save time, to omit notating the base. That is,
>
> $$\log_{10} x = \log x$$
>
> When the base is not shown, it is assumed to be 10.

Common logarithms of powers of 10 are simple to evaluate. We need only recognize that $\log 10 = \log_{10} 10 = 1$ and apply the third property of logarithms: $\log_b x^r = r \log_b x$.

$$\log 1{,}000 = \log 10^3 \ \ = 3 \log 10 \ \ = 3(1) \ \ \ = 3$$
$$\log 100 \ \ = \log 10^2 \ \ = 2 \log 10 \ \ = 2(1) \ \ \ = 2$$
$$\log 10 \ \ \ = \log 10^1 \ \ = 1 \log 10 \ \ = 1(1) \ \ \ = 1$$
$$\log 1 \ \ \ \ = \log 10^0 \ \ = 0 \log 10 \ \ = 0(1) \ \ \ = 0$$
$$\log 0.1 \ \ \ = \log 10^{-1} = -1 \log 10 = -1(1) = -1$$
$$\log 0.01 \ \ = \log 10^{-2} = -2 \log 10 = -2(1) = -2$$
$$\log 0.001 = \log 10^{-3} = -3 \log 10 = -3(1) = -3$$

To find common logarithms of numbers that are not powers of 10, we use a calculator with a ⟨ log ⟩ key or a table of logarithms. Finding common logarithms on a calculator is simply a matter of entering the number and pressing the ⟨ log ⟩ key. We will assume the use of a calculator for the rest of this appendix.

Check the following logarithms to be sure you know how to use your calculator:

$$\log 7.02 = 0.8463$$
$$\log 1.39 = 0.1430$$
$$\log 6.00 = 0.7782$$
$$\log 9.99 = 0.9996$$

▶ **EXAMPLE 1**   Use a calculator to find log 2,760.

Solution $\log 2{,}760 = 3.4409$

To work this problem on a calculator, we simply enter the number 2,760 and press the key labeled ⟨ log ⟩.

$$2{,}760 \ \boxed{\log}$$

The 3 in the answer is called the *characteristic,* and its main function is to keep track of the decimal point. The decimal part of this logarithm is called the *mantissa.* ◀

▶ **EXAMPLE 2**   Find log 0.0391.

Solution $\log 0.0391 = -1.4078$ ◀

▶ **EXAMPLE 3** Find log 0.00523.

Solution log 0.00523 = −2.2815 ◀

▶ **EXAMPLE 4** Find x if log x = 3.8774.

Solution We are looking for the number whose logarithm is 3.8774. On a calculator, we enter 3.8774 and press the key labeled $\boxed{10^x}$. (Sometimes it is the inverse of the $\boxed{\log}$ key.) The result is 7,540 to four significant digits.

$$\text{If} \quad \log x = 3.8774$$
$$\text{then} \quad x = 10^{3.8774}$$
$$x = 7{,}540$$

The number 7,540 is called the *antilogarithm* or just *antilog* of 3.8774. That is, 7,540 is the number whose logarithm is 3.8774. ◀

▶ **EXAMPLE 5** Find x if log x = −2.4179.

Solution We enter 2.4179, change it to a negative number with $\boxed{+/-}$, then press the $\boxed{10^x}$ key. The result is 0.00382.

$$\text{If} \quad \log x = -2.4179$$
$$\text{then} \quad x = 10^{-2.4179}$$
$$= 0.00382$$

The antilog of −2.4179 is 0.00382. That is, the logarithm of 0.00382 is −2.4179. ◀

In Section A.1, we found that the magnitude M of an earthquake that produces a shockwave T times larger than the smallest shockwave that can be measured on a seismograph is given by the formula

$$M = \log_{10} T$$

We can rewrite this formula using our shorthand notation for common logarithms as

$$M = \log T$$

▶ **EXAMPLE 6** The San Francisco earthquake of 1906 measured 8.3 on the Richter scale. The San Fernando earthquake of 1971 measured 6.6 on the Richter scale. Find T for each earthquake and then give some indication of how much stronger the 1906 earthquake was than the 1971 earthquake.

Solution For the 1906 earthquake:

$$\text{If } \log T = 8.3, \text{ then } T = 2.00 \times 10^8$$

For the 1971 earthquake:

$$\text{If } \log T = 6.6, \text{ then } T = 3.98 \times 10^6$$

Dividing the two values of T and rounding our answer to the nearest whole number, we have

$$\frac{2.00 \times 10^8}{3.98 \times 10^6} = 50$$

The shockwave for the 1906 earthquake was approximately 50 times as large as the shockwave for the 1971 earthquake. ◀

If you purchase a new car for P dollars and t years later it is worth W dollars, then the annual rate of depreciation, r, for that car can be found from the formula

$$\log (1 - r) = \frac{1}{t} \log \frac{W}{P}$$

▶ **EXAMPLE 7** Find the annual rate of depreciation on a car that is purchased for $P = \$9,000$ and sold 4 years later for $W = \$4,500$.

Solution Using the formula just given with $t = 4$, $W = 4,500$, and $P = 9,000$ and doing the calculations on a calculator, we have

$$\log (1 - r) = \frac{1}{4} \log \frac{4,500}{9,000}$$
$$= 0.25 \log 0.5$$
$$= 0.25(-0.3010) \qquad 0.5 \boxed{\log}$$
$$\log (1 - r) = -0.0753 \qquad \text{Multiplication}$$

Now to find $1 - r$, we must find the antilog of -0.0753. Using a calculator, we first enter 0.0753, press the $\boxed{+/-}$ key, and then press the $\boxed{10^x}$ key. Doing so gives the antilog of -0.0753 as 0.841. Here is the rest of the problem:

$$1 - r = 0.841$$
$$-r = -0.159 \qquad \text{Add } -1 \text{ to each side}$$
$$r = 0.159 \text{ or } 15.9\%$$

The annual rate of depreciation is 15.9% on a car that is purchased for $9,000 and sold 4 years later for $4,500. (We should mention here that 15.9% is actually the *average* annual rate of depreciation because cars usually lose more of their value during the first year than they do in later years.) ◀

Natural Logarithms

The next kind of logarithms we want to discuss are *natural logarithms.* In order to give a definition for natural logarithms, we need to talk about a special number that is denoted by the letter e. The number e is a number like π. It is irrational and occurs in many formulas that describe the world around us. Like π, it can be approximated with a decimal number. Whereas π is approximately 3.1416, e is approximately 2.7183. (If you have a calculator with a key labeled e^x, press 1 and then the e^x key to see a more accurate approximation to e.) We cannot give a more precise definition of the number e without using some of the topics taught in calculus. For the work we are going to do with the number e, we need know only that it is an irrational number that is approximately 2.7183. (If this bothers you, try to think of the last time you saw a precise definition for the number π. Even though you may not know the definition of π, you are still able to work problems that use the number π, simply by knowing that it is an irrational number that is approximately 3.1416.) Here is our definition for natural logarithms.

> **DEFINITION**   A *natural logarithm* is a logarithm with a base of e. The natural logarithm of x is denoted by $\ln x$. That is,
>
> $$\ln x = \log_e x$$

Since the number e is an irrational number that is approximately 2.7183, we can assume that all our properties of exponents and logarithms hold for expressions with a base of e. Here are some examples intended to make you more familiar with the number e and natural logarithms.

▶ **EXAMPLES 8–14**   Simplify each of the following expressions.

   8. $e^0 = 1$
   9. $e^1 = e$
   10. $\ln e = 1$ In exponential form, $e^1 = e$
   11. $\ln 1 = 0$ In exponential form, $e^0 = 1$
   12. $\ln e^3 = 3$
   13. $\ln e^{-4} = -4$
   14. $\ln e^t = t$ ◀

▶ **EXAMPLE 15**   Use the properties of logarithms to expand the expression $\ln Ae^{5t}$.

   Solution   Since the properties of logarithms hold for natural logarithms, we have

$$\ln Ae^{5t} = \ln A + \ln e^{5t}$$
$$= \ln A + 5t \ln e$$
$$= \ln A + 5t \qquad \text{Because } \ln e = 1 \qquad ◀$$

▶ **EXAMPLE 16**   If $\ln 2 = 0.6931$ and $\ln 3 = 1.0986$, find

   a. $\ln 6$ **b.** $\ln 0.5$ **c.** $\ln 8$

   Solution

   a. Since $6 = 2 \cdot 3$, we have

$$\ln 6 = \ln 2 \cdot 3$$
$$= \ln 2 + \ln 3$$
$$= 0.6931 + 1.0986$$
$$= 1.7917$$

b. Writing 0.5 as $\frac{1}{2}$ and applying Property 2 for logarithms gives us

$$\ln 0.5 = \ln \frac{1}{2}$$
$$= \ln 1 - \ln 2$$
$$= 0 - 0.6931$$
$$= -0.6931$$

c. Writing 8 as 2^3 and applying Property 3 for logarithms, we have

$$\ln 8 = \ln 2^3$$
$$= 3 \ln 2$$
$$= 3(0.6931)$$
$$= 2.0793$$

◀

PROBLEM SET A.3

Find the following logarithms. Round your answers to the nearest ten thousandth.

1. log 378	**2.** log 426	**3.** log 37.8	**4.** log 42,600
5. log 3,780	**6.** log 0.4260	**7.** log 0.0378	**8.** log 0.0426
9. log 37,800	**10.** log 4,900	**11.** log 600	**12.** log 900
13. log 2,010	**14.** log 10,200	**15.** log 0.00971	**16.** log 0.0312

Find x in the following equations. (Write your answers to three significant digits.)

17. $\log x = 2.8802$	**18.** $\log x = 4.8802$
19. $\log x = -2.1198$	**20.** $\log x = -3.1198$
21. $\log x = 3.1553$	**22.** $\log x = 5.5911$
23. $\log x = -5.3497$	**24.** $\log x = -1.5670$

In Problem Set A.1, we indicated that the pH of a solution is defined in terms of logarithms as

$$pH = -\log [H^+]$$

where $[H^+]$ is the concentration of the hydrogen ion in that solution.

25. Find the pH of orange juice if the concentration of the hydrogen ion in the juice is $[H^+] = 6.5 \times 10^{-4}$.

26. Find the pH of milk if the concentration of the hydrogen ion in milk is $[H^+] = 1.88 \times 10^{-6}$.

27. Find the concentration of hydrogen ion in a glass of wine if the pH is 4.75.

28. Find the concentration of hydrogen ion in a bottle of vinegar if the pH is 5.75.

Find the relative size T of the shockwave of earthquakes with the following magnitudes, as measured on the Richter scale.

29. 5.5	**30.** 6.6	**31.** 8.3	**32.** 8.7

33. How much larger is the shockwave of an earthquake that measures 6.5 on the Richter scale than one that measures 5.5 on the same scale?

34. How much larger is the shockwave of an earthquake that measures 8.5 on the Richter scale than one that measures 5.5 on the same scale?

In the section we just covered, we found that the annual rate of depreciation, r, on a car that is purchased for P dollars and is worth W dollars t years later can be found from the formula

$$\log (1 - r) = \frac{1}{t} \log \frac{W}{P}$$

35. Find the annual rate of depreciation on a car that is purchased for $9,000 and sold 5 years later for $4,500.

36. Find the annual rate of depreciation on a car that is purchased for $9,000 and sold 4 years later for $3,000.

37. Find the annual rate of depreciation on a car that is purchased for $7,550 and sold 5 years later for $5,750.

38. Find the annual rate of depreciation on a car that is purchased for $7,550 and sold 3 years later for $5,750.

Simplify each of the following expressions.

39. $\ln e$ **40.** $\ln 1$ **41.** $\ln e^5$ **42.** $\ln e^{-3}$

43. $\ln e^x$ **44.** $\ln e^y$

Use the properties of logarithms to expand each of the following expressions.

45. $\ln 10e^{3t}$ **46.** $\ln 10e^{4t}$ **47.** $\ln Ae^{-2t}$ **48.** $\ln Ae^{-3t}$

If $\ln 2 = 0.6931$, $\ln 3 = 1.0986$, and $\ln 5 = 1.6094$, find each of the following.

49. $\ln 15$ **50.** $\ln 10$ **51.** $\ln \frac{1}{3}$ **52.** $\ln \frac{1}{5}$

53. $\ln 9$ **54.** $\ln 25$ **55.** $\ln 16$ **56.** $\ln 81$

SECTION A.4   # Exponential Equations and Change of Base

Logarithms are important in solving equations in which the variable appears as an exponent. The equation

$$5^x = 12$$

is an example of one such equation. Equations of this form are called *exponential equations*. Since the quantities 5^x and 12 are equal, so are their common logarithms. We begin our solution by taking the logarithm of both sides:

$$\log 5^x = \log 12$$

We now apply Property 3 for logarithms, $\log x^r = r \log x$, to turn x from an exponent into a coefficient:

$$x \log 5 = \log 12$$

Dividing both sides by $\log 5$ gives us

$$x = \frac{\log 12}{\log 5}$$

If we want a decimal approximation to the solution, we can find $\log 12$ and $\log 5$ on a calculator and divide:

$$x = \frac{1.0792}{0.6990}$$

$$= 1.544$$

The complete problem looks like this:

$$5^x = 12$$
$$\log 5^x = \log 12$$
$$x \log 5 = \log 12$$
$$x = \frac{\log 12}{\log 5}$$
$$= \frac{1.0792}{0.6990}$$
$$= 1.544$$

Here is another example of solving an exponential equation using logarithms.

▶ **EXAMPLE 1**   Solve for x: $25^{2x+1} = 15$.

Solution   Taking the logarithm of both sides and then writing the exponent $(2x + 1)$ as a coefficient, we proceed as follows:

$$25^{2x+1} = 15$$
$$\log 25^{2x+1} = \log 15 \qquad \text{Take the log of both sides}$$
$$(2x + 1) \log 25 = \log 15 \qquad \text{Property 3}$$
$$2x + 1 = \frac{\log 15}{\log 25} \qquad \text{Divide by } \log 25$$
$$2x = \frac{\log 15}{\log 25} - 1 \qquad \text{Add } -1 \text{ to both sides}$$
$$x = \frac{1}{2}\left(\frac{\log 15}{\log 25} - 1\right) \qquad \text{Multiply both sides by } \tfrac{1}{2}$$

Using a calculator, we can write a decimal approximation to the answer:

$$x = \frac{1}{2}\left(\frac{1.1761}{1.3979} - 1\right)$$

$$= \frac{1}{2}(0.8413 - 1)$$

$$= \frac{1}{2}(-0.1587)$$

$$= -0.079 \qquad \blacktriangleleft$$

If you invest P dollars in an account with an annual interest rate r that is compounded n times a year, then t years later the amount of money in that account will be

$$A = P\left(1 + \frac{r}{n}\right)^{nt}$$

▶ **EXAMPLE 2**   If \$5,000 is placed in an account with an annual interest rate of 12% compounded twice a year, how much money will be in the account 10 years later?

Solution   Substituting $P = 5,000$, $r = 0.12$, $n = 2$, and $t = 10$ into the formula above, we have

$$A = 5,000\left(1 + \frac{0.12}{2}\right)^{2 \cdot 10}$$

$$= 5,000(1.06)^{20}$$

To evaluate this last expression on a calculator, we would use the following sequence:

$$1.06 \quad \boxed{y^x} \quad 20 \quad \boxed{\times} \quad 5,000 \quad \boxed{=}$$

giving 16,035.68 as the result.

We could also use logarithms to solve this problem. With logarithms, we take the common logarithm of each side of our last equation to obtain

$$\log A = \log [5,000(1.06)^{20}]$$
$$\log A = \log 5,000 + 20 \log 1.06$$
$$\log A = 4.2050873$$
$$A = 16,035.68$$

The original amount, \$5,000, will become \$16,035.68 in 10 years if invested at 12% interest compounded twice a year. ◀

▶ **EXAMPLE 3**   How long does it take for \$5,000 to double if it is deposited in an account that yields 5% interest compounded once a year?

Solution   Substituting $P = 5,000$, $r = 0.05$, $n = 1$, and $A = 10,000$ into our formula, we have

$$10,000 = 5,000(1 + 0.05)^t$$
$$10,000 = 5,000(1.05)^t$$
$$2 = (1.05)^t \qquad \text{Divide by 5,000}$$

This is an exponential equation. We solve by taking the logarithm of both sides:

$$\log 2 = \log (1.05)^t$$
$$\log 2 = t \log (1.05)$$
$$0.3010 = 0.0212t$$

Dividing both sides by 0.0212, we have

$$t = 14.2$$

It takes a little over 14 years for $5,000 to double if it earns 5% interest per year.

◀

There is a fourth property of logarithms we have not yet considered. This last property allows us to change from one base to another and is therefore called the *change-of-base property*.

Property 4 (Change of Base)

If a and b are both positive numbers other than 1, and if $x > 0$, then

$$\log_a x = \frac{\log_b x}{\log_b a}$$

$$\uparrow \qquad\qquad \uparrow$$
$$\text{Base } a \qquad \text{Base } b$$

The logarithm on the left side has a base of a, while both logarithms on the right side have a base of b. This allows us to change from base a to any other base b that is a positive number other than 1. Here is a proof of Property 4 for logarithms.

PROOF   We begin by writing the identity

$$a^{\log_a x} = x$$

Taking the logarithm base b of both sides and writing the exponent $\log_a x$ as a coefficient, we have

$$\log_b a^{\log_a x} = \log_b x$$
$$\log_a x \log_b a = \log_b x$$

Dividing both sides by $\log_b a$, we have the desired result:

$$\frac{\log_a x \log_b a}{\log_b a} = \frac{\log_b x}{\log_b a}$$

$$\log_a x = \frac{\log_b x}{\log_b a}$$

We can use this property to find logarithms we could not otherwise compute on our calculators—that is, logarithms with bases other than 10 or e. The next example illustrates the use of this property.

▶ **EXAMPLE 4**   Find $\log_8 24$.

Solution   Since we do not have base-8 logarithms on our calculators, we can change this expression to an equivalent expression that contains only base-10 logarithms:

$$\log_8 24 = \frac{\log 24}{\log 8}$$

Don't be confused. We did not just drop the base, we changed to base 10. We could have written the last line like this:

$$\log_8 24 = \frac{\log_{10} 24}{\log_{10} 8}$$

From our calculators, we write

$$\log_8 24 = \frac{1.3802}{0.9031}$$

$$= 1.528$$

Here is the complete calculator solution to Example 4:

$$24 \;\boxed{\log}\; \boxed{\div}\; 8 \;\boxed{\log}\; \boxed{=}$$ ◀

▶ **EXAMPLE 5**   Suppose the population in a small city is 32,000 in the beginning of 1993 and that the city council assumes that the population size t years later can be estimated by the equation

$$P = 32{,}000e^{0.05t}$$

Approximately when will the city have a population of 50,000?

Solution   We substitute 50,000 for P in the equation and solve for t:

$$50{,}000 = 32{,}000e^{0.05t}$$

$$1.56 = e^{0.05t} \qquad \frac{50{,}000}{32{,}000} \text{ is approximately } 1.56$$

To solve this equation for t, we can take the natural logarithm of each side:

$$\ln 1.56 = \ln e^{0.05t}$$

$$\ln 1.56 = 0.05t \ln e \qquad \text{Property 3 for logarithms}$$

$$\ln 1.56 = 0.05t \qquad \text{Because } \ln e = 1$$

$$t = \frac{\ln 1.56}{0.05} \qquad \text{Divide each side by 0.05}$$

$$= \frac{0.4447}{0.05}$$

$$= 8.89 \text{ years}$$

We can estimate that the population will reach 50,000 toward the end of 2001.

◄

PROBLEM SET A.4

Solve each exponential equation. Use calculators to write the answer in decimal form.

1. $3^x = 5$ **2.** $4^x = 3$ **3.** $5^x = 3$ **4.** $3^x = 4$

5. $5^{-x} = 12$ **6.** $7^{-x} = 8$ **7.** $4^{x-1} = 4$ **8.** $3^{x-1} = 9$

9. $3^{2x+1} = 2$ **10.** $2^{2x+1} = 3$ **11.** $6^{5-2x} = 4$ **12.** $9^{7-3x} = 5$

13. If \$5,000 is placed in an account with an annual interest rate of 12% compounded once a year, how much money will be in the account 10 years later?

14. If \$5,000 is placed in an account with an annual interest rate of 12% compounded four times a year, how much money will be in the account 10 years later?

15. If \$200 is placed in an account with an annual interest rate of 8% compounded twice a year, how much money will be in the account 10 years later?

16. If \$200 is placed in an account with an annual interest rate of 8% compounded once a year, how much money will be in the account 10 years later?

17. How long will it take for \$500 to double if it is invested at 6% annual interest compounded twice a year?

18. How long will it take for \$500 to double if it is invested at 6% annual interest compounded 12 times a year?

19. How long will it take for \$1,000 to triple if it is invested at 12% annual interest compounded six times a year?

20. How long will it take for \$1,000 to become \$4,000 if it is invested at 12% annual interest compounded six times a year?

Use the change-of-base property and a calculator to find a decimal approximation to each of the following logarithms.

21. $\log_8 16$ **22.** $\log_9 27$ **23.** $\log_{16} 8$ **24.** $\log_{27} 9$

25. $\log_7 15$ **26.** $\log_3 12$ **27.** $\log_{15} 7$ **28.** $\log_{12} 3$

Find a decimal approximation to each of the following natural logarithms.

29. ln 345   **30.** ln 3,450   **31.** ln 0.345   **32.** ln 0.0345

33. ln 10   **34.** ln 100   **35.** ln 45,000   **36.** ln 450,000

37. Suppose that the population in a small city is 32,000 in 1993 and that the city council assumes that the population size t years later can be estimated by the equation

$$P = 32,000e^{0.05t}$$

Approximately when will the city have a population of 64,000?

38. Suppose the population of a city is given by the equation

$$P = 100,000e^{0.05t}$$

where t is the number of years from the present time. How large is the population now? (Now corresponds to a certain value of t. Once you realize what that value of t is, the problem becomes very simple.)

39. Suppose the population of a city is given by the equation

$$P = 15,000e^{0.04t}$$

where t is the number of years from the present time. How long will it take for the population to reach 45,000?

40. Suppose the population of a city is given by the equation

$$P = 15,000e^{0.08t}$$

where t is the number of years from the present time. How long will it take for the population to reach 45,000?

41. Solve the formula $A = Pe^{rt}$ for t.

42. Solve the formula $A = Pe^{-rt}$ for t.

43. Solve the formula $A = P2^{-kt}$ for t.

44. Solve the formula $A = P2^{kt}$ for t.

45. Solve the formula $A = P(1 - r)^t$ for t.

46. Solve the formula $A = P(1 + r)^t$ for t.

Answers to Odd-Numbered
Exercises and Chapter Tests

I I I I I I I I I I

CHAPTER 1

Problem Set 1.1

1. Acute, complement is $80°$, supplement is $170°$ **3.** Acute, complement is $45°$, supplement is $135°$
5. Obtuse, complement is $-30°$ [because $120° + (-30°) = 90°$]; supplement is $60°$
7. We can't tell if x is acute or obtuse (or neither), complement is $90° - x$, supplement is $180° - x$ **9.** $60°$
11. $45°$ **13.** $50°$ (Look at it in terms of the big triangle ABC.)   **15.** Complementary   **17.** $65°$ **19.** $180°$

21. Using the proportion $\dfrac{x}{360°} = \dfrac{4 \text{ hrs}}{12 \text{ hrs}}$, we find x is $120°$.   **23.** $60°$

25. 5 (This triangle is called a $3-4-5$ right triangle. You will see it again.)
Note Whenever the three sides in a right triangle are whole numbers, those three numbers are called a Pythagorean triple.
27. 15   **29.** 5   **31.** $3\sqrt{2}$ (Note that this must be a $45°-45°-90°$ triangle.)
33. 4 (This is a $30°-60°-90°$ triangle.)   **35.** Find x by solving the equation $(\sqrt{10})^2 = (x + 2)^2 + x^2$ to get $x = 1$.
37. $\sqrt{41}$ **39.** 6   **41.** 22.5 ft   **43.** Longest side 2, third side $\sqrt{3}$ **45.** Shortest side 4, third side $4\sqrt{3}$

47. Shortest side $6/\sqrt{3} = 2\sqrt{3}$, longest side $4\sqrt{3}$ **49.** 40 ft   **51.** $\dfrac{176}{\sqrt{3}} = \dfrac{176\sqrt{3}}{3}$ ft^2 = 101.6 ft^2 **53.** $4\sqrt{2}/5$

55. 8   **57.** $4/\sqrt{2} = 2\sqrt{2}$ **59.** 1,414 ft   **61. a.** $\sqrt{2}$ in   **b.** $\sqrt{3}$ in   **63. a.** $x\sqrt{2}$ **b.** $x\sqrt{3}$

65. $\sqrt{3}$ ft

Problem Set 1.2

1.–11.   (odd)

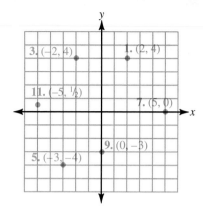

13.

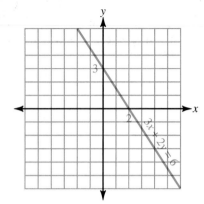

15.

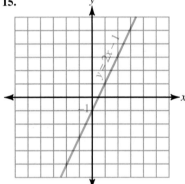

17.

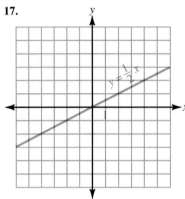

19.

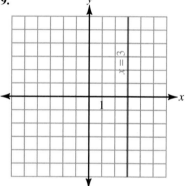

21. $x^2 + y^2 = 25$

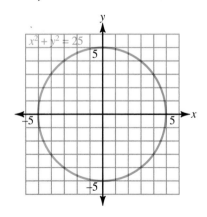

23. $x^2 + y^2 = 5$

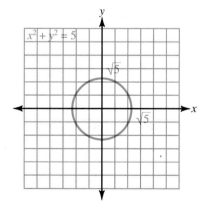

25. (5, 0) and (0, 5)   **27.** $\left(-\dfrac{\sqrt{2}}{2}, -\dfrac{\sqrt{2}}{2}\right)$ and $\left(\dfrac{\sqrt{2}}{2}, \dfrac{\sqrt{2}}{2}\right)$ **29.** 5   **31.** 13   **33.** $\sqrt{61}$ **35.** $\sqrt{130}$

37. 5   **39.** $-1, 3$ **41.** 1.3 mi

43. homeplate: (0, 0); first base: (60, 0); second base: (60, 60); third base: (0, 60)   **45.** QII and QIII

47. QI and QII   **49.** QIII   **51.**

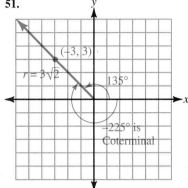

53.

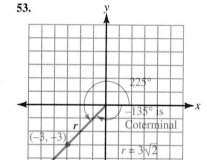

55.

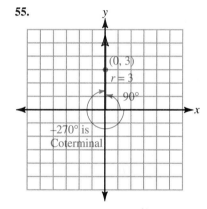

57.

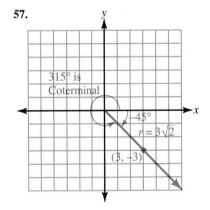

59.

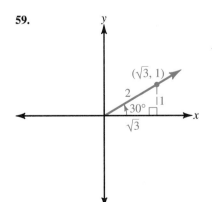

61.

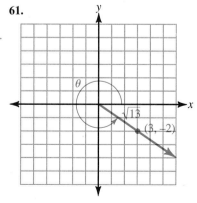

63.

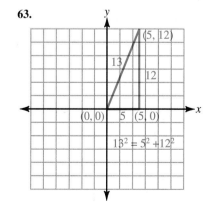

Problem Set 1.3

	$\sin\theta$	$\cos\theta$	$\tan\theta$	$\cot\theta$	$\sec\theta$	$\csc\theta$
1.	$\dfrac{4}{5}$	$\dfrac{3}{5}$	$\dfrac{4}{3}$	$\dfrac{3}{4}$	$\dfrac{5}{3}$	$\dfrac{5}{4}$
3.	$\dfrac{4}{5}$	$-\dfrac{3}{5}$	$-\dfrac{4}{3}$	$-\dfrac{3}{4}$	$-\dfrac{5}{3}$	$\dfrac{5}{4}$
5.	$\dfrac{12}{13}$	$-\dfrac{5}{13}$	$-\dfrac{12}{5}$	$-\dfrac{5}{12}$	$-\dfrac{13}{5}$	$\dfrac{13}{12}$
7.	$-\dfrac{2}{\sqrt{5}}$	$-\dfrac{1}{\sqrt{5}}$	2	$\dfrac{1}{2}$	$-\sqrt{5}$	$-\dfrac{\sqrt{5}}{2}$
9.	$\dfrac{b}{\sqrt{a^2+b^2}}$	$\dfrac{a}{\sqrt{a^2+b^2}}$	$\dfrac{b}{a}$	$\dfrac{a}{b}$	$\dfrac{\sqrt{a^2+b^2}}{a}$	$\dfrac{\sqrt{a^2+b^2}}{b}$
11.	0	-1	0	undefined	-1	undefined
13.	$-\dfrac{1}{2}$	$\dfrac{\sqrt{3}}{2}$	$-\dfrac{1}{\sqrt{3}}$	$-\sqrt{3}$	$\dfrac{2}{\sqrt{3}}$	-2
15.	$\dfrac{2}{3}$	$-\dfrac{\sqrt{5}}{3}$	$-\dfrac{2}{\sqrt{5}}$	$-\dfrac{\sqrt{5}}{2}$	$-\dfrac{3}{\sqrt{5}}$	$\dfrac{3}{2}$
17.	$\dfrac{4}{5}$	$\dfrac{3}{5}$	$\dfrac{4}{3}$	$\dfrac{3}{4}$	$\dfrac{5}{3}$	$\dfrac{5}{4}$
19.	$-\dfrac{12}{13}$	$\dfrac{5}{13}$	$-\dfrac{12}{5}$	$-\dfrac{5}{12}$	$\dfrac{13}{5}$	$-\dfrac{13}{12}$

21. $\sin\theta = \dfrac{7.02}{11.7} = 0.6$, $\cos\theta = \dfrac{9.36}{11.7} = 0.8$

23.

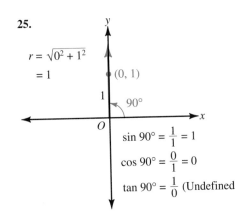

$r = \sqrt{(-1)^2 + 1^2}$
$= \sqrt{2}$

$(-1, 1)$

$r = \sqrt{2}$ $135°$

$\sin 135° = \dfrac{1}{\sqrt{2}}$

$\cos 135° = -\dfrac{1}{\sqrt{2}}$

$\tan 135° = -1$

25.

$r = \sqrt{0^2 + 1^2}$
$= 1$

$(0, 1)$

$90°$

$\sin 90° = \dfrac{1}{1} = 1$

$\cos 90° = \dfrac{0}{1} = 0$

$\tan 90° = \dfrac{1}{0}$ (Undefined)

27.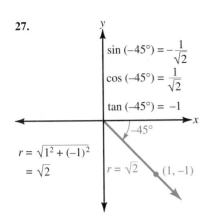

29.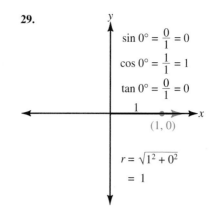

31. QI, QIV because $\cos \theta = x/r$ and x is positive in QI and QIV. (Remember, r is always positive.)

33. QIII, QIV   **35.** QI, QIII   **37.** QIII   **39.** QI (both positive), QIV (both negative)

	$\sin \theta$	$\cos \theta$	$\tan \theta$	$\cot \theta$	$\sec \theta$	$\csc \theta$
41.	$\dfrac{12}{13}$	$\dfrac{5}{13}$	$\dfrac{12}{5}$	$\dfrac{5}{12}$	$\dfrac{13}{5}$	$\dfrac{13}{12}$
43.	$\dfrac{8}{17}$	$\dfrac{15}{17}$	$\dfrac{8}{15}$	$\dfrac{15}{8}$	$\dfrac{17}{15}$	$\dfrac{17}{8}$
45.	$-\dfrac{7}{25}$	$\dfrac{24}{25}$	$-\dfrac{7}{24}$	$-\dfrac{24}{7}$	$\dfrac{25}{24}$	$-\dfrac{25}{7}$
47.	$\dfrac{3}{5}$	$\dfrac{4}{5}$	$\dfrac{3}{4}$	$\dfrac{4}{3}$	$\dfrac{5}{4}$	$\dfrac{5}{3}$
49.	$-\dfrac{20}{29}$	$-\dfrac{21}{29}$	$\dfrac{20}{21}$	$\dfrac{21}{20}$	$-\dfrac{29}{21}$	$-\dfrac{29}{20}$
51.	$-\dfrac{12}{13}$	$\dfrac{5}{13}$	$-\dfrac{12}{5}$	$-\dfrac{5}{12}$	$\dfrac{13}{5}$	$-\dfrac{13}{12}$
53.	$\dfrac{2}{\sqrt{5}}$	$\dfrac{1}{\sqrt{5}}$	2	$\dfrac{1}{2}$	$\sqrt{5}$	$\dfrac{\sqrt{5}}{2}$
55.	$-\dfrac{1}{2}$	$\dfrac{\sqrt{3}}{2}$	$-\dfrac{1}{\sqrt{3}}$	$-\sqrt{3}$	$\dfrac{2}{\sqrt{3}}$	-2
57.	$\dfrac{2}{\sqrt{5}}$	$\dfrac{1}{\sqrt{5}}$	2	$\dfrac{1}{2}$	$\sqrt{5}$	$\dfrac{\sqrt{5}}{2}$
59.	$\dfrac{a}{\sqrt{a^2+b^2}}$	$\dfrac{b}{\sqrt{a^2+b^2}}$	$\dfrac{a}{b}$	$\dfrac{b}{a}$	$\dfrac{\sqrt{a^2+b^2}}{b}$	$\dfrac{\sqrt{a^2+b^2}}{a}$

61. $90°$ **63.** $225°$

65. A point on the terminal side of θ is $(1, 2)$ because it satisfies the equation $y = 2x$; $r = \sqrt{1^2 + 2^2} = \sqrt{5}$; $\sin \theta = 2/\sqrt{5}$; $\cos \theta = 1/\sqrt{5}$.

67. Find a point in QII that is on the terminal side of $y = -3x$ to obtain $\sin \theta = 3/\sqrt{10}$ and $\tan \theta = -3$.

69.

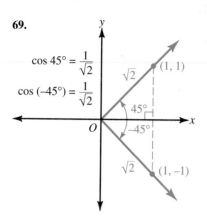

$$\cos 45° = \frac{1}{\sqrt{2}}$$

$$\cos (-45°) = \frac{1}{\sqrt{2}}$$

71. ±4

Problem Set 1.4

1. 1/7   **3.** −3/2   **5.** −√2   **7.** 1/x (x ≠ 0)   **9.** 5/4   **11.** −1/2   **13.** 1/a   **15.** −3/4
17. 12/5   **19.** 1/2   **21.** 8   **23.** 12/5   **25.** −13/5   **27.** 4/5   **29.** −2√2/3   **31.** −3/5
33. 1/2   **35.** −2/√5   **37.** 1/2√2   **39.** −17/15   **41.** 29/20

	sin θ	cos θ	tan θ	cot θ	sec θ	csc θ
43.	$\dfrac{5}{13}$	$\dfrac{12}{13}$	$\dfrac{5}{12}$	$\dfrac{12}{5}$	$\dfrac{13}{12}$	$\dfrac{13}{5}$
45.	$-\dfrac{1}{2}$	$\dfrac{\sqrt{3}}{2}$	$-\dfrac{1}{\sqrt{3}}$	$-\sqrt{3}$	$\dfrac{2}{\sqrt{3}}$	-2
47.	$\dfrac{\sqrt{3}}{2}$	$\dfrac{1}{2}$	$\sqrt{3}$	$\dfrac{1}{\sqrt{3}}$	2	$\dfrac{2}{\sqrt{3}}$
49.	$-\dfrac{3}{\sqrt{13}}$	$\dfrac{2}{\sqrt{13}}$	$-\dfrac{3}{2}$	$-\dfrac{2}{3}$	$\dfrac{\sqrt{13}}{2}$	$-\dfrac{\sqrt{13}}{3}$
51.	$-\dfrac{2\sqrt{2}}{3}$	$-\dfrac{1}{3}$	$2\sqrt{2}$	$\dfrac{1}{2\sqrt{2}}$	-3	$-\dfrac{3}{2\sqrt{2}}$
53.	$\dfrac{1}{a}$	$\dfrac{\sqrt{a^2-1}}{a}$	$\dfrac{1}{\sqrt{a^2-1}}$	$\sqrt{a^2-1}$	$\dfrac{a}{\sqrt{a^2-1}}$	a
55.	0.23	0.97	0.24	4.17	1.03	4.35
57.	0.59	−0.81	−0.73	−1.37	−1.24	1.69

Your answers for 55 and 57 may differ from the answers here in the hundredths column, if you found the reciprocal of an unrounded number.

Note   As Problems 59 through 62 indicate, the slope of a line through the origin is the same as the tangent of the angle the line makes with the positive *x*-axis.

59. 3   **61.** *m*

Problem Set 1.5

1. $\cos\theta = \pm\sqrt{1 - \sin^2\theta}$ **3.** $\cot\theta = \pm\dfrac{\sqrt{1 - \sin^2\theta}}{\sin\theta}$ **5.** $\sec\theta = 1/\cos\theta$ **7.** $\tan\theta = \pm\dfrac{\sqrt{1 - \cos^2\theta}}{\cos\theta}$

9. $\csc\theta\cot\theta = \dfrac{1}{\sin\theta}\cdot\dfrac{\cos\theta}{\sin\theta} = \dfrac{\cos\theta}{\sin^2\theta}$ **11.** $1/\cos\theta$ **13.** $\dfrac{\sec\theta}{\csc\theta} = \dfrac{1/\cos\theta}{1/\sin\theta} = \dfrac{\sin\theta}{\cos\theta}$

15. $\dfrac{\sec\theta}{\tan\theta} = \dfrac{1/\cos\theta}{\sin\theta/\cos\theta} = \dfrac{1}{\sin\theta}$ **17.** $\dfrac{\tan\theta}{\cot\theta} = \dfrac{\sin\theta/\cos\theta}{\cos\theta/\sin\theta} = \dfrac{\sin^2\theta}{\cos^2\theta}$ **19.** $\sin^2\theta$ **21.** $\dfrac{\sin\theta + 1}{\cos\theta}$

23. $2\cos\theta$ **25.** $\cos\theta$ **27.** $\dfrac{\sin\theta}{\cos\theta} + \dfrac{1}{\sin\theta} = \dfrac{\sin\theta}{\cos\theta}\cdot\dfrac{\sin\theta}{\sin\theta} + \dfrac{1}{\sin\theta}\cdot\dfrac{\cos\theta}{\cos\theta} = \dfrac{\sin^2\theta + \cos\theta}{\sin\theta\cos\theta}$

29. $\dfrac{\cos\theta - \sin\theta}{\sin\theta\cos\theta}$ **31.** $\dfrac{\sin\theta\cos\theta + 1}{\cos\theta}$ **33.** $\dfrac{1 - \sin^2\theta}{\sin\theta} = \dfrac{\cos^2\theta}{\sin\theta}$ **35.** $\sin^2\theta + 7\sin\theta + 12$

37. $8\cos^2\theta + 2\cos\theta - 15$ **39.** $1 - \sin^2\theta = \cos^2\theta$ **41.** $1 - \tan^2\theta$

43. $\sin^2\theta - 2\sin\theta\cos\theta + \cos^2\theta = 1 - 2\sin\theta\cos\theta$ **45.** $\sin^2\theta - 8\sin\theta + 16$

47. $\cos\theta\tan\theta = \cos\theta\cdot\dfrac{\sin\theta}{\cos\theta} = \sin\theta$ **51.** $\dfrac{\sin\theta}{\csc\theta} = \dfrac{\sin\theta}{\dfrac{1}{\sin\theta}}$

$$= \sin\theta\cdot\dfrac{\sin\theta}{1}$$

$$= \sin^2\theta$$

59. $\sin\theta\tan\theta + \cos\theta = \sin\theta\cdot\dfrac{\sin\theta}{\cos\theta} + \cos\theta$ **63.** $\csc\theta - \sin\theta = \dfrac{1}{\sin\theta} - \sin\theta = \dfrac{1}{\sin\theta} - \dfrac{\sin^2\theta}{\sin\theta}$

$$= \dfrac{\sin^2\theta}{\cos\theta} + \cos\theta \qquad\qquad\qquad = \dfrac{1 - \sin^2\theta}{\sin\theta}$$

$$= \dfrac{\sin^2\theta + \cos^2\theta}{\cos\theta} \qquad\qquad\qquad = \dfrac{\cos^2\theta}{\sin\theta}$$

$$= \dfrac{1}{\cos\theta}$$

$$= \sec\theta$$

71. $\dfrac{\cos\theta}{\sec\theta} + \dfrac{\sin\theta}{\csc\theta} = \dfrac{\cos\theta}{\dfrac{1}{\cos\theta}} + \dfrac{\sin\theta}{\dfrac{1}{\sin\theta}}$ **75.** $\sin\theta(\sec\theta + \csc\theta) = \sin\theta\cdot\sec\theta + \sin\theta\cdot\csc\theta$

$$= \cos^2\theta + \sin^2\theta \qquad\qquad\qquad\qquad = \sin\theta\cdot\dfrac{1}{\cos\theta} + \sin\theta\cdot\dfrac{1}{\sin\theta}$$

$$= 1 \qquad\qquad\qquad\qquad\qquad\qquad = \dfrac{\sin\theta}{\cos\theta} + \dfrac{\sin\theta}{\sin\theta}$$

$$= \tan\theta + 1$$

81. $2|\sec\theta|$ **83.** $3|\cos\theta|$ **85.** $4|\sec\theta|$

Chapter 1 Test

1. $3\sqrt{3}$ **2.** 3 **3.** $h = 5\sqrt{3}, r = 5\sqrt{6}, y = 5, x = 10$ **4.** $x = 6, h = 3\sqrt{3}, s = 3\sqrt{3}, r = 3\sqrt{6}$

5. $\sqrt{52} = 7.21$ to the nearest hundredth   **6.** $90°$ **7.** $5/2$ and $5\sqrt{3}/2$

8.

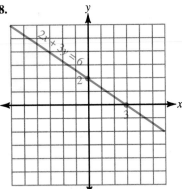

9. 13   **10.** $\sqrt{a^2 + b^2}$ **11.** $-5, 1$

12. $\sin 90° = 1$, $\cos 90° = 0$, $\tan 90°$ is undefined   **13.** $\sin(-45°) = -1/\sqrt{2}$, $\cos(-45°) = 1/\sqrt{2}$, $\tan(-45°) = -1$

14. QIV   **15.** QII

	$\sin \theta$	$\cos \theta$	$\tan \theta$	$\cot \theta$	$\sec \theta$	$\csc \theta$
16.	$\dfrac{4}{5}$	$-\dfrac{3}{5}$	$-\dfrac{4}{3}$	$-\dfrac{3}{4}$	$-\dfrac{5}{3}$	$\dfrac{5}{4}$
17.	$-\dfrac{1}{\sqrt{10}}$	$-\dfrac{3}{\sqrt{10}}$	$\dfrac{1}{3}$	3	$-\dfrac{\sqrt{10}}{3}$	$-\sqrt{10}$
18.	$\dfrac{1}{2}$	$-\dfrac{\sqrt{3}}{2}$	$-\dfrac{1}{\sqrt{3}}$	$-\sqrt{3}$	$-\dfrac{2}{\sqrt{3}}$	2
19.	$-\dfrac{12}{13}$	$-\dfrac{5}{13}$	$\dfrac{12}{5}$	$\dfrac{5}{12}$	$-\dfrac{13}{5}$	$-\dfrac{13}{12}$

20. $\sin \theta = -2/\sqrt{5}$ $\cos \theta = 1/\sqrt{5}$ **21.** $-4/3$ **22.** $-1/2$ **23.** $1/27$

24. $\cos \theta = 1/3$, $\sin \theta = -2\sqrt{2}/3$, $\tan \theta = -2\sqrt{2}$ **25.** $\cos \theta = \sqrt{a^2 - 1}/a$, $\csc \theta = a$, $\cot \theta = \sqrt{a^2 - 1}$

26. $\sin^2 \theta - 4 \sin \theta - 21$ **27.** $1 - 2 \sin \theta \cos \theta$ **28.** $\cos^2 \theta/\sin \theta$

CHAPTER 2

Problem Set 2.1

	$\sin A$	$\cos A$	$\tan A$	$\cot A$	$\sec A$	$\csc A$
1.	$\dfrac{4}{5}$	$\dfrac{3}{5}$	$\dfrac{4}{3}$	$\dfrac{3}{4}$	$\dfrac{5}{3}$	$\dfrac{5}{4}$
3.	$\dfrac{2}{\sqrt{5}}$	$\dfrac{1}{\sqrt{5}}$	2	$\dfrac{1}{2}$	$\sqrt{5}$	$\dfrac{\sqrt{5}}{2}$
5.	$\dfrac{2}{3}$	$\dfrac{\sqrt{5}}{3}$	$\dfrac{2}{\sqrt{5}}$	$\dfrac{\sqrt{5}}{2}$	$\dfrac{3}{\sqrt{5}}$	$\dfrac{3}{2}$

	$\sin A$	$\cos A$	$\tan A$	$\sin B$	$\cos B$	$\tan B$
7.	$\dfrac{5}{6}$	$\dfrac{\sqrt{11}}{6}$	$\dfrac{5}{\sqrt{11}}$	$\dfrac{\sqrt{11}}{6}$	$\dfrac{5}{6}$	$\dfrac{\sqrt{11}}{5}$

9. $\dfrac{1}{\sqrt{2}}$ $\dfrac{1}{\sqrt{2}}$ 1 $\dfrac{1}{\sqrt{2}}$ $\dfrac{1}{\sqrt{2}}$ 1

11. $\dfrac{3}{5}$ $\dfrac{4}{5}$ $\dfrac{3}{4}$ $\dfrac{4}{5}$ $\dfrac{3}{5}$ $\dfrac{4}{3}$

13. $\dfrac{\sqrt{3}}{2}$ $\dfrac{1}{2}$ $\sqrt{3}$ $\dfrac{1}{2}$ $\dfrac{\sqrt{3}}{2}$ $\dfrac{1}{\sqrt{3}}$

15. $90° - 10° = 80°$ **17.** $90° - 8° = 82°$ **19.** $17°$ **21.** $90° - x$ **23.** x **25.** $4\left(\dfrac{1}{2}\right) = 2$

27. 3   **29.** $1/2\sqrt{2}$ or $\sqrt{2}/4$ **31.** $\left(\dfrac{1 + \sqrt{3}}{2}\right)^2 = \dfrac{1 + 2\sqrt{3} + 3}{4} = \dfrac{4 + 2\sqrt{3}}{4} = \dfrac{2 + \sqrt{3}}{2}$ **33.** 0

35. $4 + 2\sqrt{3}$ **37.** 1   **39.** $2\sqrt{3}$ **41.** $-3\sqrt{3}/2$ **43.** $\sqrt{2}$ **45.** $\sec 30° = \dfrac{1}{\cos 30°} = \dfrac{1}{\sqrt{3}/2} = \dfrac{2}{\sqrt{3}}$

47. $2/\sqrt{3}$ **49.** 1   **51.** $\sqrt{2}$ **53.** 0.8660   **55.** 0.8660   **57.** 0.7071   **59.** 0.7071

	sin A	cos A	sin B	cos B
61.	0.60	0.80	0.80	0.60
63.	0.96	0.28	0.28	0.96

65. $\dfrac{1}{\sqrt{3}}, \dfrac{\sqrt{2}}{\sqrt{3}}$ **67.** $\dfrac{1}{\sqrt{3}}, \dfrac{\sqrt{2}}{\sqrt{3}}$ **69.** 5   **71.** $-1, 3$

73.

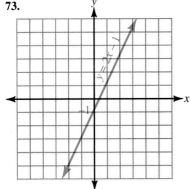

75. See Example 6, Section 1.2

77. $225°$ **79.** $60°$

Problem Set 2.2

1. $64° 9'$ **3.** $89° 40'$ **5.** $106° 49'$ **7.** $\begin{array}{r} 89° 60' \\ -34° 12' \\ \hline 55° 48' \end{array}$ **9.** $59° 43'$ **11.** $\begin{array}{r} 76° 24' = \\ -22° 34' \\ \hline \end{array}$ $\begin{array}{r} 75° 84' \\ -22° 34' \\ \hline 53° 50' \end{array}$

13. $39° 50'$ **15.** $35° 24'$ (*Calculator:* $0.4 \boxed{\times} 60 \boxed{=}$)   **17.** $16° 15'$ **19.** $92° 33'$ **21.** $19° 54'$

23. $45.2°$ (*Calculator:* $12 \boxed{\div} 60 \boxed{+} 45 \boxed{=}$)   **25.** $62.6°$

27. $17.33°$ **29.** $48.45°$ **31.** 0.4571   **33.** 0.9511   **35.** 21.3634   **37.** 1.6643   **39.** 1.5003

41. 4.0906   **43.** 0.9100   **45.** 0.9083   **47.** 0.8355   **49.** 0.0297   **51.** 1.6772   **53.** 1.4370

55. $12.3°$ **57.** $34.5°$ **59.** $78.9°$ **61.** $11.1°$ **63.** $33.3°$ **65.** $55.5°$ **67.** $44° 44'$

69. $65° 43'$ **71.** $50° 5'$ **73.** $14° 41'$ **75.** $10° 10'$ **77.** $9° 9'$ **79.** 0.3907   **81.** 1.2134

83. 0.0787   **85.** 1   **87.** 1

89. You should get an error message. The sine of an angle can never exceed 1.
91. You get an error message; tan 90° is undefined.
93. tan 86° = 14, tan 87° = 19, tan 88° = 29, tan 89° = 57, tan 89.9° = 573
95. sin θ = −2/√13, cos θ = 3/√13, tan θ = −2/3   **97.** sin 90° = 1, cos 90° = 0, tan 90° is undefined
99. sin θ = −12/13, tan θ = 12/5, cot θ = 5/12, sec θ = −13/5, csc θ = −13/12   **101.** QII

Problem Set 2.3

1. 10 ft   **3.** 39 m   **5.** 2.13 ft   **7.** 8.535 yd   **9.** 32°   **11.** 30°   **13.** 59.20°
15. $B = 65°$, $a = 10$ m, $b = 22$ m   **17.** $B = 57.4°$, $b = 67.9$ inches, $c = 80.6$ inches
19. $B = 79° 18'$, $a = 1.121$ cm, $c = 6.037$ cm   **21.** $A = 14°$, $a = 1.4$ ft, $b = 5.6$ ft
23. $A = 63° 30'$, $a = 650$ mm, $c = 726$ mm   **25.** $A = 66.55°$, $b = 2.356$ mi, $c = 5.921$ mi
27. $A = 23°$, $B = 67°$, $c = 95$ ft   **29.** $A = 42.8°$, $B = 47.2°$, $b = 2.97$ cm
31. $A = 61.42°$, $B = 28.58°$, $a = 22.41$ inches   **33.** 49°   **35.** 11   **37.** 36   **39.** $x = 79$, $h = 40$
41. 42°   **43.** $x = 7.4$, $y = 6.2$ **45.** $h = 18$, $x = 11$ **47.** 15   **49.** 35.3°   **51.** 35.3°   **53.** 30.7 ft
55. 201.5 ft   **57.** **a.** 159.8 ft   **b.** 195.8 ft   **c.** 40.8 ft   **59.** 1/4   **61.** $2\sqrt{2}/3$ **63.** $-\sqrt{21}/5$

	sin θ	cos θ	tan θ	cot θ	sec θ	csc θ
65.	$\dfrac{\sqrt{3}}{2}$	$-\dfrac{1}{2}$	$-\sqrt{3}$	$-\dfrac{1}{\sqrt{3}}$	-2	$\dfrac{2}{\sqrt{3}}$
67.	$-\dfrac{\sqrt{3}}{2}$	$-\dfrac{1}{2}$	$\sqrt{3}$	$\dfrac{1}{\sqrt{3}}$	-2	$-\dfrac{2}{\sqrt{3}}$

Problem Set 2.4

1. 39 cm, 69°   **3.** 78.4°   **5.** 39 ft   **7.** 36.6°   **9.** 55.1°   **11.** 61 cm
13. **a.** 800 ft   **b.** 200 ft   **c.** 14.0°   **15.** 31 mi, N 78° E   **17.** 39 mi   **19.** 63.4 mi N, 48.0 mi W
21. 6.4 ft   **23.** 161 ft   **25.** 78.9 ft   **27.** 26 ft   **29.** 4,000 mi   **31.** 6.2 mi
33. $\theta_1 = 45.00°$, $\theta_2 = 35.26°$, $\theta_3 = 30.00°$ **35.** $1 - 2 \sin \theta \cos \theta$

Problem Set 2.5

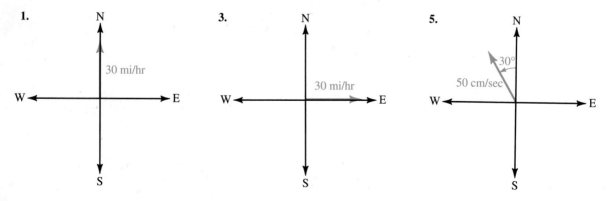

7.

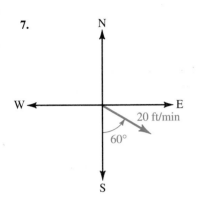

'9.

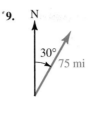

11.

13. 12.4 mph at N 74.8° E   **15.** 198 mph at N 20.5° E   **17.** 2.6 mph   **19.** 49.6 mi, N 85.8° E
21. 6.8 mi   **23.** $|\mathbf{V}_x| = 12.6$ $|\mathbf{V}_y| = 5.66$ **25.** $|\mathbf{V}_x| = 339$ $|\mathbf{V}_y| = 248$ **27.** $|\mathbf{V}_x| = 55$ $|\mathbf{V}_y| = 32$
29. $|\mathbf{V}| = 43.6$ **31.** $|\mathbf{V}| = 5.9$ **33.** both are 850 ft/sec   **35.** 2,550 ft
37. 38.1 ft/sec at an elevation of 23.2°   **39.** 100 km south, 90.3 km east   **41.** 245 mi north, 145 mi east
43. $|\mathbf{H}| = 42.0$ lbs $|\mathbf{T}| = 59.$⁵ lbs   **45.** $|\mathbf{N}| = 9.7$ lbs $|\mathbf{F}| = 2.6$ lbs   **47.** $|\mathbf{F}| = 33.1$ lbs
49. $\sin 135° = 1/\sqrt{2}$, $\cos 135° = -1/\sqrt{2}$, $\tan 135° = -1$ **51.** $\sin\theta = 2/\sqrt{5}$, $\cos\theta = 1/\sqrt{5}$ **53.** $x = \pm 6$

Chapter 2 Test

	sin A	cos A	tan A	sin B	cos B	tan B
1.	$\dfrac{1}{\sqrt{5}}$	$\dfrac{2}{\sqrt{5}}$	$\dfrac{1}{2}$	$\dfrac{2}{\sqrt{5}}$	$\dfrac{1}{\sqrt{5}}$	2
2.	$\dfrac{\sqrt{3}}{2}$	$\dfrac{1}{2}$	$\sqrt{3}$	$\dfrac{1}{2}$	$\dfrac{\sqrt{3}}{2}$	$\dfrac{1}{\sqrt{3}}$
3.	$\dfrac{3}{5}$	$\dfrac{4}{5}$	$\dfrac{3}{4}$	$\dfrac{4}{5}$	$\dfrac{3}{5}$	$\dfrac{4}{3}$
4.	$\dfrac{5}{13}$	$\dfrac{12}{13}$	$\dfrac{5}{12}$	$\dfrac{12}{13}$	$\dfrac{5}{13}$	$\dfrac{12}{5}$

5. 76°   **6.** 17°   **7.** 5/4   **8.** 2   **9.** 0   **10.** $\sqrt{3}/2$ **11.** 73° 10′   **12.** 9°43′   **13.** 73° 12′
14. 16° 27′   **15.** 2.8°   **16.** 79.5°   **17.** 0.4120   **18.** 0.7902   **19.** 2.0353   **20.** 0.3378
21. 4.7° or 4° 40′   **22.** 58.7° or 58° 40′   **23.** 71.2° or 71° 10′   **24.** 13.3° or 13° 20′   **25.** 4   **26.** 2
27. $A = 33.2°$, $B = 56.8°$, $c = 124$ **28.** $A = 30.3°$, $B = 59.7°$, $b = 41.5$ **29.** $A = 65.1°$, $a = 657$, $c = 724$
30. $B = 54° 30′$, $a = 0.268$, $b = 0.376$ **31.** 86 cm   **32.** 5.8 ft   **33.** 70 ft   **34.** $|\mathbf{V}_x| = 4.3$ $|\mathbf{V}_y| = 2.5$
35. 70°   **36.** $|\mathbf{V}_x| = 380$ ft/sec $|\mathbf{V}_y| = 710$ ft/sec   **37.** 60 mi S, 104 mi E
38. 249 mph at 118.0° clockwise from due north   **39.** $|\mathbf{H}| = 45.6$ lbs   **40.** $|\mathbf{F}| = 8.6$ lbs

CHAPTER 3

Problem Set 3.1

1. 30°   **3.** 36.6°   **5.** 48.3°   **7.** 15° 10′   **9.** 60°   **11.** 60°   **13.** $-1/\sqrt{2}$ **15.** $\sqrt{3}/2$
17. -1 **19.** $-1/2$ **21.** -2 **23.** 2   **25.** 1/2   **27.** $-1/\sqrt{3}$ **29.** 0.9744   **31.** -4.8901
33. -0.7427 **35.** 1.5032   **37.** 1.7321   **39.** -1.7263 **41.** 0.7071   **43.** 0.2711   **45.** 1.4225

47. -0.8660 **49.** $198.0°$ **51.** $140.0°$ **53.** $210.5°$ **55.** $74.7°$ **57.** $105.2°$ **59.** $314.3°$
61. $156.4°$ **63.** $126.4°$ **65.** $236.0°$ **67.** $240°$ **69.** $135°$ **71.** $300°$ **73.** $240°$ **75.** $120°$
77. $135°$ **79.** $315°$ **81.** Complement $20°$, supplement $110°$
83. Complement $90° - x$, supplement $180° - x$ **85.** Side opposite $30°$ is 5, side opposite $60°$ is $5\sqrt{3}$ **87.** 1/4
89. 1

Problem Set 3.2

1. 3   **3.** 1/2   **5.** 3π **7.** 2   **9.** $\theta = s/r = 450/4{,}000 = 0.1125$ radians

	Angle In Radians		Reference Angle	
	Exact Value	Approximation	In degrees	In radians
11.	$\dfrac{\pi}{6}$	≈ 0.52	$30° =$	$\dfrac{\pi}{6}$
13.	$\dfrac{\pi}{2}$	≈ 1.57	$90° =$	$\dfrac{\pi}{2}$
15.	$\dfrac{13\pi}{9}$	≈ 4.54	$80° =$	$\dfrac{4\pi}{9}$
17.	$-\dfrac{5\pi}{6}$	≈ -2.62	$30° =$	$\dfrac{\pi}{6}$
19.	$\dfrac{7\pi}{3}$	≈ 7.33	$60° =$	$\dfrac{\pi}{3}$
21.	$-\dfrac{3\pi}{4}$	≈ -2.36	$45° =$	$\dfrac{\pi}{4}$

23. 2.11   **25.** 0.000291
27. Use the answer to Problem 25 to write $\theta = 1' = 0.000291$ radians, then substitute this value of θ into $\theta = s/r$ along with $r = 4{,}000$ and solve for s. $s = 1.16$ mi
29. $60°$, $\hat{\theta} = 60° = \pi/3$ **31.** $120°$, $\hat{\theta} = 60° = \pi/3$ **33.** $-210°$, $\hat{\theta} = 30° = \pi/6$
35. $300°$, $\hat{\theta} = 60° = \pi/3$ **37.** $720°$, $\hat{\theta} = 0° = 0$ **39.** $15°$, $\hat{\theta} = 15° = \pi/12$ **41.** $57.3°$ **43.** $74.5°$
45. $43.0°$ **47.** $286.5°$ **49.** $-\sqrt{3}/2$ **51.** $1/\sqrt{3}$ **53.** -2 **55.** $\sqrt{2}$ **57.** $-2\sqrt{2}$ **59.** $-1/\sqrt{2}$
61. $\sqrt{3}$ **63.** $\sqrt{3}/2$ **65.** 0   **67.** $\sqrt{3}/2$ **69.** -2 **71.** $(0, 0)\ \left(\dfrac{\pi}{4}, \dfrac{1}{\sqrt{2}}\right)\left(\dfrac{\pi}{2}, 1\right)\left(\dfrac{3\pi}{4}, \dfrac{1}{\sqrt{2}}\right)(\pi, 0)$

73. $(0, 0)\ \left(\dfrac{\pi}{2}, 2\right)(\pi, 0)\left(\dfrac{3\pi}{2}, -2\right)(2\pi, 0)$ **75.** $(0, 0)\ \left(\dfrac{\pi}{4}, 1\right)\left(\dfrac{\pi}{2}, 0\right)\left(\dfrac{3\pi}{4}, -1\right)(\pi, 0)$

77. $\left(\dfrac{\pi}{2}, 0\right)(\pi, 1)\left(\dfrac{3\pi}{2}, 0\right)(2\pi, -1)\left(\dfrac{5\pi}{2}, 0\right)$ **79.** $\left(-\dfrac{\pi}{4}, 0\right)(0, 3)\left(\dfrac{\pi}{4}, 0\right)\left(\dfrac{\pi}{2}, -3\right)\left(\dfrac{3\pi}{4}, 0\right)$

	$\sin\theta$	$\cos\theta$	$\tan\theta$	$\cot\theta$	$\sec\theta$	$\csc\theta$	
81.	$-\dfrac{3}{\sqrt{10}}$	$\dfrac{1}{\sqrt{10}}$	-3	$-\dfrac{1}{3}$	$\sqrt{10}$	$-\dfrac{\sqrt{10}}{3}$	
83.	$\dfrac{n}{r}$	$\dfrac{m}{r}$	$\dfrac{n}{m}$	$\dfrac{m}{n}$	$\dfrac{r}{m}$	$\dfrac{r}{n}$	where $r = \sqrt{m^2 + n^2}$
85.	$\dfrac{1}{2}$	$-\dfrac{\sqrt{3}}{2}$	$-\dfrac{1}{\sqrt{3}}$	$-\sqrt{3}$	$-\dfrac{2}{\sqrt{3}}$	2	
87.	$\dfrac{2}{\sqrt{5}}$	$\dfrac{1}{\sqrt{5}}$	2	$\dfrac{1}{2}$	$\sqrt{5}$	$\dfrac{\sqrt{5}}{2}$	

Problem Set 3.3

	$\sin\theta$	$\cos\theta$	$\tan\theta$	$\cot\theta$	$\sec\theta$	$\csc\theta$
1. $\theta=150°$	$\dfrac{1}{2}$	$-\dfrac{\sqrt{3}}{2}$	$-\dfrac{1}{\sqrt{3}}$	$-\sqrt{3}$	$-\dfrac{2}{\sqrt{3}}$	2
3. $\theta=\dfrac{11\pi}{6}$	$-\dfrac{1}{2}$	$\dfrac{\sqrt{3}}{2}$	$-\dfrac{1}{\sqrt{3}}$	$-\sqrt{3}$	$\dfrac{2}{\sqrt{3}}$	-2
5. $\theta=180°$	0	-1	0	undefined	-1	undefined
7. $\theta=\dfrac{3\pi}{4}$	$\dfrac{1}{\sqrt{2}}$	$-\dfrac{1}{\sqrt{2}}$	-1	-1	$-\sqrt{2}$	$\sqrt{2}$

9. $\cos(-60°)=\cos 60°=1/2$ **11.** $\cos(-5\pi/6)=\cos 5\pi/6=-\sqrt{3}/2$ **13.** $\sin(-30°)=-\sin 30°=-1/2$
15. $\sin(-3\pi/4)=-\sin 3\pi/4=-1/\sqrt{2}$
17. On the unit circle, we locate all points with a y-coordinate of $1/2$. The angles associated with these points are $\pi/6$ and $5\pi/6$.
19. $5\pi/6,\,7\pi/6$ **21.** Look for points for which $y/x=-\sqrt{3}$. The angles associated with these points are $2\pi/3$ and $5\pi/3$.
23. $\sin\theta=-2/\sqrt{5},\ \cos\theta=1/\sqrt{5},\ \tan\theta=-2$ **25.** $\sin(-\theta)=-\sin\theta=-(-1/3)=1/3$

27. $\sin\left(2\pi+\dfrac{\pi}{2}\right)=\sin\dfrac{\pi}{2}=1$ **29.** $\sin\left(2\pi+\dfrac{\pi}{6}\right)=\sin\dfrac{\pi}{6}=\dfrac{1}{2}$ **31.** $\dfrac{5\pi}{2}=2\pi+\dfrac{\pi}{2}$, so $\sin\dfrac{5\pi}{2}=1$

33. $\dfrac{13\pi}{6}=2\pi+\dfrac{\pi}{6}$, so $\sin\dfrac{13\pi}{6}=\dfrac{1}{2}$ **35.**

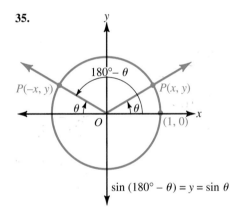

$\sin(180°-\theta)=y=\sin\theta$

37. $\tan(-\theta)=\dfrac{\sin(-\theta)}{\cos(-\theta)}=\dfrac{-\sin\theta}{\cos\theta}=-\dfrac{\sin\theta}{\cos\theta}=-\tan\theta$ **45.**

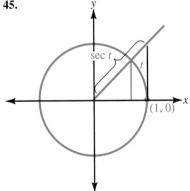

47. $B = 48°, a = 24, b = 27$ **49.** $A = 68°, a = 790, c = 850$ **51.** $A = 33.1°, B = 56.9°, c = 37.5$
53. $A = 44.7°, B = 45.3°, b = 4.41$

Problem Set 3.4

1. 6 inches   **3.** 2.25 ft   **5.** 2π cm ≈ 6.28 cm   **7.** $4\pi/3$ mm ≈ 4.19 mm
9. $40\pi/3$ inches ≈ 41.9 inches   **11.** 5.03 cm   **13.** $1,400\pi$ mi $\approx 4,400$ mi   **15.** $4\pi/9$ ft ≈ 1.40 ft

17. 2,100 mi   **19.** $\dfrac{125\pi}{6} = 65.4$ ft to the nearest tenth   **21.** 480.0 ft to the nearest tenth

23. **a.** 103.1 ft to the nearest tenth   **b.** 361.0 ft to the nearest tenth   **c.** 490.0 ft to the nearest tenth   **25.** 0.5 ft
27. 3 inches   **29.** 4 cm   **31.** 1 m   **33.** $16/5\pi$ km ≈ 1.02 km   **35.** 9 cm^2 **37.** 19.2 inches^2
39. $9\pi/10$ m$^2 \approx 2.83$ m^2 **41.** $25\pi/24$ m$^2 \approx 3.27$ m^2 **43.** 4 inches^2 **45.** 2 cm
47. $4/\sqrt{3}$ inches ≈ 2.31 inches   **49.** 900π ft$^2 \approx 2,830$ ft^2 **51.** 60°   **53.** 2.3 ft   **55.** 74.0°   **57.** 62.3 ft

Problem Set 3.5

1. 1.5 ft/min   **3.** 3 cm/sec   **5.** 15 mph   **7.** 80 ft   **9.** 22.5 mi
11. 7 mi (first change 20 min to 1/3 hr)   **13.** $2\pi/15$ rad/sec ≈ 0.419 rad/sec   **15.** 4 rad/min
17. 8/3 rad/sec ≈ 2.67 rad/sec   **19.** 37.5π rad/hr ≈ 118 rad/hr
21. $d = 100 \tan \frac{1}{2}\pi t$; when $t = 1/2, d = 100$ ft; when $t = 3/2, d = -100$ ft; when $t = 1, d$ is undefined because the light
rays are parallel to the wall.   **23.** 40 inches   **25.** 180π m ≈ 565 m   **27.** 4,500 ft
29. 20π rad/min ≈ 62.8 rad/min   **31.** $200\pi/3$ rad/min ≈ 209 rad/min   **33.** 11.6π rad/min ≈ 36.4 rad/min
35. 10 inches/sec   **37.** 0.5 rad/sec   **39.** 80π ft/min ≈ 251 ft/min   **41.** $\pi/12$ rad/hr ≈ 0.262 rad/hr
43. 10π ft ≈ 31.4 ft   **45.** $33,000/\pi$ rpm $\approx 10,500$ rpm
47. 2.5 inches from the center $v = 1,800\pi$ inches/min $\approx 5,655$ inches/min; 1.25 inches from the center $v =$
900π inches/min $\approx 2,827$ inches/min   **49.** 0.47 mi/hr to the nearest hundredth
51. 576π cm/day $\approx 1,810$ cm/day   **53.** 889 rad/min (53,300 rad/hr)   **55.** 10.4 mi/hr at N 24.1° W
57. 182 mph at 54.5° clockwise from due north   **59.** $|\mathbf{V}_x| = 54$ ft/sec, $|\mathbf{V}_y| = 41$ ft/sec   **61.** 46.2 mi S, 71.9 mi W

Chapter 3 Test

1. 55°   **2.** 62.2°   **3.** 50° 20′   **4.** 45°   **5.** -1.1918 **6.** -2.1445 **7.** 1.1964   **8.** 1.2639
9. -1.2991 **10.** -6.5121 **11.** 174°   **12.** 241° 30′ or 241.5°   **13.** 226°   **14.** 310° 20′ or 310.3°
15. $-1/\sqrt{2}$ **16.** $-1/\sqrt{2}$ **17.** $-1/\sqrt{3}$ **18.** $2/\sqrt{3}$ **19.** $25\pi/18$ **20.** $-13\pi/6$ **21.** 240°
22. 105°   **23.** $\sqrt{3}/2$ **24.** $-1/2$ **25.** $-2\sqrt{2}$ **26.** 1   **27.** $-2/\sqrt{3}$ **28.** 2   **29.** 0

30. $2\sqrt{2}$ **31.** $\cot(-\theta) = \dfrac{\cos(-\theta)}{\sin(-\theta)} = \dfrac{\cos\theta}{-\sin\theta} = -\cot\theta$

32. First use odd and even functions to write everything in terms of θ instead of $-\theta$.   **33.** 2π m ≈ 6.28 m
34. 2π ft ≈ 6.28 ft   **35.** 4 cm   **36.** 3/8 cm = 0.375 cm   **37.** 4π inches$^2 \approx 12.6$ inches^2 **38.** 10.8 cm^2
39. 2π cm   **40.** 8 inches^2 **41.** 90 ft   **42.** 3,960 ft   **43.** 72 inches   **44.** 120π ft ≈ 377 ft
45. 12π rad/min ≈ 37.7 rad/min   **46.** 4π rad/min ≈ 12.6 rad/min   **47.** 0.5 rad/sec   **48.** 5/3 rad/sec
49. 80π ft/min ≈ 251 ft/min   **50.** 20π ft/min ≈ 62.8 ft/min
51. 4 rad/sec for the 6-cm pulley and 3 rad/sec for the 8-cm pulley   **52.** $2,700\pi$ ft/min $\approx 8,480$ ft/min

CHAPTER 4

Problem Set 4.1

If you want graphs that look similar to the ones here, use graph paper on which each square is 1/8 inch on each side. Then let two squares equal one unit. This way you can let the number π be approximately 6 squares. Here is an example:

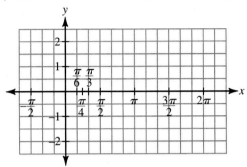

1.

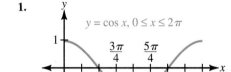

3.

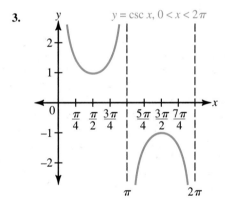

5.

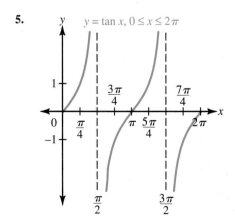

7. $\pi/3, 5\pi/3$ **9.** $3\pi/2$ **11.** $3\pi/4, 5\pi/4$

13. $\pi/4,\ 5\pi/4$ **15.** $\pi/6,\ 11\pi/6$

17.

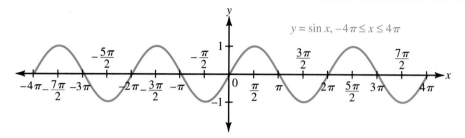

$y = \sin x,\ -4\pi \le x \le 4\pi$

19.

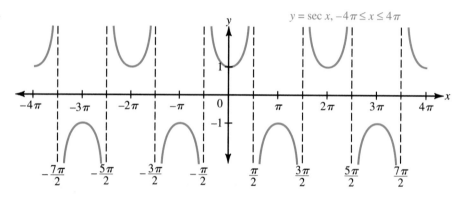

$y = \sec x,\ -4\pi \le x \le 4\pi$

21.

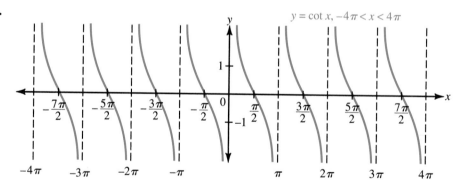

$y = \cot x,\ -4\pi < x < 4\pi$

23. $\pi/3,\ 2\pi/3,\ 7\pi/3,\ 8\pi/3$

25. $\pi,\ 3\pi$ **27.** $7\pi/6,\ 11\pi/6,\ 19\pi/6,\ 23\pi/6$ **29.** $\pi/4,\ 5\pi/4,\ 9\pi/4,\ 13\pi/4$ **31.** $\pi/4,\ 3\pi/4,\ 9\pi/4,\ 11\pi/4$

33. $\pi/2 + k\pi$ **35.** $\pi/2 + 2k\pi$ **37.** $k\pi$ **39.** Amplitude = 3, period = π **41.** Amplitude = 2, period = 2

43. Amplitude = 3, period = π **45.** After you have tried the problem yourself, look at Example 1 in Section 4.2.

47. After you have tried the problem yourself, look at Example 4 in Section 4.2.   **55.** $60°$ **57.** $45°$

59. $120°$ **61.** $330°$

Problem Set 4.2

1.

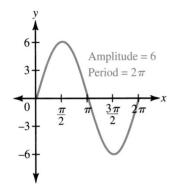

Amplitude = 6
Period = 2π

3.

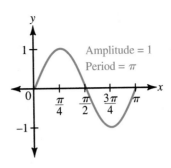

Amplitude = 1
Period = π

5.

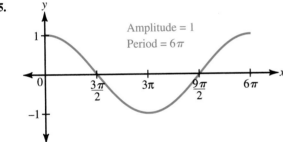

Amplitude = 1
Period = 6π

7.

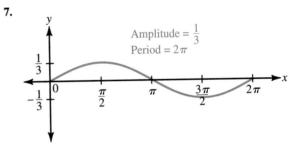

Amplitude = $\frac{1}{3}$
Period = 2π

9.

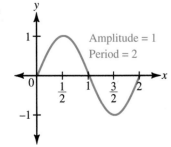

Amplitude = 1
Period = 2

11.

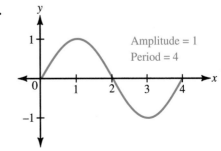

Amplitude = 1
Period = 4

13.

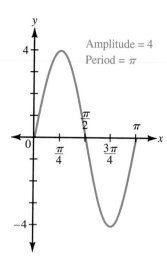

Amplitude = 4
Period = π

15.

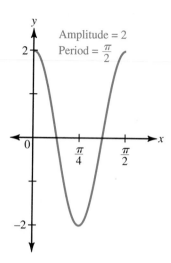

Amplitude = 2
Period = $\dfrac{\pi}{2}$

17.

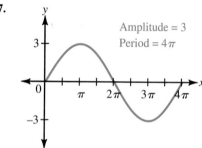

Amplitude = 3
Period = 4π

19.

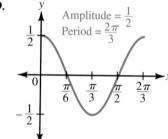

Amplitude = $\dfrac{1}{2}$
Period = $\dfrac{2\pi}{3}$

21.

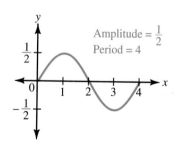

Amplitude = $\dfrac{1}{2}$
Period = 4

23.

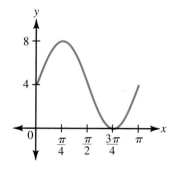

25.

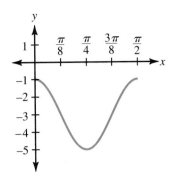

27.

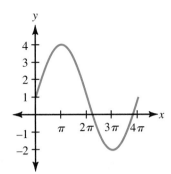

29.

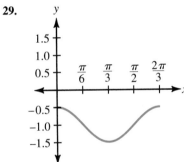

31.

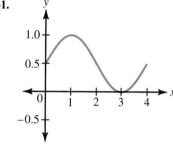

33.

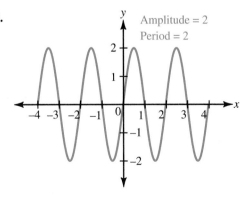

Amplitude = 2
Period = 2

35.

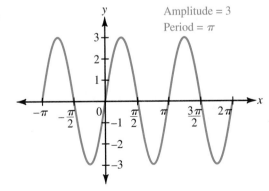

Amplitude = 3
Period = π

37.

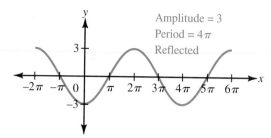

Amplitude = 3
Period = 4π
Reflected

39.

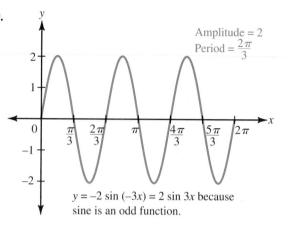

Amplitude = 2
Period = $\dfrac{2\pi}{3}$

$y = -2 \sin(-3x) = 2 \sin 3x$ because
sine is an odd function.

41.   Maximum value of I is 20 amperes; one complete cycle takes $2\pi/120\pi = 1/60$ seconds.

43.

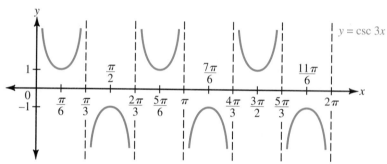

$y = \csc 3x$

45.

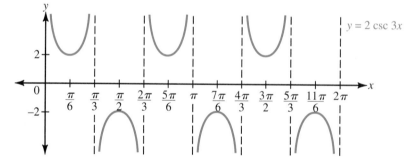

$y = 2 \csc 3x$

47.

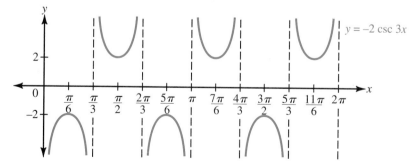

$y = -2 \csc 3x$

49.

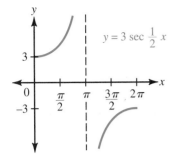

$y = 3 \sec \frac{1}{2} x$

51.

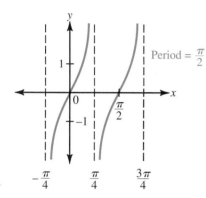

Period $= \frac{\pi}{2}$

53.

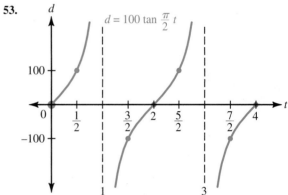

$d = 100 \tan \frac{\pi}{2} t$

55.

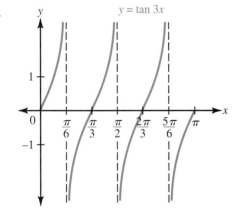

$y = \tan 3x$

57.

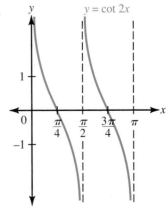

$y = \cot 2x$

59.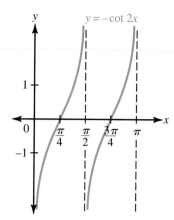
$y = -\cot 2x$

61.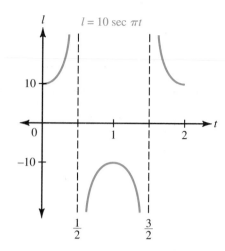
$l = 10 \sec \pi t$

63. 0   **65.** 1   **67.** $\sqrt{3}/2$ **69.** 3/2   **71.** $\pi/4$ **73.** $\pi/3$ **75.** $5\pi/6$ **77.** $5\pi/4$

Problem Set 4.3

1.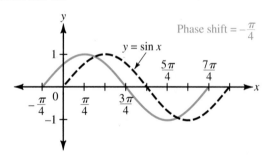
Phase shift $= -\frac{\pi}{4}$
$y = \sin x$

3.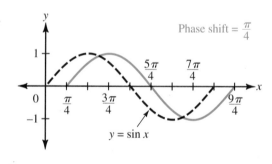
Phase shift $= \frac{\pi}{4}$
$y = \sin x$

5.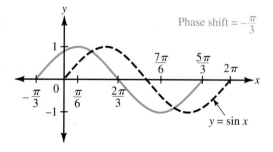
Phase shift $= -\frac{\pi}{3}$
$y = \sin x$

7.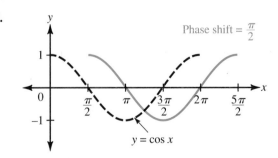
Phase shift $= \frac{\pi}{2}$
$y = \cos x$

9.

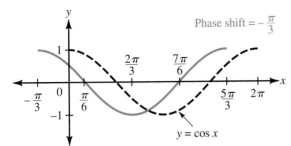

Phase shift $= -\dfrac{\pi}{3}$

$y = \cos x$

11.

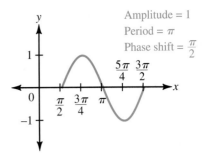

Amplitude $= 1$
Period $= \pi$
Phase shift $= \dfrac{\pi}{2}$

13.

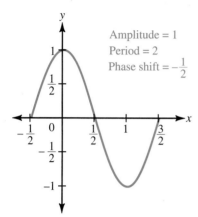

Amplitude $= 1$
Period $= 2$
Phase shift $= -\dfrac{1}{2}$

15.

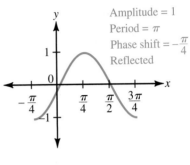

Amplitude $= 1$
Period $= \pi$
Phase shift $= -\dfrac{\pi}{4}$
Reflected

17.

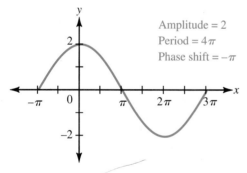

Amplitude $= 2$
Period $= 4\pi$
Phase shift $= -\pi$

19.

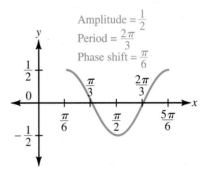

Amplitude $= \dfrac{1}{2}$
Period $= \dfrac{2\pi}{3}$
Phase shift $= \dfrac{\pi}{6}$

21.

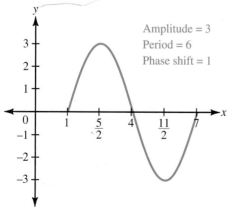

Amplitude $= 3$
Period $= 6$
Phase shift $= 1$

23.

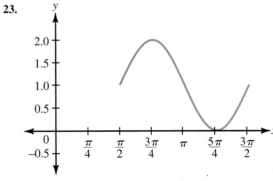

25.

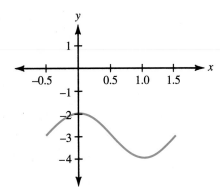

27.

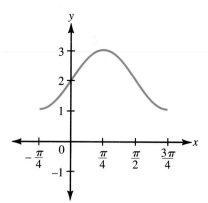

29.

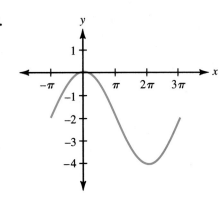

31.

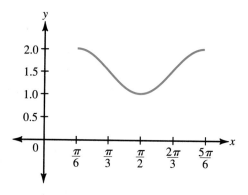

33.

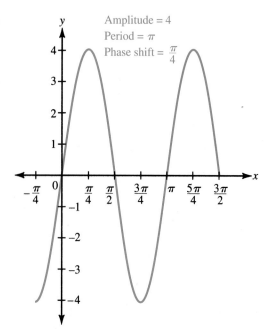

Amplitude = 4
Period = π
Phase shift = $\dfrac{\pi}{4}$

35.

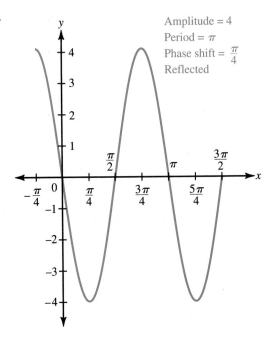

Amplitude = 4
Period = π
Phase shift = $\dfrac{\pi}{4}$
Reflected

37.

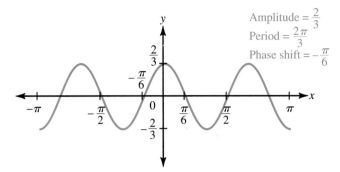

Amplitude $= \dfrac{2}{3}$

Period $= \dfrac{2\pi}{3}$

Phase shift $= -\dfrac{\pi}{6}$

39.

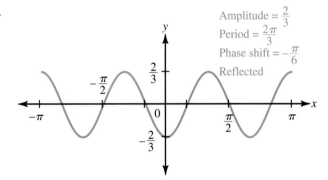

Amplitude $= \dfrac{2}{3}$

Period $= \dfrac{2\pi}{3}$

Phase shift $= -\dfrac{\pi}{6}$

Reflected

41.

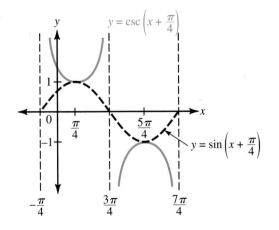

$y = \csc\left(x + \dfrac{\pi}{4}\right)$

$y = \sin\left(x + \dfrac{\pi}{4}\right)$

43.

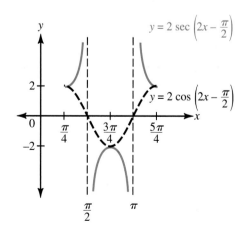

$y = 2\sec\left(2x - \dfrac{\pi}{2}\right)$

$y = 2\cos\left(2x - \dfrac{\pi}{2}\right)$

45.

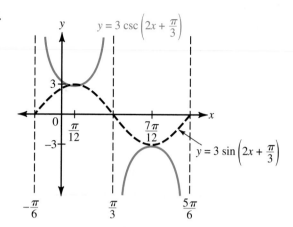

47.

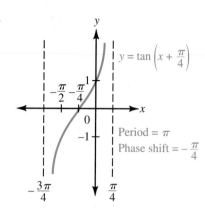

49.

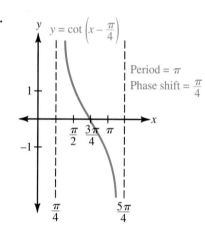

51.

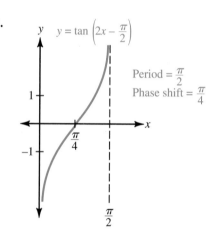

53. $5\pi/3$ cm   **55.** $2.6\pi \approx 8.2$ cm   **57.** 2/3 ft or 8 inches   **59.** 8π inches^2

Problem Set 4.4

1. $y = \frac{1}{2}x + 1$ **3.** $y = 2x - 3$ **5.** $y = \sin x$ **7.** $y = 3\cos x$ **9.** $y = -3\cos x$ **11.** $y = \sin 3x$
13. $y = \sin \frac{1}{3}x$ **15.** $y = 2\cos 3x$ **17.** $y = 4\sin \pi x$ **19.** $y = -4\sin \pi x$ **21.** $y = 2 - 4\sin \pi x$
23. $y = 3\cos\left(2x + \frac{\pi}{2}\right)$ **25.** $y = -2\cos\left(3x + \frac{\pi}{2}\right)$ **27.** $y = -3 + 3\cos\left(2x + \frac{\pi}{2}\right)$

29. $y = 2 - 2\cos\left(3x + \dfrac{\pi}{2}\right)$ **31.**

t	h
0 min	12 ft
1.875 min	40.8 ft
3.75 min	110.5 ft
5.625 min	180.2 ft
7.5 min	209 ft
9.375 min	180.2 ft
11.25 min	110.5 ft
13.125 min	40.8 ft
15 min	12 ft

$$h = 110.5 - 98.5\cos\dfrac{2\pi}{15}t$$

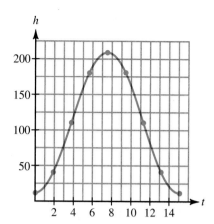

33. $39°$ **35.** $56°$ **37.** $84°$ **39.** $131.8°$ **41.** $205.5°$ **43.** $281.8°$

Problem Set 4.5

1.

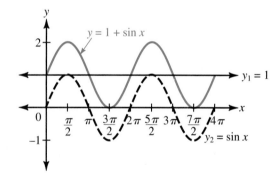

3.

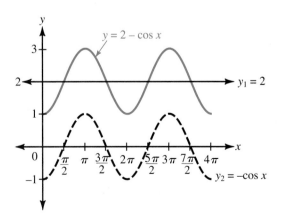

5.

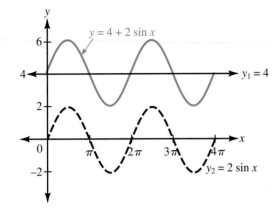

7.

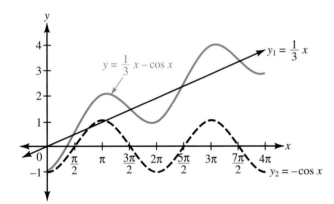

9.

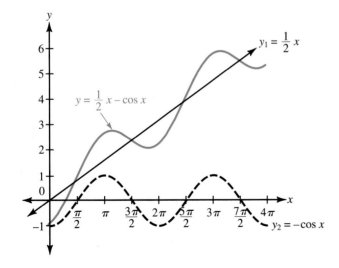

11.

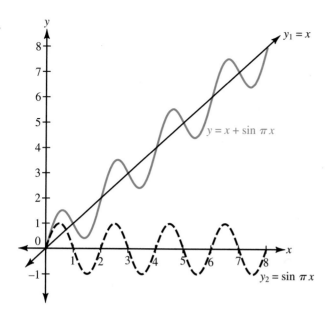

13.

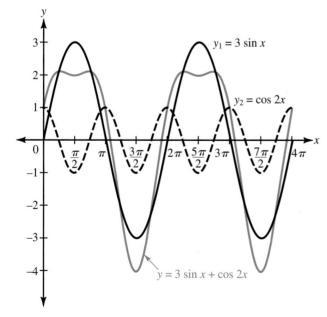

15.

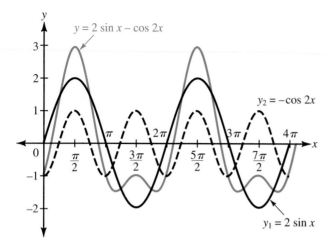

17.

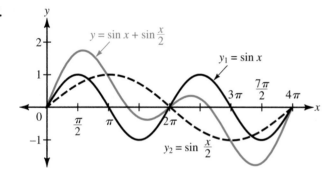

19.

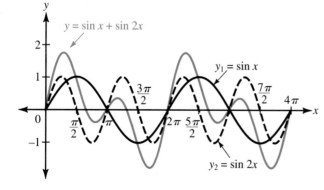

21.

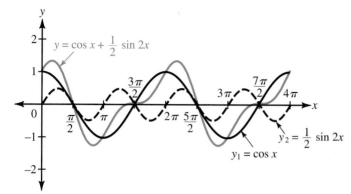

23.

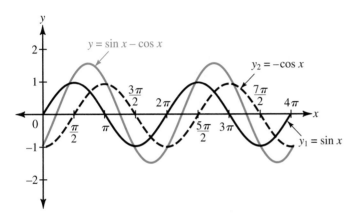

25.

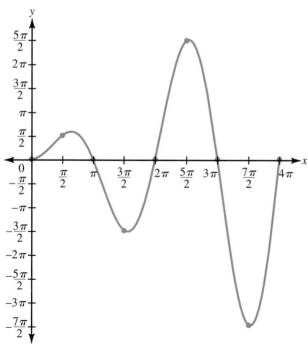

27. 1/4 ft/sec   **29.** 1,200 ft   **31.** 180 m
33. 4π rad/sec   **35.** 40π cm

Problem Set 4.6

1.

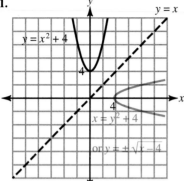

3.

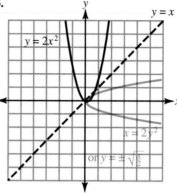

5.

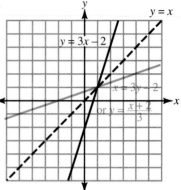

7.

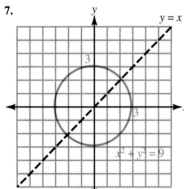

9.

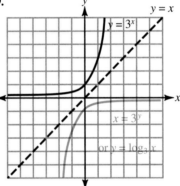

11.

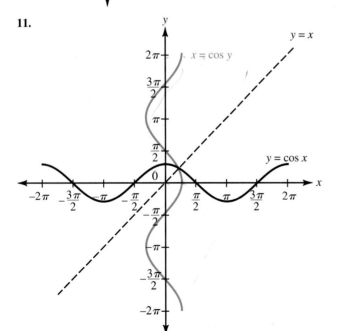

13. We have not included the graph of $y = \tan x$ with the graph of $x = \tan y$ because placing them both on the same coordinate system makes the diagram too complicated. It is best to graph $y = \tan x$ lightly in pencil and then reflect that graph about the line $y = x$ to get the graph of $x = \tan y$.

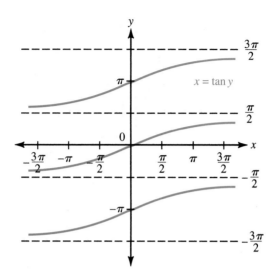

15. $-1 \le x \le 1$ **17.** $0 \le y \le \pi$ **19.** $\pi/3$ **21.** π **23.** $\pi/4$ **25.** $3\pi/4$ **27.** $-\pi/6$ **29.** $\pi/$
29. $\pi/3$ **31.** 0   **33.** $-\pi/6$ **35.** $2\pi/3$ **37.** $\pi/6$ **39.** $9.8°$ **41.** $147.4°$ **43.** $20.8°$
45. $74.3°$ **47.** $117.8°$ **49.** $-70.0°$ **51.** $-50.0°$ **53.** $4/5$ **55.** $3/4$ **57.** $\sqrt{5}$ **59.** $\sqrt{3}/2$

61. 2   **63.** 3/5   **65.** 1/2   **67.** 1/2   **69.** x **71.** $\sqrt{1 - x^2}$ **73.** $\dfrac{x}{\sqrt{x^2 + 1}}$ **75.** $\dfrac{\sqrt{x^2 - 1}}{x}$

77. x

79.

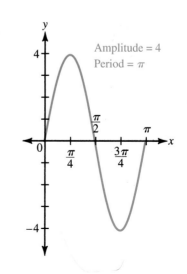

Amplitude = 4
Period = π

81.

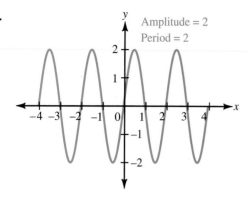

Amplitude = 2
Period = 2

83.

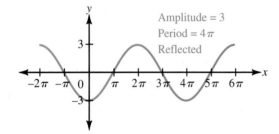

Amplitude = 3
Period = 4π
Reflected

85.

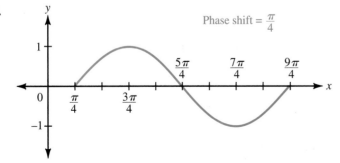

Phase shift = $\frac{\pi}{4}$

87.

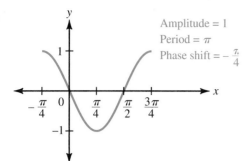

Amplitude = 1
Period = π
Phase shift = $-\dfrac{\pi}{4}$

89.

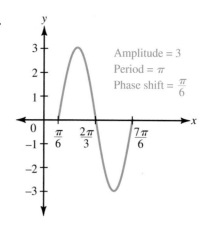

Amplitude = 3
Period = π
Phase shift = $\dfrac{\pi}{6}$

Chapter 4 Test

1.

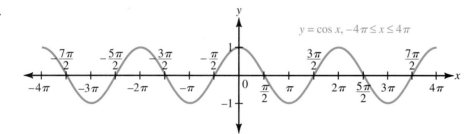

$y = \sin x,\ -4\pi \le x \le 4\pi$

2.

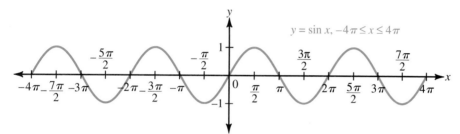

$y = \cos x,\ -4\pi \le x \le 4\pi$

3.

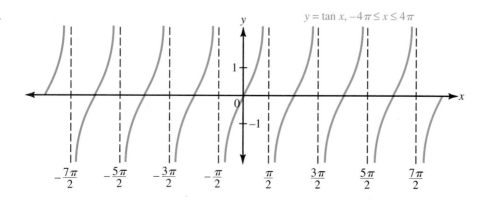

$y = \tan x,\ -4\pi \le x \le 4\pi$

4.

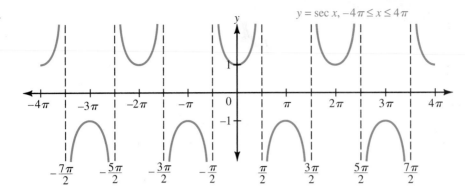

$y = \sec x,\ -4\pi \le x \le 4\pi$

5. 4   **6.** 8   **7.** $-3\pi, -\pi, \pi, 3\pi$ **8.** $-11\pi/3, -7\pi/3, -5\pi/3, -\pi/3, \pi/3, 5\pi/3, 7\pi/3, 11\pi/3$

9.

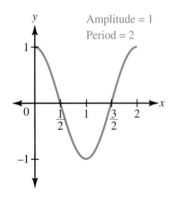

Amplitude = 1
Period = 2

10.

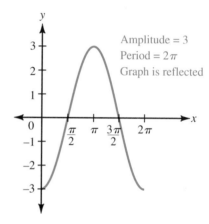

Amplitude = 3
Period = 2π
Graph is reflected

11.

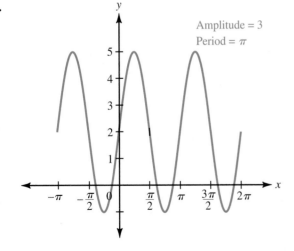

Amplitude = 3
Period = π

12.

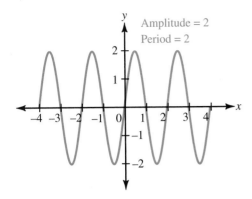

Amplitude = 2
Period = 2

13.

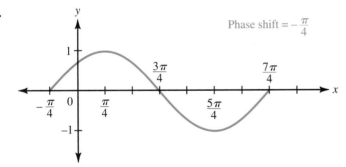

Phase shift = $-\dfrac{\pi}{4}$

14.

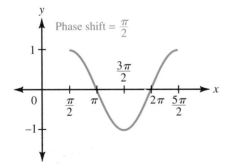

Phase shift = $\dfrac{\pi}{2}$

15.

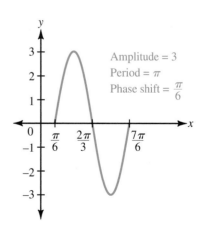

Amplitude = 3
Period = π
Phase shift = $\dfrac{\pi}{6}$

16.

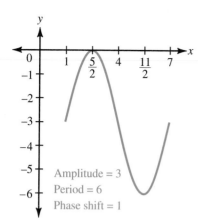

Amplitude = 3
Period = 6
Phase shift = 1

17.

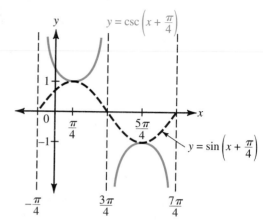

$y = \csc\left(x + \frac{\pi}{4}\right)$

$y = \sin\left(x + \frac{\pi}{4}\right)$

18.

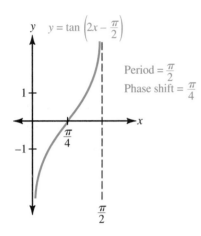

$y = \tan\left(2x - \frac{\pi}{2}\right)$

Period = $\frac{\pi}{2}$
Phase shift = $\frac{\pi}{4}$

19.

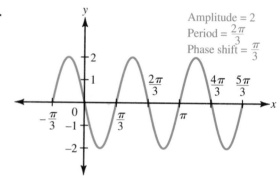

Amplitude = 2
Period = $\frac{2\pi}{3}$
Phase shift = $\frac{\pi}{3}$

20.

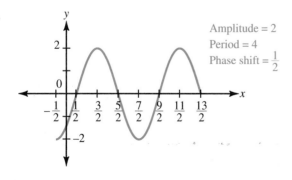

Amplitude = 2
Period = 4
Phase shift = $\frac{1}{2}$

21. $y = 2\sin\left(\frac{1}{2}x + \frac{\pi}{2}\right)$

22. $y = \frac{1}{2} + \frac{1}{2}\sin\frac{\pi}{2}x$

23.

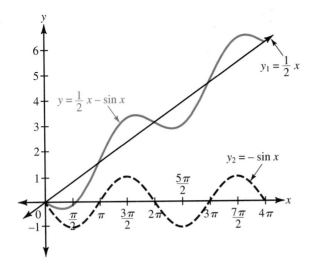

24.

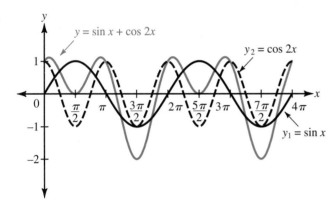

25.

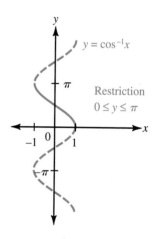

26.

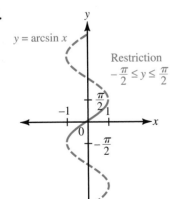

27. $\pi/4$ **28.** $5\pi/6$ **29.** $-\pi/4$ **30.** $\pi/2$ **31.** $36.4°$

$\pi/6$

32. $-39.7°$ **33.** $134.3°$ **34.** $-12.5°$ **35.** $\sqrt{5}/2$ **36.** $3/\sqrt{13}$ **37.** $\sqrt{1-x^2}$ **38.** $\dfrac{x}{\sqrt{1-x^2}}$

CHAPTER 5

Problem Set 5.1

For some of the problems in the beginning of this problem set we will give the complete proof. Remember, however, that there is often more than one way to prove an identity. You may have a correct proof even if it doesn't match the one you find here. As the problem set progresses, we will give hints on how to begin the proof instead of the complete proof.

1. $\cos \theta \tan \theta = \cos \theta \cdot \dfrac{\sin \theta}{\cos \theta}$

$\quad = \sin \theta$

9. $\cos x(\csc x + \tan x) = \cos x \csc x + \cos x \tan x$

$\quad = \cos x \cdot \dfrac{1}{\sin x} + \cos x \cdot \dfrac{\sin x}{\cos x}$

$\quad = \dfrac{\cos x}{\sin x} + \sin x$

$\quad = \cot x + \sin x$

17. $\dfrac{\cos^4 t - \sin^4 t}{\sin^2 t} = \dfrac{(\cos^2 t + \sin^2 t)(\cos^2 t - \sin^2 t)}{\sin^2 t}$

$\quad = \dfrac{\cos^2 t - \sin^2 t}{\sin^2 t}$

$\quad = \dfrac{\cos^2 t}{\sin^2 t} - \dfrac{\sin^2 t}{\sin^2 t}$

$\quad = \cot^2 t - 1$

19. Write the numerator on the right side as $1 - \sin^2 \theta$ and then factor it.
25. Factor the left side and then write in terms of sines and cosines.
27. Change the left side to sines and cosines and then add the resulting fractions.   **33.** See Example 6 in this section.
37. Rewrite the left side in terms of cosine and then simplify.   **81.** $\sin (30° + 60°) = \sin 90° = 1$

$\sin 30° + \sin 60° = \dfrac{1}{2} + \dfrac{\sqrt{3}}{2} \neq 1$

83. $\cos A = 4/5$, $\tan A = 3/4$ **85.** $\sqrt{3}/2$ **87.** $\sqrt{3}/2$ **89.** $15°$ **91.** $105°$

Problem Set 5.2

1. $\dfrac{\sqrt{6} - \sqrt{2}}{4}$ **3.** $\dfrac{\sqrt{6} - \sqrt{2}}{\sqrt{6} + \sqrt{2}}$ or $\dfrac{\sqrt{3} - 1}{\sqrt{3} + 1}$ **5.** $\dfrac{\sqrt{6} + \sqrt{2}}{4}$ **7.** $\dfrac{\sqrt{2} - \sqrt{6}}{4}$

9. $\sin (x + 2\pi) = \sin x \cos 2\pi + \cos x \sin 2\pi$

$\quad = \sin x(1) + \cos x(0)$

$\quad = \sin x$

For problems 11–19, proceed as in Problem 9. Expand the left side and simplify.

21. $\sin 5x$ **23.** $\cos 6x$ **25.** $\cos 90° = 0$

27.

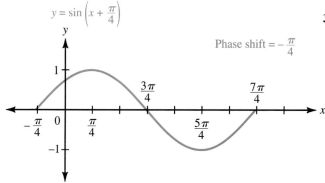

$y = \sin 2x$

29.

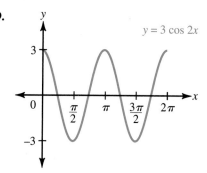

$y = 3 \cos 2x$

31.

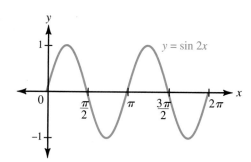

$y = \sin\left(x + \frac{\pi}{4}\right)$

Phase shift $= -\frac{\pi}{4}$

33.

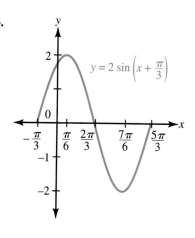

$y = 2 \sin\left(x + \frac{\pi}{3}\right)$

35. $-16/65,\ 63/65,\ -16/63,$ QIV   **37.** 2, 1/2, QI   **39.** 1   **41.** $\sin 2x = 2 \sin x \cos x$

63.

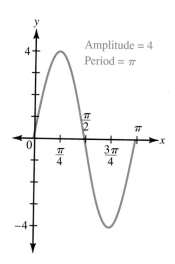

Amplitude $= 4$
Period $= \pi$

65.

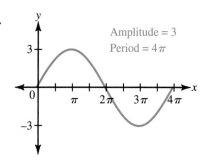

Amplitude $= 3$
Period $= 4\pi$

67.

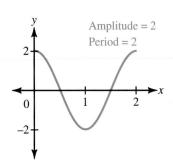

Amplitude = 2
Period = 2

69. See the solution to Problem 43 in Problem Set 4.2.

71.

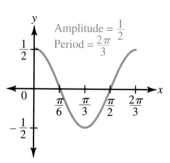

Amplitude = $\frac{1}{2}$
Period = $\frac{2\pi}{3}$

73.

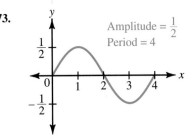

Amplitude = $\frac{1}{2}$
Period = 4

Problem Set 5.3

1. 24/25   **3.** 24/7   **5.** −4/5   **7.** 4/3   **9.** 120/169   **11.** 169/120   **13.** 3/5   **15.** 5/3

17. $y = 4 - 8 \sin^2 x$
$= 4(1 - 2 \sin^2 x)$
$= 4 \cos 2x$

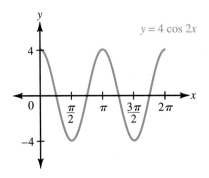

$y = 4 \cos 2x$

19. $y = 3 \cos 2x$ The graph will look like the graph for Problem 17 but with an amplitude of 3 instead of 4.
21. $y = \cos 4x$ The graph will be a cosine curve with amplitude 1 and period $\pi/2$.   **27.** 24/7

29. $\sin 30° = 1/2$ **31.** $\cos 150° = -\sqrt{3}/2$ **33.** $\frac{1}{2} \sin \frac{\pi}{6} = \frac{1}{4}$ **35.** $\frac{1}{2} \tan 45° = \frac{1}{2}$

59. $\frac{1}{2} \left(\tan^{-1} \frac{x}{5} - \frac{5x}{x^2 + 25} \right)$ **61.** $\frac{1}{2} \left(\sin^{-1} \frac{x}{3} - \frac{x\sqrt{9 - x^2}}{9} \right)$

63.

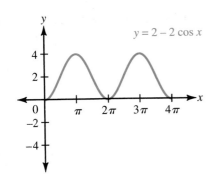

65.

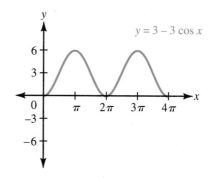

67. See the solution to Problem 21 in Problem Set 4.5.

69.

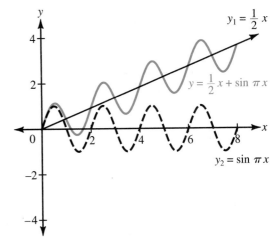

Problem Set 5.4

1. $1/2$ **3.** 2 **5.** $-1/\sqrt{10}$ **7.** $-\sqrt{10}$ **9.** $\sqrt{\dfrac{3 + 2\sqrt{2}}{6}}$ **11.** $-\sqrt{\dfrac{3 - 2\sqrt{2}}{6}}$ **13.** $-3 - 2\sqrt{2}$

15. $2/\sqrt{5}$ **17.** $-7/25$ **19.** $-25/7$ **21.** $3/\sqrt{10}$ **23.** $7/25$ **25.** 0

27. See the solution to Problem 63 in Problem Set 5.3.   **29.**

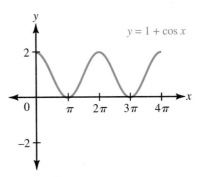

31. $\dfrac{\sqrt{2 + \sqrt{3}}}{2}$ **33.** $\dfrac{\sqrt{2 + \sqrt{3}}}{2}$ **35.** $-\dfrac{\sqrt{2 - \sqrt{3}}}{2}$ **49.** $3/5$ **51.** $1/\sqrt{5}$ **53.** $x/\sqrt{x^2 + 1}$

55. $x/\sqrt{1 - x^2}$ **57.** See Figure 4 in Section 4.6.

Problem Set 5.5

1. $-1/\sqrt{5}$ **3.** $(2\sqrt{3} - 1)/2\sqrt{5}$ **5.** $4/5$ **7.** $x/\sqrt{1 - x^2}$ **9.** $2x\sqrt{1 - x^2}$ **11.** $2x^2 - 1$

15. $5(\sin 8x + \sin 2x)$ **17.** $\dfrac{1}{2}(\cos 10x + \cos 6x)$ **19.** $\dfrac{1}{2}(\sin 90° + \sin 30°) = \dfrac{1}{2}\left(1 + \dfrac{1}{2}\right) = \dfrac{3}{4}$

21. $\dfrac{1}{2}(\cos 2\pi - \cos 6\pi) = \dfrac{1}{2}(1 - 1) = 0$ **25.** $2 \sin 5x \cos 2x$ **27.** $2 \cos 30° \cos 15° = \sqrt{3} \cos 15°$

29. $2 \cos \dfrac{\pi}{3} \sin \dfrac{\pi}{4} = 1/\sqrt{2}$

37.

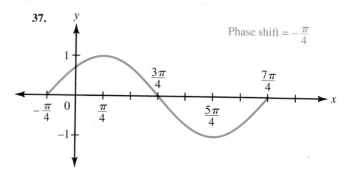

Phase shift $= -\dfrac{\pi}{4}$

39.

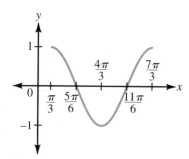

41.

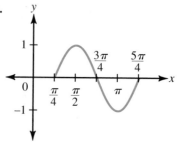

43.

Amplitude $= \dfrac{1}{2}$

Period $= \dfrac{2\pi}{3}$

Phase shift $= \dfrac{\pi}{6}$

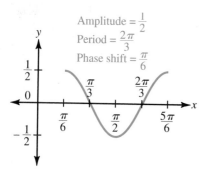

45.

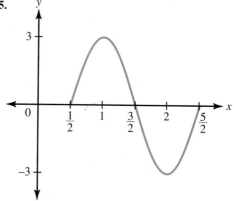

Chapter 5 Test

13. $63/65$ **14.** $-56/65$ **15.** $-119/169$ **16.** $-120/169$ **17.** $1/\sqrt{10}$ **18.** $-3/\sqrt{10}$

19. $\dfrac{\sqrt{6} + \sqrt{2}}{4}$ **20.** $\dfrac{\sqrt{6} + \sqrt{2}}{4}$ **21.** $\dfrac{\sqrt{6} - \sqrt{2}}{\sqrt{6} + \sqrt{2}}$ or $\dfrac{\sqrt{3} - 1}{\sqrt{3} + 1}$ **22.** $\dfrac{\sqrt{6} + \sqrt{2}}{\sqrt{6} - \sqrt{2}}$ or $\dfrac{\sqrt{3} + 1}{\sqrt{3} - 1}$ **23.** $\cos 9x$

24. $\sin 90° = 1$ **25.** $3/5, -\sqrt{\dfrac{5 - 2\sqrt{5}}{10}}$ **26.** $3/5, \sqrt{\dfrac{10 - \sqrt{10}}{20}}$ **27.** 1   **28.** $\pm\sqrt{3}/2$ **29.** $11/5\sqrt{5}$

30. $11/5\sqrt{5}$ **31.** $1 - 2x^2$ **32.** $2x\sqrt{1 - x^2}$ **33.** $\dfrac{1}{2}(\cos 2x - \cos 10x)$ **34.** $2 \cos 45° \cos(-30°) = \sqrt{6}/2$

CHAPTER 6

Problem Set 6.1

1. $30°, 150°$ **3.** $30°, 330°$ **5.** $135°, 315°$ **7.** $\pi/3, 2\pi/3$ **9.** $\pi/6, 11\pi/6$ **11.** $3\pi/2$
13. $48.6°, 131.4°$ **15.** $\varnothing$ **17.** $228.6°, 311.4°$ **19.** $\pi/2, \pi/6, 5\pi/6$ **21.** $0, \pi/4, \pi, 5\pi/4$
23. $0, 2\pi/3, \pi, 4\pi/3$ **25.** $\pi/2, 7\pi/6, 11\pi/6$ **27.** $120°, 150°, 210°, 240°$ **29.** $0°, 60°, 180°, 120°$
31. $120°, 240°$ **33.** $201.5°, 338.5°$ **35.** $51.8°, 308.2°$ **37.** $17.0°, 163°$ **39.** $30° + 360°k, 150° + 360°k$
41. $\pi/3 + 2k\pi, 2\pi/3 + 2k\pi$ **43.** $3\pi/2 + 2k\pi$ **45.** $48.6° + 360°k, 131.4° + 360°k$
47. $40° + 180°k, 190° + 180°k$ **49.** $120°k, 40° + 120°k$ **51.** $35° + 90°k, 65° + 90°k$
53. $42° + 72°k, 60° + 72°k$ **55.** $h = -16t^2 + 750t$ **57.** $h = -16(2)^2 + 750(2) = 1{,}436$ ft   **59.** $15.7°$
61. $\sin 2A = 2 \sin A \cos A$ **63.** $\cos 2A = 2 \cos^2 A - 1$

65. $\sin \theta \cos 45° + \cos \theta \sin 45° = \dfrac{1}{\sqrt{2}} \sin \theta + \dfrac{1}{\sqrt{2}} \cos \theta$ **67.** $\dfrac{\sqrt{6} + \sqrt{2}}{4}$

Problem Set 6.2

1. $30°, 330°$ **3.** $225°, 315°$ **5.** $45°, 135°, 225°, 315°$ **7.** $30°, 150°$ **9.** $30°, 90°, 150°, 270°$
11. $60°, 180°, 300°$ **13.** $7\pi/6, 3\pi/2, 11\pi/6$ **15.** $0, 2\pi/3, 4\pi/3$ **17.** $\pi/2, 7\pi/6, 11\pi/6$ **19.** $\pi/3, 5\pi/3$
21. $2\pi/3, 4\pi/3$ **23.** $\pi/4$ **25.** $30°, 90°$ **27.** $60°, 180°$ **29.** $60°, 300°$ **31.** $120°, 180°$
33. $210°, 330°$ **35.** $36.9°, 48.2°, 311.8°, 323.1°$ **37.** $36.9°, 143.1°, 216.9°, 323.1°$

39. $225° + 360°k, 315° + 360°k$ **41.** $\dfrac{\pi}{4} + 2k\pi$ **43.** $120° + 360°k, 180° + 360°k$ **47.** $68.5°, 291.5°$

49. $218.2°, 321.8°$ **51.** $73.0°, 287.0°$ **53.** $\sqrt{\dfrac{3 - \sqrt{5}}{6}}$ **55.** $\sqrt{\dfrac{6}{3 - \sqrt{5}}}$ **57.** $\sqrt{\dfrac{3 - \sqrt{5}}{3 + \sqrt{5}}}$ or $\dfrac{3 - \sqrt{5}}{2}$

59. See the solution to Problem 63 in Problem Set 5.3.   **61.** $\dfrac{\sqrt{2 - \sqrt{2}}}{2}$

Problem Set 6.3

1. $30°, 60°, 210°, 240°$ **3.** $67.5°, 157.5°, 247.5°, 337.5°$ **5.** $60°, 180°, 300°$ **7.** $\pi/8, 3\pi/8, 9\pi/8, 11\pi/8$
9. $\pi/3, \pi, 5\pi/3$ **11.** $\pi/6, 2\pi/3, 7\pi/6, 5\pi/3$ **13.** $15° + 180°k, 75° + 180°k$
15. $30° + 120°k, 90° + 120°k$ **17.** $6° + 36°k, 12° + 36°k$
19. $\pi/18, 5\pi/18, 13\pi/18, 17\pi/18, 25\pi/18, 29\pi/18$ **21.** $5\pi/18, 7\pi/18, 17\pi/18, 19\pi/18, 29\pi/18, 31\pi/18$

23. $\dfrac{\pi}{10} + \dfrac{2k\pi}{5}$ **25.** $\dfrac{\pi}{8} + \dfrac{k\pi}{2}, \dfrac{3\pi}{8} + \dfrac{k\pi}{2}$ **27.** $\dfrac{\pi}{5} + \dfrac{2k\pi}{5}$ **29.** $10° + 120°k, 50° + 120°k, 90° + 120°k$

31. $60° + 180°k, 90° + 180°k, 120° + 180°k$ **33.** $20° + 60°k, 40° + 60°k$ **35.** $0°, 270°$ **37.** $180°, 270°$

39. $96.8°, 173.2°, 276.8°, 353.2°$ **41.** $27.4°, 92.6°, 147.4°, 212.6°, 267.4°, 332.6°$

43. $50.4°, 84.6°, 140.4°, 174.6°, 230.4°, 264.6°, 320.4°, 354.6°$ **45.** 4.0 min and 16.0 min   **47.** 6

49. 1/4 second (and every second after that)   **51.** 1/12   **59.** $-4\sqrt{2}/9$

61. $\sqrt{\dfrac{3 - 2\sqrt{2}}{6}}$ **63.** $\dfrac{4 - 6\sqrt{2}}{15}$ **65.** $\dfrac{15}{4 - 6\sqrt{2}}$

Problem Set 6.4

1.

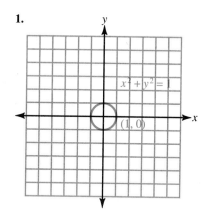

3.

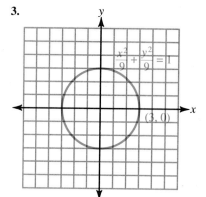

5.

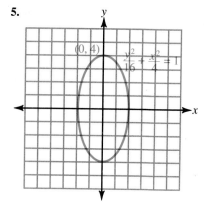

7.

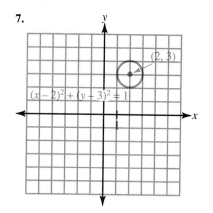

9.

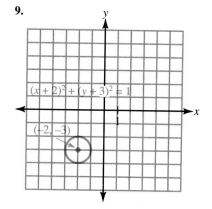

11.

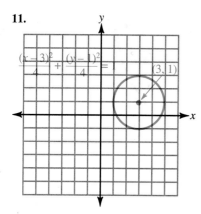

$$\frac{(x-3)^2}{4} + \frac{(y-1)^2}{4} = 1$$

(3, 1)

13.

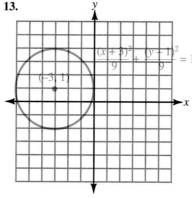

$$\frac{(x+3)^2}{9} + \frac{(y-1)^2}{9} = 1$$

(−3, 1)

15. $x^2 - y^2 = 1$

17. $\dfrac{x^2}{9} - \dfrac{y^2}{9} = 1$

19. $\dfrac{(y-4)^2}{9} - \dfrac{(x-2)^2}{9} = 1$

21. $x = 1 - 2y^2$ **23.** $y = x$ **25.** $2x = 3y$

27. $x = 98.5 \sin \theta = 98.5 \sin \dfrac{2\pi}{15} t$; $y = 110.5 - 98.5 \cos \theta = 110.5 - 98.5 \cos \dfrac{2\pi}{15} t$ **29.** $\sqrt{3}/2$ **31.** $\dfrac{\sqrt{15}+3}{4\sqrt{10}}$

33. $\sqrt{1-x^2}$ **35.** $\dfrac{1-x^2}{1+x^2}$ **37.** $4(\sin 5x + \sin x)$

Chapter 6 Test

1. $30°, 150°$ **2.** $150°, 330°$ **3.** $30°, 90°, 150°, 270°$ **4.** $0°, 60°, 180°, 300°$ **5.** $45°, 135°, 225°, 315°$
6. $90°, 210°, 330°$ **7.** $180°$ **8.** $0°, 240°$ **9.** $48.6°, 131.4°, 210°, 330°$
10. $95° + 120°k, 115° + 120°k$ where $k = 0, 1, 2$ **11.** $0°, 90°$ **12.** $90°, 180°$
13. $2k\pi, \dfrac{\pi}{3} + 2k\pi, \dfrac{5\pi}{3} + 2k\pi$ **14.** $\dfrac{\pi}{6} + 2k\pi, \dfrac{7\pi}{6} + 2k\pi$ **15.** $\dfrac{\pi}{2} + \dfrac{2k\pi}{3}$ **16.** $\dfrac{\pi}{8} + \dfrac{k\pi}{2}$

17. $90°, 203.6°, 336.4°$ **18.** $111.5°, 248.5°$ **19.**

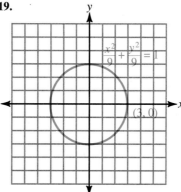

$$\frac{x^2}{9} + \frac{y^2}{9} = 1$$

(3, 0)

20.

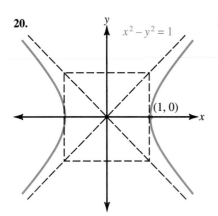

$x^2 - y^2 = 1$

$(1, 0)$

21.

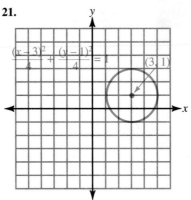

$$\frac{(x-3)^2}{4} + \frac{(y-1)^2}{4} = 1$$

$(3, 1)$

22.

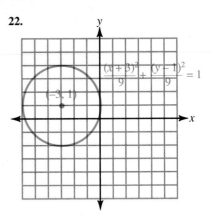

$$\frac{(x+3)^2}{9} + \frac{(y-1)^2}{9} = 1$$

$(-3, 1)$

CHAPTER 7

Problem Set 7.1

1. 16 cm   **3.** 71 inches   **5.** 140 yd   **7.** $C = 70°$, $c = 44$ km   **9.** $C = 80°$, $a = 11$ cm
11. $C = 66.1°$, $b = 302$ inches, $c = 291$ inches   **13.** $C = 39°$, $a = 7.8$ m, $b = 11$ m
15. $B = 16°$, $b = 1.39$ ft, $c = 4.36$ ft   **17.** $A = 141.8°$, $b = 118$ cm, $c = 214$ cm
19. $\sin B = 5$, which is impossible   **21.** 11   **23.** 20   **25.** 209 ft   **27.** 273 ft   **29.** 5,900 ft
31. 42 ft   **33.** 14 mi, 9.3 mi   **35.** $|\mathbf{CB}| = 341$ lbs $|\mathbf{CA}| = 345$ lbs   **37.** $|\mathbf{CA}| = 3{,}243$ $|\mathbf{CB}| = 3{,}193$
39. $45°, 135°$ **41.** $90°, 270°$ **43.** $30°, 90°, 150°$ **45.** $54.1°, 305.9°$

47. $\dfrac{\pi}{6} + 2k\pi, \dfrac{3\pi}{2} + 2k\pi, \dfrac{5\pi}{6} + 2k\pi$ **49.** $47.6°, 132.4°$ **51.** $75.2°, 104.8°$

Problem Set 7.2

1. $\sin B = 2$ is impossible   **3.** $B = 35.3$ is the only possibility for B **5.** $B = 77°$ or $B' = 103°$
7. $B = 54°$, $C = 88°$, $c = 67$ ft or $B' = 126°$, $C' = 16°$, $c' = 18$ ft   **9.** $B = 28.1°$, $C = 39.7°$, $c = 30.2$ cm
11. $B = 34°\,50'$, $A = 117°\,20'$, $a = 660$ m or $B' = 145°\,10'$, $A' = 7°$, $a' = 90.6$ m
13. $C = 26°\,20'$, $A = 108°\,30'$, $a = 2.39$ inches   **15.** no solution   **17.** no solution
19. $B = 26.8°$, $A = 126.4°$, $a = 65.7$ km   **21.** 15 ft or 38 ft   **23.** 310 mph or 360 mph
25. Yes, it makes an angle of $88°$ with the ground.   **27.** $30°, 150°, 210°, 330°$ **29.** $90°, 270°$

31. $41.8°, 48.6°, 131.4°, 138.2°$ **33.** $\dfrac{\pi}{2} + 2k\pi, \dfrac{7\pi}{6} + 2k\pi, \dfrac{11\pi}{6} + 2k\pi$ **35.** $\dfrac{3\pi}{4} + 2k\pi, \dfrac{7\pi}{4} + 2k\pi$

Problem Set 7.3

1. 100 inches   **3.** $C = 69°$ **5.** 9.4 m   **7.** $A = 128°$ **9.** $A = 15.6°$, $C = 12.9°$, $b = 727$ m
11. $A = 39°$, $B = 57°$, $C = 84°$ **13.** $B = 114°\,10'$, $C = 22°\,30'$, $a = 0.694$ km
15. $A = 55.4°$, $B = 45.5°$, $C = 79.1°$ **17.** $a^2 = b^2 + c^2 - 2bc \cos 90° = b^2 + c^2 - 2bc(0) = b^2 + c^2$
19. 24 inches   **21.** 130 mi   **23.** 462 mi   **25.** 190 mph with heading $153°$

27. 18.5 mph at 81.3° from due north   **29.** $\dfrac{\pi}{18} + \dfrac{2k\pi}{3}, \dfrac{5\pi}{18} + \dfrac{2k\pi}{3}$ **31.** $\dfrac{\pi}{12} + \dfrac{k\pi}{3}, \dfrac{\pi}{4} + \dfrac{k\pi}{3}$

33. $20° + 120°k, 100° + 120°k$ **35.** $45° + 60°k$ **37.** $0°, 90°$

Problem Set 7.4

The answers to problems 1 through 21 have been rounded to three significant digits.
1. $1,520 \text{ cm}^2$ **3.** 342 m^2 **5.** 0.123 km^2 **7.** 26.3 m^2 **9.** $28,300 \text{ inches}^2$ **11.** 2.09 ft^2
13. $1,410 \text{ inches}^2$ **15.** 15.0 yd^2 **17.** 8.15 ft^2 **19.** 156 inches^2 **21.** 14.3 cm
23. See the solution to Problem 1 in Problem Set 6.4.   **25.** See the solution to Problem 11 in Problem Set 6.4.

27. $\dfrac{y^2}{9} - \dfrac{x^2}{4} = 1$ **29.** $y = 1 - 2x^2$

Problem Set 7.5

1. $\mathbf{V} = 4\mathbf{i} + 4\mathbf{j}$ **3.** $\mathbf{V} = 2\mathbf{i} + 5\mathbf{j}$ **5.** $\mathbf{V} = -3\mathbf{i} + 6\mathbf{j}$ **7.** $\mathbf{V} = 4\mathbf{i} - 5\mathbf{j}$ **9.** $\mathbf{V} = -\mathbf{i} - 5\mathbf{j}$
11. $|\mathbf{V}| = 5$ **13.** $|\mathbf{U}| = 13$ **15.** $|\mathbf{W}| = \sqrt{5}$ **17.** $\sqrt{61}$ **19.** 2 **21.** $\sqrt{29}$
23. $2\mathbf{i}, 2\mathbf{j}, 5\mathbf{i} + \mathbf{j}, 0$ **25.** $6\mathbf{i} - 8\mathbf{j}, 6\mathbf{i} + 8\mathbf{j}, 18\mathbf{i} - 16\mathbf{j}, 0$
27. $7\mathbf{i} + 7\mathbf{j}, -3\mathbf{i} + 3\mathbf{j}, 16\mathbf{i} + 19\mathbf{j}, 20$ **29.** 48 **31.** -369 **33.** $\theta = 90.0°$ **35.** $\theta = 81.1°$
37. $\theta = 111.3°$ **39.** $\langle 1, 0 \rangle \bullet \langle 0, 1 \rangle = 0$ **41.** $\langle -1, 0 \rangle \bullet \langle 0, 1 \rangle = 0$
43. $\langle a, b \rangle \bullet \langle -b, a \rangle = -ab + ab = 0$ **47.** $|\mathbf{N}| = 9.7 \text{ lbs}$ $|\mathbf{F}| = 2.6 \text{ lbs}$ **49.** $|\mathbf{H}| = 45.6 \text{ lbs}$
51. $|\mathbf{T_1}| = 16 \text{ lbs}$ $|\mathbf{T_2}| = 20 \text{ lbs}$

Chapter 7 Test

1. 6.7 inches   **2.** 4.3 inches   **3.** $C = 78.4°, a = 26.5 \text{ cm}, b = 38.3 \text{ cm}$
4. $B = 49.2°, a = 18.8 \text{ cm}, c = 43.2 \text{ cm}$ **5.** $\sin B = 3.0311$, which is impossible
6. $B = 29°$ is the only possibility for B **7.** $B = 71°, C = 58°, c = 7.1 \text{ ft}$ or $B' = 109°, C' = 20°, c' = 2.9 \text{ ft}$
8. $B = 59°, C = 95°, c = 11 \text{ ft}$ or $B' = 121°, C' = 33°, c' = 6.0 \text{ ft}$ **9.** 11 cm   **10.** 19 cm   **11.** $95.7°$
12. $69.5°$ **13.** $A = 43°, B = 18°, c = 8.1 \text{ cm}$ **14.** $B = 34°, C = 111°, a = 3.8 \text{ m}$ **15.** $51°$ **16.** 59 ft
17. 410 ft   **18.** 14.1 m   **19.** 145 mi   **20.** 4.2 mi, S 75° W   **21.** 90 ft   **22.** 300 mph or 387 mph
23. 65 ft   **24.** 390 mph at N 89.2° W   **25.** 13   **26.** $-5\mathbf{i} + 41\mathbf{j}$ **27.** $35\mathbf{i} + 31\mathbf{j}$ **28.** $\sqrt{117} \approx 10.8$
29. -8 **30.** $98.6°$ **31.** 498 cm^2 **32.** 307 cm^2 **33.** 52 cm^2 **34.** 52 cm^2 **35.** 17 km^2
36. 52 km^2

CHAPTER 8

Problem Set 8.1

1. $4i$ **3.** $11i$ **5.** $3i\sqrt{2}$ **7.** $2i\sqrt{2}$ **9.** -6 **11.** -3 **13.** $x = 2/3, y = -1/2$
15. $x = 2/5, y = -4$ **17.** $x = -2$ or $3, y = \pm 3$ **19.** $x = \pi/4$ or $5\pi/4, y = \pi/2$
21. $x = \pi/2, y = \pi/4$ or $5\pi/4$ **23.** $10 - 2i$ **25.** $2 + 6i$ **27.** $3 - 13i$ **29.** $5 \cos x - 3i \sin y$
31. $2 + 2i$ **33.** $12 + 2i$ **35.** 1 **37.** -1 **39.** 1 **41.** i **43.** $-48 - 18i$ **45.** $10 - 10i$

47. $5 + 12i$ **49.** 41 **51.** 53 **53.** $-28 + 4i$ **55.** -6 **57.** $\dfrac{1}{5} + \dfrac{3}{5}i$ **59.** $-\dfrac{5}{13} + \dfrac{12}{13}i$

61. $-2 - 5i$ **63.** $\dfrac{4}{61} + \dfrac{17}{61}i$ **65.** 13   **67.** $-7 + 22i$ **69.** $10 - 3i$ **71.** $16 + 20i$ **73.** $x^2 + 9$

79. $x = 4 + 2i, y = 4 - 2i;\ \ x = 4 - 2i, y = 4 + 2i$

81. $x = 1 + i\sqrt{3}, y = 2 - 2i\sqrt{3};\ \ x = 1 - i\sqrt{3}, y = 2 + 2i\sqrt{3}$ **85.** yes   **87.** $\sin\theta = -\frac{4}{5}, \cos\theta = \frac{3}{5}$

89. $\sin\theta = \dfrac{b}{\sqrt{a^2 + b^2}}, \cos\theta = \dfrac{a}{\sqrt{a^2 + b^2}}$ **91.** $135°$ **93.** $B = 69.6°, C = 37.3°, a = 248$ cm

95. $A = 40.5°, B = 61.3°, C = 78.2°$

Problem Set 8.2

1.
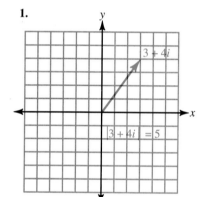

3.
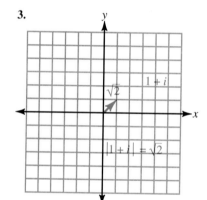

5.
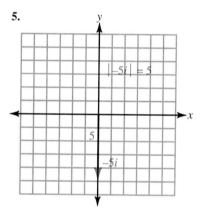

7.
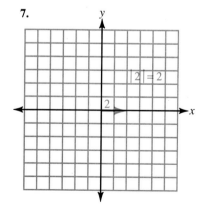

9.
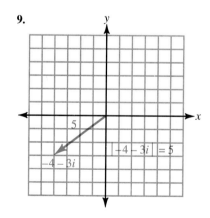

11.
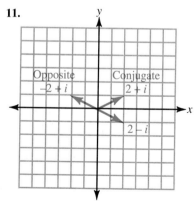

13.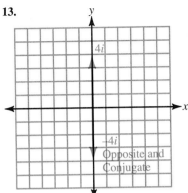

15.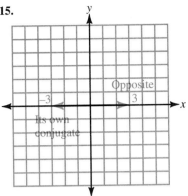

17.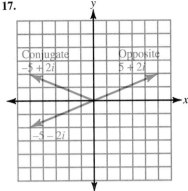

19. $\sqrt{3} + i$ **21.** $-2 + 2i\sqrt{3}$ **23.** $-\dfrac{\sqrt{3}}{2} - \dfrac{1}{2}i$ **25.** $\dfrac{\sqrt{2}}{2} - \dfrac{\sqrt{2}}{2}i$ **27.** $9.78 + 2.08i$

29. $-79.86 + 60.18i$ **31.** $-0.91 - 0.42i$ **33.** $9.51 - 3.09i$ **35.** $\sqrt{2}(\cos 135° + i \sin 135°)$
37. $\sqrt{2}(\cos 315° + i \sin 315°)$ **39.** $3\sqrt{2}(\cos 45° + i \sin 45°)$ **41.** $8(\cos 90° + i \sin 90°)$
43. $9(\cos 180° + i \sin 180°)$ **45.** $4(\cos 120° + i \sin 120°)$ **47.** $5(\cos 53.13° + i \sin 53.13°)$
49. $29(\cos 46.40° + i \sin 46.40°)$ **51.** $25(\cos 286.26° + i \sin 286.26°)$ **53.** $5\sqrt{5}(\cos 10.30° + i \sin 10.30°)$
55. $(2i)(3i) = [2(\cos 90° + i \sin 90°)][3(\cos 90° + i \sin 90°)]$
$\qquad = 2 \cdot 3(\cos 90° \cos 90° + i \cos 90° \sin 90° + i \sin 90° \cos 90° + i^2 \sin 90° \sin 90°)$
$\qquad = 6[0 \cdot 0 + i \cdot 0 \cdot 1 + i \cdot 1 \cdot 0 + (-1)(1)(1)]$
$\qquad = 6(-1)$
$\qquad = -6$

57. $2(\cos 30° + i \sin 30°) = 2\left(\dfrac{\sqrt{3}}{2} + i\dfrac{1}{2}\right) = \sqrt{3} + i;\quad 2[\cos(-30°) + i \sin(-30°)] = 2\left(\dfrac{\sqrt{3}}{2} - i\dfrac{1}{2}\right) = \sqrt{3} - i$

59. If $z = \cos \theta + i \sin \theta$, then $|z| = \sqrt{\cos^2 \theta + \sin^2 \theta} = \sqrt{1} = 1$ **61.** $\cos 75° = \cos(45° + 30°)$
$\qquad\qquad\qquad\qquad\qquad\qquad\qquad\qquad\qquad\qquad\qquad\qquad\qquad = \cos 45° \cos 30° - \sin 45° \sin 30°$
$\qquad\qquad\qquad\qquad\qquad\qquad\qquad\qquad\qquad\qquad\qquad\qquad\qquad = \dfrac{\sqrt{6} - \sqrt{2}}{4}$

63. $\sin(A + B) = \sin A \cos B + \cos A \sin B$ **65.** $\sin(30° + 90°) = \sin 120° = \sqrt{3}/2$
$\qquad\qquad = \dfrac{3}{5} \cdot \dfrac{12}{13} + \dfrac{4}{5} \cdot \dfrac{5}{13} = \dfrac{56}{65}$

67. $\cos(18° + 32°) = \cos 50°$ **69.** No triangle exists.   **71.** $B = 62.7°$ or $B = 117.3°$

Problem Set 8.3

1. $12(\cos 50° + i \sin 50°)$ **3.** $56(\cos 157° + i \sin 157°)$ **5.** $4(\cos 180° + i \sin 180°)$
7. $2(\cos 180° + i \sin 180°) = -2$ **9.** $4(\cos 210° + i \sin 210°) = -2\sqrt{3} - 2i$ **11.** $12(\cos 360° + i \sin 360°) = 12$
13. $4\sqrt{2}(\cos 135° + i \sin 135°) = -4 + 4i$ **15.** $10(\cos 240° + i \sin 240°) = -5 - 5i\sqrt{3}$

17. $64(\cos 60° + i \sin 60°) = 32 + 32i\sqrt{3}$ **19.** $\cos 120° + i \sin 120° = -\dfrac{1}{2} + \dfrac{\sqrt{3}}{2}i$

21. $81(\cos 240° + i \sin 240°) = -\dfrac{81}{2} - \dfrac{81\sqrt{3}}{2}i$ **23.** $32(\cos 450° + i \sin 450°) = 32i$

25. $(1 + i)^4 = [\sqrt{2}(\cos 45° + i \sin 45°)]^4 = 4(\cos 180° + i \sin 180°) = -4$ **27.** $-8 - 8i\sqrt{3}$ **29.** $-8i$

31. $16 + 16i$ **33.** $4(\cos 35° + i \sin 35°)$ **35.** $1.5(\cos 19° + i \sin 19°)$ **37.** $0.5(\cos 60° + i \sin 60°)$

39. $2(\cos 0° + i \sin 0°) = 2$ **41.** $\cos(-60°) + i \sin(-60°) = \dfrac{1}{2} - \dfrac{\sqrt{3}}{2}i$

43. $2[\cos(-270°) + i \sin(-270°)] = 2i$ **45.** $2[\cos(-180°) + i \sin(-180°)] = -2$ **47.** $-4 - 4i$ **49.** 8

51. $[2(\cos 60° + i \sin 60°)]^2 - 2[2(\cos 60° + i \sin 60°)] + 4 = 0$

$$4(\cos 120° + i \sin 120°) - 4(\cos 60° + i \sin 60°) + 4 = 0$$
$$-2 + 2i\sqrt{3} - 2 - 2i\sqrt{3} + 4 = 0$$
$$0 = 0$$

53. $w^4 = [2(\cos 15° + i \sin 15°)]^4 = 16(\cos 60° + i \sin 60°) = 8 + 8i\sqrt{3}$

55. $(1 + i)^{-1} = [\sqrt{2}(\cos 45° + i \sin 45°)]^{-1}$ **57.** $\dfrac{\sqrt{3}}{4} + \dfrac{1}{4}i$ **59.** $-7/9$ **61.** $\sqrt{6}/3$ **63.** $\sqrt{6}/2$

$$= \sqrt{2}^{-1}[\cos(-45°) + i \sin(-45°)]$$
$$= \frac{1}{\sqrt{2}}\left(\frac{1}{\sqrt{2}} - \frac{1}{\sqrt{2}}i\right) = \frac{1}{2} - \frac{1}{2}i$$

65. $4\sqrt{2}/7$ **67.** 6.0 mi   **69.** $103°$ at 160 mph

Problem Set 8.4

1.

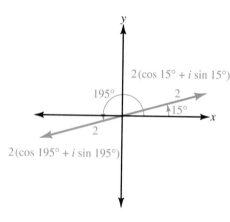

3.

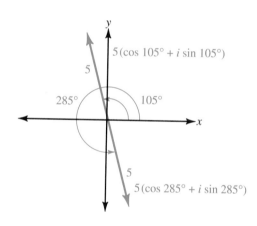

5.

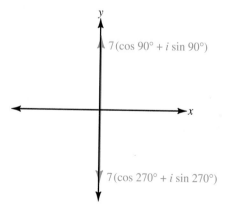

7. $\sqrt{3} + i, -\sqrt{3} - i$ **9.** $\sqrt{2} + i\sqrt{2}, -\sqrt{2} - i\sqrt{2}$ **11.** $5i, -5i$ **13.** $\dfrac{\sqrt{6}}{2} + \dfrac{\sqrt{2}}{2}i, -\dfrac{\sqrt{6}}{2} - \dfrac{\sqrt{2}}{2}i$

15. $2(\cos 70° + i \sin 70°), 2(\cos 190° + i \sin 190°), 2(\cos 310° + i \sin 310°)$
17. $2(\cos 10° + i \sin 10°), 2(\cos 130° + i \sin 130°), 2(\cos 250° + i \sin 250°)$
19. $3(\cos 60° + i \sin 60°), 3(\cos 180° + i \sin 180°), 3(\cos 300° + i \sin 300°)$

21. $4(\cos 30° + i \sin 30°), 4(\cos 150° + i \sin 150°), 4(\cos 270° + i \sin 270°)$ **23.** $3, -\dfrac{3}{2} + \dfrac{3\sqrt{3}}{2}i, -\dfrac{3}{2} - \dfrac{3\sqrt{3}}{2}i$

25. $2, -2, 2i, -2i$ **27.** $\sqrt{3} + i, -1 + i\sqrt{3}, -\sqrt{3} - i, 1 - i\sqrt{3}$ **29.**

$$10(\cos 3° + i \sin 3°) \approx 9.99 + 0.52i$$
$$10(\cos 75° + i \sin 75°) \approx 2.59 + 9.66i$$
$$10(\cos 147° + i \sin 147°) \approx -8.39 + 5.45i$$
$$10(\cos 219° + i \sin 219°) \approx -7.77 - 6.29i$$
$$10(\cos 291° + i \sin 291°) \approx 3.58 - 9.34i$$

31.

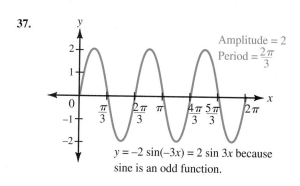

cos 90° + i sin 90°
cos 150° + i sin 150°
cos 30° + i sin 30°
cos 210° + i sin 210°
cos 330° + i sin 330°
cos 270° + i sin 270°

33. $\sqrt{2}(\cos \theta + i \sin \theta)$ where $\theta = 30°, 150°, 210°, 330°$
35. $\sqrt[4]{2}(\cos \theta + i \sin \theta)$ where $\theta = 67.5°, 112.5°. 247.5°, 292.5°$

37.

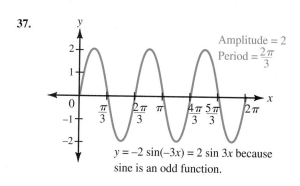

Amplitude = 2
Period = $\dfrac{2\pi}{3}$

$y = -2 \sin(-3x) = 2 \sin 3x$ because sine is an odd function.

39.

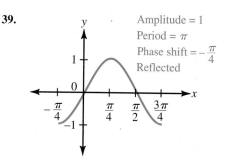

Amplitude = 1
Period = π
Phase shift = $-\dfrac{\pi}{4}$
Reflected

41.

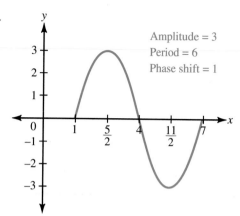

Amplitude = 3
Period = 6
Phase shift = 1

43. 4.23 cm² **45.** 3.8 ft²

Problem Set 8.5

1.–11. (odd)

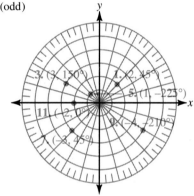

13. $(2, -300°), (-2, 240°), (-2, -120°)$

15. $(5, -225°), (-5, 315°), (-5, -45°)$ **17.** $(-3, -330°), (3, -150°), (3, 210°)$ **19.** $(1, \sqrt{3})$
21. $(0, -3)$ **23.** $(-1, -1)$ **25.** $(-6, -2\sqrt{3})$ **27.** $(3\sqrt{2}, 135°)$ **29.** $(4, 150°)$ **31.** $(2, 0°)$
33. $(2, 210°)$ **35.** $(5, 53.1°)$ **37.** $(\sqrt{5}, 116.6°)$ **39.** $(\sqrt{13}, 236.3°)$ **41.** $x^2 + y^2 = 9$
43. $x^2 + y^2 = 6y$ **45.** $(x^2 + y^2)^2 = 8xy$ **47.** $x + y = 3$ **49.** $r(\cos \theta - \sin \theta) = 5$

51. $r^2 = 4$ **53.** $r = 6 \cos \theta$ **55.** $\theta = 45°$ or $\cos \theta = \sin \theta$ **57.** y

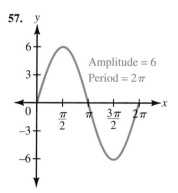

Amplitude = 6
Period = 2π

59.

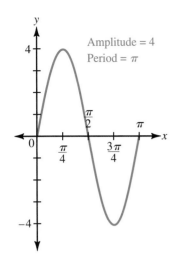

Amplitude = 4
Period = π

61.

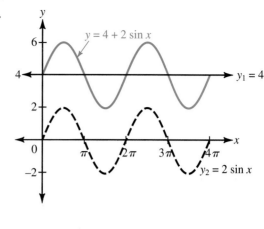

$y = 4 + 2 \sin x$
$y_1 = 4$
$y_2 = 2 \sin x$

Problem Set 8.6

1.

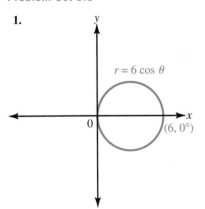

$r = 6 \cos \theta$
$(6, 0°)$

3.

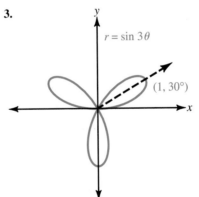

$r = \sin 3\theta$
$(1, 30°)$

5.

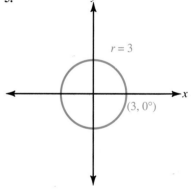

$r = 3$
$(3, 0°)$

7.

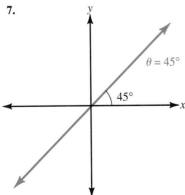

$\theta = 45°$
$45°$

9.

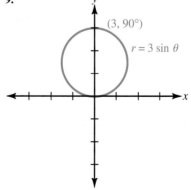

$(3, 90°)$
$r = 3 \sin \theta$

11.

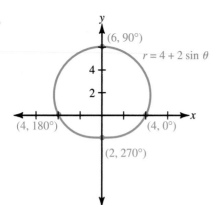

13.

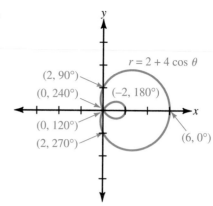

15.

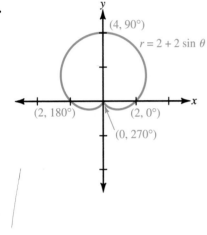

17.

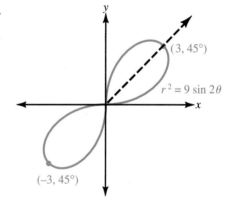

19.

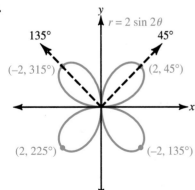

21.

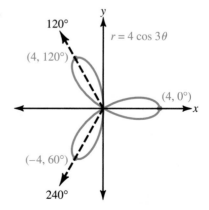

23.

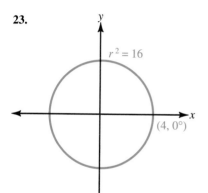

$r^2 = 16$

$(4, 0°)$

25.

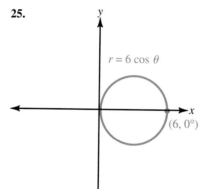

$r = 6 \cos \theta$

$(6, 0°)$

27.

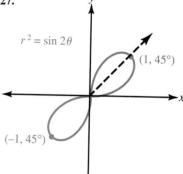

$r^2 = \sin 2\theta$

$(1, 45°)$

$(-1, 45°)$

29.

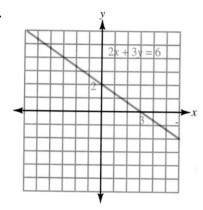

$2x + 3y = 6$

2

3

31.

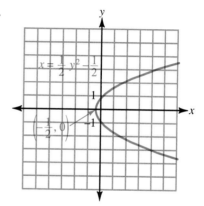

$x = \frac{1}{2}y^2 - \frac{1}{2}$

1

1

$\left(-\frac{1}{2}, 0\right)$

-1

33.

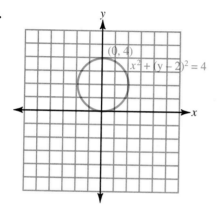

$(0, 4)$

$x^2 + (y - 2)^2 = 4$

35.

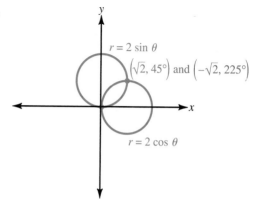

$r = 2 \sin \theta$

$\left(\sqrt{2}, 45°\right)$ and $\left(-\sqrt{2}, 225°\right)$

$r = 2 \cos \theta$

37.

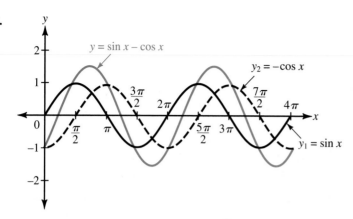

39.

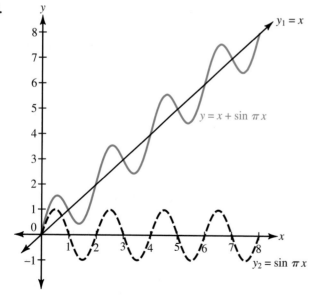

41.

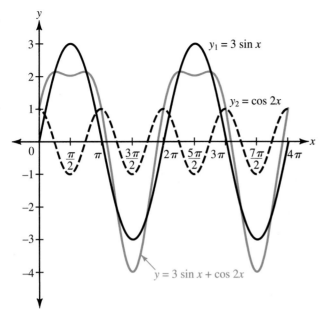

Chapter 8 Test

1. $5i$ **2.** $2i\sqrt{3}$ **3.** $x = 2, y = 2$ **4.** $x = -2$ or $5, y = 2$ **5.** $7 - 6i$ **6.** $8 - 2i$ **7.** 1 **8.** i

9. 89 **10.** $-16 + 30i$ **11.** $-2 - \dfrac{5}{2}i$ **12.** $\dfrac{11}{61} + \dfrac{60}{61}i$ **13.** **a.** 5 **b.** $-3 - 4i$ **c.** $3 - 4i$

14. **a.** 5 **b.** $-3 + 4i$ **c.** $3 + 4i$ **15.** **a.** 8 **b.** $-8i$ **c.** $-8i$ **16.** **a.** 4 **b.** 4 **c.** -4

17. $4\sqrt{3} - 4i$ **18.** $-\sqrt{2} + i\sqrt{2}$ **19.** $2\sqrt{2}(\cos 45° + i \sin 45°)$ **20.** $2(\cos 150° + i \sin 150°)$

21. $5(\cos 90° + i \sin 90°)$ **22.** $3(\cos 180° + i \sin 180°)$ **23.** $15(\cos 65° + i \sin 65°)$

24. $5(\cos 30° + i \sin 30°)$ **25.** $32(\cos 50° + i \sin 50°)$ **26.** $81(\cos 80° + i \sin 80°)$

27. $7(\cos 25° + i \sin 25°), 7(\cos 205° + i \sin 205°)$ **28.** $\sqrt{2}(\cos \theta + i \sin \theta)$ where $\theta = 15°, 105°, 195°, 285°$

29. $x = \sqrt{2}(\cos \theta + i \sin \theta)$ where $\theta = 15°, 165°, 195°, 345°$ **30.** $x = \cos \theta + i \sin \theta$ where $\theta = 60°, 180°, 300°$

31. $(-4, 45°), (4, -135°); (-2\sqrt{2}, -2\sqrt{2})$ **32.** $(6, 240°), (6, -120°); (-3, -3\sqrt{3})$ **33.** $(3\sqrt{2}, 135°)$

34. $(5, 90°)$ **35.** $x^2 + y^2 = 6y$ **36.** $(x^2 + y^2)^{3/2} = 2xy$ **37.** $r(\cos \theta + \sin \theta) = 2$ **38.** $r = 8 \sin \theta$

39.

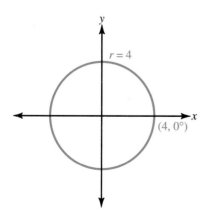

40.

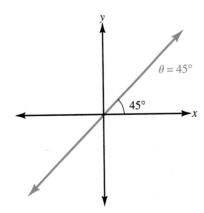

41.

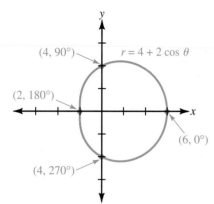

42.

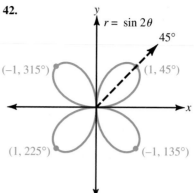

APPENDIX A

Problem Set A.1

1. $\log_2 16 = 4$ **3.** $\log_{10} 0.01 = -2$ **5.** $\log_2 \frac{1}{32} = -5$ **7.** $\log_{1/2} 8 = -3$ **9.** $10^2 = 100$ **11.** $8^0 = 1$
13. $10^{-3} = 0.001$ **15.** $5^{-2} = \frac{1}{25}$ **17.** 9   **19.** $\frac{1}{3}$ **21.** 2

23.

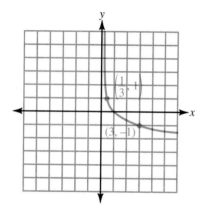

25.

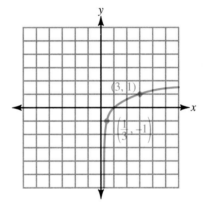

27. 4   **29.** $\frac{3}{2}$ **31.** 3   **33.** 1   **35.** 0   **37.** 0   **39.** $\frac{1}{2}$ **41.** 7   **43.** 10^{-6} **45.** 2
47. 10^8 times as large

Problem Set A.2

1. $\log_3 4 + \log_3 x$ **3.** $\log_6 5 - \log_6 x$ **5.** $5 \log_2 y$ **7.** $\frac{1}{3} \log_9 z$ **9.** $2 \log_6 x + 3 \log_6 y$
11. $\frac{1}{2} \log_5 x + 4 \log_5 y$ **13.** $\log_b x + \log_b y - \log_b z$ **15.** $\log_{10} 4 - \log_{10} x - \log_{10} y$
17. $2 \log_{10} x + \log_{10} y - \frac{1}{2} \log_{10} z$ **19.** $3 \log_{10} x + \frac{1}{2} \log_{10} y - 4 \log_{10} z$ **21.** $\frac{2}{3} \log_b x + \frac{1}{3} \log_b y - \frac{4}{3} \log_b z$

23. $\log_b xz$ **25.** $\log_3 \dfrac{x^2}{y^3}$ **27.** $\log_{10} \sqrt{x}\sqrt[3]{y}$ **29.** $\log_2 \dfrac{x^3\sqrt{y}}{z}$ **31.** $\log_2 \dfrac{\sqrt{x}}{y^3 z^4}$ **33.** $\log_{10} \dfrac{x^{3/2}}{y^{3/4} z^{4/5}}$

35. $\frac{2}{3}$ **37.** 18   **39.** Possible solutions -1 and 3; only 3 checks   **41.** 3
43. Possible solutions -2 and 4; only 4 checks   **45.** Possible solutions -1 and 4; only 4 checks
47. Possible solutions $-\frac{5}{2}$ and $\frac{5}{3}$; only $\frac{5}{3}$ checks   **49.** 2.52 min

Problem Set A.3

1. 2.5775   **3.** 1.5775   **5.** 3.5775   **7.** -1.4225 **9.** 4.5775   **11.** 2.7782   **13.** 3.3032
15. -2.0128 **17.** 759   **19.** 0.00759   **21.** 1,430   **23.** 0.00000447   **25.** Approximately 3.2
27. 1.78×10^{-5} **29.** 3.16×10^5 **31.** 2×10^8 **33.** 10 times as large   **35.** 12.9%   **37.** 5.3%
39. 1   **41.** 5   **43.** x **45.** $\ln 10 + 3t$ **47.** $\ln A - 2t$ **49.** 2.7080   **51.** -1.0986
53. 2.1972   **55.** 2.7724

Problem Set A.4

1. 1.4650   **3.** 0.6826   **5.** -1.5440 **7.** 2.0000   **9.** -0.1845 **11.** 2.1131   **13.** \$15,529.24
15. \$438.22   **17.** 11.7 years   **19.** 9.25 years   **21.** 1.3333   **23.** 0.7500   **25.** 1.3917   **27.** 0.7186
29. 5.8435   **31.** -1.0642 **33.** 2.3026   **35.** 10.7144   **37.** 13.9 years later   **39.** 27.5 years

41. $t = \dfrac{1}{r} \ln \dfrac{A}{P}$ **43.** $t = \dfrac{1}{k} \cdot \dfrac{\log P - \log A}{\log 2}$ **45.** $t = \dfrac{\log A - \log P}{\log(1 - r)}$

Index

| | | | | | | | | |

COMPLEX NUMBERS

Operations on Complex Numbers in Standard Form [8.1]

If $z_1 = a_1 + b_1 i$ and $z_2 = a_2 + b_2 i$ are two complex numbers in standard form, then the following definitions and operations apply.

Addition

$$z_1 + z_2 = (a_1 + a_2) + (b_1 + b_2)i$$

Add real parts; add imaginary parts.

Subtraction

$$z_1 - z_2 = (a_1 - a_2) + (b_1 - b_2)i$$

Subtract real parts; subtract imaginary parts.

Multiplication

$$z_1 z_2 = (a_1 a_2 - b_1 b_2) + (a_1 b_2 + a_2 b_1)i$$

In actual practice, simply multiply as you would multiply two binomials.

Conjugates

The conjugate of $a + bi$ is $a - bi$. Their product is the real number $a^2 + b^2$.

Division

Multiply the numerator and denominator of the quotient by the conjugate of the denominator.

Graph of a Complex Number [8.2]

The graph of the complex number $x + yi$ is a vector (arrow) that extends from the origin out to the point (x, y).

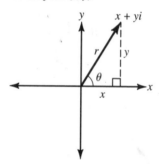

Absolute Value of a Complex Number [8.2]

The *absolute value* (or *modulus*) of the complex number $z = x + yi$ is the distance from the origin to the point (x, y).

$$r = |z| = |x + yi| = \sqrt{x^2 + y^2}$$

Argument of a Complex Number [8.2]

The argument of the complex number $z = x + yi$ is the smallest positive angle from the positive x-axis to the graph of z.

$$\sin \theta = \frac{y}{r}, \ \cos \theta = \frac{x}{r}, \ \text{and } \tan \theta = \frac{y}{x}$$

Trigonometric Form of a Complex Number [8.2]

The complex number $z = x + yi$ is written in trigonometric form when it is written as

$$z = r(\cos \theta + i \sin \theta)$$

where r is the absolute value of z and θ is the argument of z.

Products and Quotients in Trigonometric Form [8.3]

If $z_1 = r_1(\cos \theta_1 + i \sin \theta_1)$ and $z_2 = r_2(\cos \theta_2 + i \sin \theta_2)$, then

$$z_1 z_2 = r_1 r_2 [\cos(\theta_1 + \theta_2) + i \sin(\theta_1 + \theta_2)]$$

$$\frac{z_1}{z_2} = \frac{r_1}{r_2} [\cos(\theta_1 - \theta_2) + i \sin(\theta_1 - \theta_2)]$$

DeMoivre's Theorem [8.3]

If $z = r(\cos \theta + i \sin \theta)$ is a complex number in trigonometric form and n is an integer, then

$$z^n = r^n(\cos n\theta + i \sin n\theta)$$

Roots of a Complex Number [8.4]

The n nth roots of the complex number $z = r(\cos \theta + i \sin \theta)$ are given by the formula

$$w_k = r^{1/n}\left(\cos \frac{\theta + 360°k}{n} + i \sin \frac{\theta + 360°k}{n}\right)$$

where $k = 0, 1, 2, \ldots, n - 1$. That is, the n nth roots are

$$w_0 = r^{1/n}\left(\cos \frac{\theta}{n} + i \sin \frac{\theta}{n}\right)$$

$$w_1 = r^{1/n}\left(\cos \frac{\theta + 360°}{n} + i \sin \frac{\theta + 360°}{n}\right)$$

$$w_2 = r^{1/n}\left(\cos \frac{\theta + 720°}{n} + i \sin \frac{\theta + 720°}{n}\right)$$

$$\vdots$$

$$w_{n-1} = r^{1/n}\left(\cos \frac{\theta + 360°(n - 1)}{n} + i \sin \frac{\theta + 360°(n - 1)}{n}\right)$$